Yankee Science in the Making

Science and Engineering in New England from Colonial Times to the Civil War

Dirk J. Struik

DOVER PUBLICATIONS, INC.
New York

Published in Canada by General Publishing Company, Ltd., 30 Lesmill Road, Don Mills, Toronto, Ontario.
Published in the United Kingdom by Constable and Company, Ltd., 3 The Lanchesters, 162–164 Fulham Palace Road, London W6 9ER.

This Dover edition, first published in 1991, is an unabridged, slightly enlarged republication of the revised edition, first published by Collier Books, New York, 1962. The original edition was published by Little, Brown and Company, Boston, 1948. For the Dover edition the author has added a new Postscript and Supplementary Bibliography. A few typographical errors have also been corrected, and a subtitle added.

Manufactured in the United States of America
Dover Publications, Inc., 31 East 2nd Street, Mineola, N.Y. 11501

Library of Congress Cataloging-in-Publication Data

Struik, Dirk Jan, 1894–
Yankee science in the making / by Dirk J. Struik.
p. cm.
"Unabridged, slightly enlarged republication of the revised edition, first published by Collier Books, New York, 1962. The original edition was published by Little, Brown and Company, Boston, 1948"—T.p. verso.
Includes bibliographical references (p.) and index.
ISBN 0-486-26927-2 (pbk.)
1. Science—United States—History. I. Title.
Q127.U6S8 1991
509.73—dc20 91-30933
CIP

To the memory of my father

HENDRIK JAN STRUIK

Preface to the First Edition

SCIENTISTS, SCHOLARS, engineers are citizens, not only in the sense that they can vote, but in the wider sense that their work contributes materially to the welfare and the ideas of society. Each of them, in his own professional way, expresses tendencies, desires, and ideals which exist among his fellow men. Understanding the science, the learning and the technology of an age means not only a knowledge of the content of the individual professions and techniques, but also an insight into the ways in which they are related to the social structure, the cultural aspirations, and the traditions of this age. And so we begin to understand that the history of an epoch includes the history of its science and technology. We must also understand that this history must combine a study of their social relations with a study of theories and techniques. History of science, if taken in the fullest meaning of the term, must include its sociology.

Sociology of science has been rarely attempted, spadework has hardly started. A promising field for a preliminary attack is the investigation of science and technology in a relatively stable community, where a certain homogeneity of the population with established traditions offers a background for a gradual unfolding of culture. Such a community was pre-Civil War New England, and this explains why the study of its literature—as shown for instance by Van Wyck Brooks's remarkable books—has been so successful. In this book we have concentrated on the republican period of this pre-Civil War New England, which is more interesting from the point of view of the historian of science.

It was a period of considerable growth in science and engineering. In 1780, when the American Academy was founded, there existed in New England only a few persons interested in science, and this interest was almost exclusively in astronomy, medicine and agronomy. Modern industry and engineering were absent. By 1860 the situation had entirely changed. There were flowering scientific institutions with outstanding men of research; large and modern industries attracted the attention of the world. We can divide this period roughly into two parts.

In the first, which we have called the Federalist period, we still find relatively few people interested in science and engineering. Bowditch continued to work in the astronomical tradition of the merchant class, Baldwin's projects mean the beginning of civil engineering. The second or Jacksonian period, however, was characterized by a mass interest in science and technical questions. In the many geological and other surveys men of science were trained. The Coast Survey and the Wilkes expedition testified to the interest of the Federal Government. Private initiative led to the foundation of the Harvard Observatory and the Scientific Schools at Harvard and Yale. New England civil engineers such as Whistler and Francis achieved international fame. The city of Lowell became a center of industrial chemistry, of textile and of hydraulical engineering. Scientists such as Agassiz, Gray and Dana reached international standing, and broke through the provincialism which had been so long characteristic of New England.

It is an instructive fact that in early republican New England most scientists came from the coastal towns, where the mercantile class had shown an early interest in mathematics, astronomy, medicine, and meteorology. Many inventors and manufacturers came from the farming towns of Connecticut and Worcester County, where the combination of "whittling boy" and "Yankee peddler" provided the incentive for technological experimentation. This difference of origin was seldom bridged in the practice of life, at any rate not in the period under consideration, though there were such notable exceptions as Jacob Bigelow. It seemed therefore fitting to end our study with the figure of William Barton Rogers, in whose lifework, the foundation of the Massachusetts Institute of Technology, we can read a successful attempt to bring theory and practice of science together. That this attempt, conceived in the atmosphere of Jeffersonian tradition, was eventually brought to fruition under the sponsorship of big corporations is a phenomenon which falls outside the time limits set by our study.

Nobody is more convinced than the author of the inadequacy of the picture he has tried to draw of the formative years of industrial New England. Some important figures have hardly been mentioned—we think of Daniel Treadwell;

important technological developments have been omitted—we think of the development of the woolen and shoe industries. Some figures such as Agassiz have perhaps received too much space as compared to others, whose work was also influential in building the structure of science in America. The author confesses to a certain prejudice in his preference for particular figures, inventions, and discoveries, partly because of the accessibility of material, partly because they appealed more to his imagination. He hopes nevertheless that his picture of scientific and technological growth can be taken as a first approximation to the truth, and also hopes that other men or women may be attracted to make this picture more adequate to reality.

The source material is scattered in books and papers, in biographies and monographs. To find the way in this printed material required considerable work, so that very little manuscript material has been used. It may be hoped that studies like this book will encourage historians and biographers to dig into the masses of manuscript material, and that we may get at least a few more adequate biographies of New England inventors, educators and scientists, such as McAlister's book on Eaton, or Cuningham's book on Dwight. There are plenty of attractive figures worth a detailed study: Hitchcock and Waterhouse, Gray and Baldwin, Folger and Whistler. Even such subjects of existing biographies as Bowditch and Agassiz still merit a more comprehensive study.

The author likes to express his appreciation of the encouragement and the help he has obtained in writing this study of New England—especially from Henry Wadsworth Longfellow Dana, I. Bernard Cohen, Harry Winner, Frank L. Hitchcock, Ruth Ramler Struik, Harold Bowditch, Angus Cameron, Gerald Thompson, and many others. He is also grateful for the assistance he obtained from many librarians, and likes to express his special appreciation of the efficiency and courtesy of the library staff of the Massachusetts Institute of Technology.

D. J. S.

THE MASSACHUSETTS INSTITUTE OF TECHNOLOGY
CAMBRIDGE, MASSACHUSETTS
June 1948

Preface to the Revised Edition

AFTER ITS first appearance in 1948, this book was republished in 1957 under the title *The Origins of American Science (New England)* (Cameron Associates, New York). The text was unchanged, but a page of errata was added. Now the book appears in a new edition under its original title, and some revisions have been made, including references to the literature that has appeared since 1948. Among this newer literature some books stand out, including E. Lurie's biography of Louis Agassiz, A. Hunter Dupree's biography of Asa Gray, and the same author's *Science in the Federal Government*. The book, *The Pursuit of Science in Revolutionary America 1735-89* by B. Hindle is a rich source of information on colonial New England as well as on other regions.

The first edition has brought to the author a number of letters from friendly readers, often with interesting new information or points of criticism. The present edition has benefited from these communications, for which the author likes to express his grateful appreciation. The book has also been the beneficiary of thoughtful and sympathetic reviews,[1] some of which have raised points which have been taken into account in this new edition of the book. The review by L. Marx contains the criticism that the discussion of the colonial epoch in Chapter 1 does not bring out "that a great many characteristics of New England science in the Nineteenth Century only make sense in terms of a more detailed delineation of the Puritan conception of science and the attitudes it left behind," so that "the essence of Emerson's critique of contemporary science," and his "awareness of a lost ideal," is missed. The point seems to be well taken, and its elaboration might have lent a deeper perspective to the way the nineteenth-century

[1] We like to mention those by C. Mabee (*Amer. Hist. Review* 54, 1948-49, pp. 617-619); M. Boas (*Archives Intern. d'Hist. des Sciences* 5, 1952, pp. 385-388); and L. Marx (*Isis* 40, 1949, pp. 62-64).

Yankee felt his science. In the long, drawn out struggle between the ancient orthodox theocracy and the more tolerant mercantile aristocracy the appreciation of the meaning of science played a significant role, as we can see by studying the works of Puritan divines and of Harvard and Yale professors. The battle between the dictates of the scientific and theological conscience, waged in the anguished breasts of Silliman, Hitchcock, and Agassiz, was indeed a continuation, in newer forms, of the great crusade against the Devil so manfully undertaken by Increase and Cotton Mather and later by Jonathan Edwards. The struggle is still with us, though usually in a more secularized form, and we witness it in the present debates among troubled men and women on the moral unneutrality of science. Even the witch hunts have remained. The subject, however, seems too large for our simple Chapter 1, and we must, for the time being, leave it to Professor Perry Miller to explain this matter in his books on the New England colonial mind.

There are more aspects of New England science with which our book (which, after all, should retain a reasonable size) has insufficiently coped. There is, for instance, the matter of cartography, a subject rather neglected in our histories of science. However, the drawing of good maps of large territories depends on good clocks, good astronomical instruments, and good tables of the ephemerides, so that cartography is intimately connected with astronomy and instrument making. More attention to cartography would have induced the author to write, for instance, an introductory chapter on New England before 1620. Such a chapter might have dwelt upon some of the following themes.

There is a belief, widely held in and around Massachusetts, that New England history began in 1620 with the landfall of the *Mayflower*. This belief pays scant attention to the fact that there was a native population long before 1620, which through its cultivation of corn and many other vegetable products—the result of millennia of careful attention to crops—was able to help the Europeans greatly in their settlement. Moreover, and more important for our subject, it is too little realized that Europeans had been visiting the North Atlantic coast ever since the time of the Cabots and the Cortereals, that is, ever since the last years of the fifteenth century. Already before 1550 we find a sketchy outline of this coast on the maps, the result of crude surveying from board ship by navigators in the service of several nations. The information gathered by these skippers, of whom Giovanni da Verrazano is best remembered, was collected and evaluated by Gerhard Mercator, whose huge world map of 1569 is also a landmark in scientific map projection. His map is graced with a curious assortment of forgotten names, but one feature stands out unmistakably, the elbow of Cape Cod, called Cabo de Arenas—Sandy Cape. The Penobscot region is indicated by the name Norombega, in such large characters that it covers a fair portion of present New England. This name, on which so much has been written, may well have originated with Verrazano, who picked up a name that sounded like it in connection with some locality East of the Hudson. This happened perhaps during his happy fortnight in the spring of 1524 with the Indians at Refugio, that is, Narragansett Bay. The name (it is spelled Oranbega on the map of 1529 drawn by Verrazano's brother) moved north on the maps until it became attached to the whole Penobscot region, a type of transfer not uncommon on ancient maps. Indeed, it was mainly through the authority of Mercator that the name America became attached to both continents, where originally it was only given to a section of the Venezuelan-Brazilian coast, described by Amerigo Vespucci. As to the name Cape Cod, it is due to Captain Bartholomew Gosnold, who spent a few weeks of 1602 at Cuttyhunk, collecting sassafras, that panacea for all sickness, including the "pox." He also is responsible for the name "Martha's Vineyard." Another landmark that appears early on the maps is the Penobscot river; its present name dates from a *Description of New England* by Captain John Smith, published in 1616.

Maps of the New England coast begin to take a less schematic character with the surveys of Samuel de Champlain, John Smith, and Adriaen Block—French, English, and Dutch, respectively. The Champlain maps of 1612 show Boston Harbor with its islands, the Kennebec, the Penobscot, and other recognizable landmarks, with the Charles River (or Mystic) flowing out of the country south of Lake Champlain. Champlain drew these maps as the result of three voyages undertaken between 1604 and 1606, taking off from his French base in the Bay of Fundy (a bay nonexistent on Mercator's map). Some of Champlain's names have been preserved, among them his "Isle des Monts Déserts," which he placed at 44½° North Latitude, which shows that Champlain knew how to use his astrolabe for measuring his distance from the equator.

John Smith sailed the New England coast in 1614, and, like Champlain, was very pleased with what he saw. "The Countrie of the Massachusetts," he wrote, using the term for the first time in print, "is the Paradise of all those parts." This can be read in Smith's *A Description of New England* (1616). Here we also meet the term "New England" for the first time, although Francis Blake had used its equivalent, "Nova Albion," in 1579 for a section of the North Pacific coast. The *Description* had a map with a record of Smith's discoveries. We now begin to feel on familiar ground: many details of the coast line can be identified, and many names sound fine: Boston, Dartmouth, Sandwich, Oxford, Ipswich. But the world of those place names is topsy turvey: all those names are now attached to other places. Three Massachusetts names alone have survived on their allotted sites, "The River Charles," "Plimouth," and "Cape Anna." "The Sandy Cape" was called Cape James, under pressure of Prince Charles, later Charles I, who had distributed his own names over Smith's map to fit his fancy. But "the Cod" won out with Captain Gosnold, and Cape Cod it has remained.

Smith was an ardent promoter of immigration, and his appeal was to merchants and shipmasters as well as to debtors and beggars. He belonged to the newer school of mercantilists for whom trade in fish promised more gold than mining. His books were widely read; the Pilgrims in 1620 landed on the spot they called Plymouth, which is indicated by the name Plymouth on Smith's map. At that time most of the Indians, whose "corne, groves, mulberries, selvage gardens" had impressed Smith as well as Champlain, conveniently had died out. The English could settle without that opposition which Champlain seems to have feared when some years before he decided to settle instead on the St. Lawrence. The settlers also met with the distrust of the Dutch, whose geographical knowledge can be studied on a map prepared by Adriaen Block.

Block sailed in 1614 from the Manhattan region and charted the southern shore of New England, perhaps coasting north as far as Massachusetts Bay. He moved up the Connecticut (his "Fresh River") to a point beyond present Hartford. His map contains a number of quaint names which haunted the standard Dutch maps of Blaeu and others throughout the seventeenth century. If the Dutch had prevailed in New England, Boston Harbor might have been known as Vos Haven ("Fox Harbor"). The name Rhode Island is a Dutch name dating from this period. There is also a Block Island, off the Connecticut coast.

Long after the settlement by Pilgrim and Puritan, maps remained very inaccurate, especially insofar as the longitude of places was concerned. Indeed, before the end of the seventeenth century the longitude of not a single location in North America had been determined by celestial methods. The first published astronomical determinations of longitude in North America (by eclipse observations) were those by Jean Deshayes at Quebec in 1686 and by William Brattle at Boston in 1694, although the men who observed the comets of 1680 and 1682 must have done some computation of position. Deshayes' latitude was only a few minutes off, but his longitude, more than a degree; Brattle was within half a minute in latitude and only a fraction of a degree in longitude. The Boston observer had better instruments than his colleague at Quebec: Brattle used a telescope, the English one presented by John Winthrop to Harvard College in 1672. During the eighteenth century more observations are taken and more accurate maps of New England and the Atlantic regions appear; among the best of the printed ones are those of Lewis Evans at Philadelphia, followed by those of Jeffries and Mitchell. The Mitchell map was used at the Paris peace conference of 1783.

Another rather neglected subject is the movement for agricultural reform and education in early republican New England. It began with agricultural societies organized by gentlemen farmers related to the merchant class, who were impressed by the results of the agricultural revolution which had taken place in Great Britain during the eighteenth century. The first such society was organized in Massachusetts in 1792. These societies were helpful in establishing the botanical garden at Harvard (1805); they also endeavored to promote the improvement of livestock, in particular the breeding of Merino sheep. In this period also begins the work of Elkanah Watson, who through the organization of country fairs (the first held in 1812 at Pittsfield, Massachusetts) tried to bridge somewhat the distance between the reform-minded gentry and the dirt farmer.

This movement to bring reform to the homestead gained impetus with the change in farming conditions due to the industrialization of the countryside, the improved transportation facilities, and the deadly competition from the newly opened western lands. Farm journals appeared, and attempts were made at agricultural education. This movement had a popular scientific aspect; in it we meet Marshall P. Wilder, long a central figure in the Massachusetts Horticultural Society (founded in 1829), Thomas G. Fessenden, popular author and editor, and Ezekiel Holmes of Gardiner, Maine. Holmes was associated with Benjamin Vaughan, and this brings us to the Kennebec experiment.

Jamaica-born Vaughan, friend of Jefferson and Priestley, settled after the Revolution on land along the Kennebec in the present state of Maine. Here, with the aid of his brother Charles, and of Robert Hallowell Gardiner, he gradually built up a center of scientific agriculture. Able craftsmen and men of culture were encouraged to settle. Augusta, the capital of Maine after its separation from Massachusetts (1820) was projected near the Gardiner settlements. In 1823 the Gardiner Lyceum was opened with a program not unlike that of Jefferson's University of Virginia: elective system, student self-government, stress on the practical side of science, no religious affiliation. However, the project was without much future, conceived in a country without a hinterland and based on ideas too advanced for its surroundings. The idyll in the pines had to disappear. Ezekiel Holmes kept the ideals alive for a while after Vaughan's death in 1835, but in 1857 the

Lyceum building was sold. Even the memory of this pioneering project has faded with the years. A recent book on the Kennebec does not even mention it.

Interest in scientific agriculture increased on the farms and in the legislature. In 1837 the Massachusetts Agricultural Survey was authorized and placed under the supervision of Henry Colman, who left a Unitarian pulpit to travel widely in the state talking to farmers and collecting material. His results were published between 1838 and 1844 in four volumes totaling 1,106 pages. Wherever in his report Colman had to deal with that new fad, agricultural chemistry, he regarded it with some scepticism. Colman also went to England, where Dickens had a crack at him.

Agricultural chemistry came to America with the published lectures of Humphrey Davy (1813), but it became the object of mass interest with the work of Liebig, of which the first English translation appeared in 1840. In New England the pioneering was done by Samuel Luther Dana, on whose work we have reported. Dana's and Liebig's theories were occasionally in conflict, Dana promoting the "humus" theory against Liebig's stress on chemicals—a debate which has continued happily ever since. At the universities Liebig won out: when Harvard and Yale opened their scientific schools they were staffed with Liebig pupils.

These examples, taken from theology, cartography, and agriculture, may show how much is still needed to gain a rounded picture of our subject. Examples illustrating this need could also be taken from other fields, such as military engineering, or the development of what used to be called "philosophical" instruments.

D. J. Struik

November 1961

Contents

PART I

BEGINNINGS

Chapter 1

The Colonial Setting

> It is a fact not generally realized that the American Philosophical Society at Philadelphia, the Royal Society of Great Britain and the Royal Institution of London, are all of them in a measure indebted for their birth and first foundation to natives or inhabitants of New England.
>
> —JACOB BIGELOW, 1816

1

IN THE YEAR of the Lord 1620 Francis Bacon, Lord Chancellor to King James of England, published his *Novum Organum*. The object of true knowledge, he claimed, was the relief of man's estate, the furtherance of man's power over nature, to alleviate human misery and to add to the comforts of life. Let man experiment, and study the ebb and flow of the sea, the motion of the planets, weight, gravity and the many forms of heat. Then true induction will give him understanding, and understanding will make him master of his fate.

Bacon drew the conclusions from a social process which was rapidly developing in the world around him. The Elizabethan period of relative peace had stimulated the growth of mercantile and industrial enterprise. Navigation had opened new seaways and new continents. The older crafts were flourishing, and new ones were established. Weavers, spinners, fullers, carders, printers, brewers, wheelwrights, blacksmiths and silversmiths increased the national income. Across the seas from Holland, France and Italy came word of new achievements in agriculture, of canal building and of mill construction, of new developments in textiles and in the art of war making. And from across the seas came also the books of Copernicus and Stevin, of Kepler and Galileo, with their new and wondrous discoveries. King James was interested especially in the great engineering works of his period, the draining of the Lincolnshire Fens and the new water supply of London. It was a period of new and hardy enterprise, stimulating invention and science, as well as a new philosophy of life.

2

In that same year 1620, on a bleak December day, the Pilgrim Fathers landed on the inhospitable shores of Plymouth and laid the foundation of the New England settlements. Several of them, their leaders, had been in Holland and seen the many crafts, the canals and windmills of that rapidly developing country. Others had stayed in England, many of them in Yorkshire, Lincolnshire and neighboring districts with their advanced workmanship and their new engineering projects. Many of them were able craftsmen. William Bradford, their leader, knew the silk business and was a fustian maker, others were wool combers, and among them were a tailor, a brewer and a smith. These Pilgrims were followed, some years later, by men and women of similar social and religious background, who spread along the rocky coast of the Atlantic Ocean and penetrated into the virginal forest. Salem was settled in 1626, Boston in 1630. There were men of many crafts among the colonists, weavers, shipbuilders, blacksmiths, printers, carpenters and tanners. Several of their leaders were men of learning. John Winthrop was "browghte up amonge boockes and learned men," as were the ministers John Cotton and John Eliot. They knew their Bible but read other books as well, and remained as much as possible in contact with their brethren overseas. In culture and crafts the new colonies of Masachusetts Bay, of New Haven, of Connecticut and other plantations were a fair reflection of the more advanced parts of England.

These men were Puritans and other nonconformists escaping from Royalist and Anglican suppression to build their own institutions under Calvinist leadership. Providence, in 1636, was settled by Roger Williams, who believed in the separation of church and state. Hartford, in 1636, was settled by Thomas Hooker, who opposed Winthrop's magisterial autocracy. Every year brought new colonists, many of them men and women of ability, craftsmen willing to try their skill at farming in the new country. Some of them were scholars, bringing their books. It is estimated that in the twenty years following the landing of the Pilgrims twenty thousand persons crossed the ocean and settled in New England.

Then, with the English Revolution, the downfall of Charles I and his execution in 1649, the emigration ceased almost entirely. Under Cromwell the Puritans and the dissenters in England enjoyed sufficient freedom of worship, and no longer felt the necessity of hazarding the perilous journey to the new colonies. Even under the Restoration there was never enough pressure to send more than a small number of individuals across the ocean. The result was that the immigrants of 1620–1640 had the country almost exclusively to themselves and their descendants. There were a few minor waves, one after 1685, when some hundreds of Huguenots came over from France after the Revocation of the Edict of Nantes by Louis XIV. In the early eighteenth century a number of Scotch-Irish settlers immigrated on grants from English land speculators. A few Negro slaves were imported from the South or on Northern slave ships. All this did not deeply affect the composition of the population. The Yankees of the American Revolution were almost without exception descendants of the early Protestant settlers, the inheritors of their tradition. Immigration on any large scale into New England was not resumed until deep into the nineteenth century, when industrialization brought representatives of new nations.

3

For six or more generations, during more than one century and a half, the original settlers and their descendants lived on their own soil in considerable seclusion from the outside world, isolated even from other British colonies. The main means of communication with other parts of the world was by sea navigation, which tended to break the provincialism of the coastal towns. Poor roads made long land travel adventurous and even dangerous, and no rivers connected New England with other parts of the colonies.

In that long period New Englanders had full opportunity to develop their own local traits and even what may be called their national character. They were, to begin with, descendants of a racially and culturally homogeneous group of immigrants, all hailing from the more economically advanced regions of Elizabethan England. Their small towns and widely scattered farms centered around their Calvinist churches. The training of the early immigrants had served them well in building their homes and providing for the other necessities of life. This training had been passed from father and mother to children and grandchildren, who applied it to the cultivation of the new country. They showed considerable ingenuity in crafts and in trading, in navigation and shipbuilding, with early touches of originality. They numbered about a million just after the Revolution.

The majority was engaged in agriculture on self-contained homesteads, often far away from the centers of population. An active minority was busy in navigation, fisheries, lumbering and the mechanical arts. The strong family ties, the predominance of self-contained farms, the small size of the towns all led to the maintenance of strong traditions in church and state. In most places the official religion was the old Puritan creed, brought from England and differing little from the original Calvinism with its predestination and its strict adherence to the Bible as the revealed word of God. The Calvinist ministers had decisive influence in matters of public policy as well as of creed. Eventually other sects, especially the Episcopalians and the Quakers, were tolerated, and sometimes became quite influential, but Calvinism set its stamp on colonial New England, as it set its stamp on the Dutch Republic and on Scotland. It was a creed of hardy and individualist workers and traders, which blessed the self-reliant man in the pursuit of his business. Calvinist church organization was democratic, with the emphasis on the community of believers; it harmonized well with the homespun democracy of the New England town government. In the eighteenth century the countryside was covered with small towns scattered along the bays and coves of the seacoast and the rivers and brooks of the hilly countryside.

These towns were all of a remarkable similarity, and some of them can still be admired in secluded spots removed from modern traffic and modern industry. They have a village green with Wren-style churches, still called meetinghouses in true Puritan fashion, and white clapboard homesteads with brick chimneys and small-paned windows looking out over old-fashioned gardens. The old houses vary in appearance; some are in saltbox style, others are gabled, the later and more prosperous ones are square mansions with hip or gambrel roofs and ornamented doorways. Along the broad village streets, around the homesteads and in the pastures, stand majestic elms, the typical trees of New England, with their slanting branches as different from the English elm as the American robin is from his namesake in England. Maples and oaks abound, and color the landscape in fall in brilliant and fantastic colors. Some relics of the past, a hitching post, a wayside inn, a gristmill or a manse, are carefully preserved. Near the church is the old cemetery, with stern and pious stones guarding the remains of the old settlers. Almost every town has lists of ancient local inventors, scholars, authors and pioneers, who have set their stamp not only on the locality, but on other parts of the Union as well. Nothing reveals better the homogeneity of old New England than the character of its towns, despite their subtle local differences, especially between seaport and inland towns.

4

The very geography of the country influenced the building of the national character. New England stretches from the Hudson River and Lake Champlain on the west to the Atlantic Ocean on the east, and northward well into the approaches of the St. Lawrence River. It is a rocky land, broken by hills and mountains that rise to more than four or five thousand feet, crowned by Mount Washington with its summit above the 6000 foot level. The country is studded with lakes and crossed by broad and impressive rivers, which drain regions by a wealth of fast-flowing brooks. Its mountains, which to the north break up into the Green Mountain chain and the White Mountains, are part of the Appalachian system which stretches from Georgia to Maine along the eastern part of the North American continent. The rock of these mountains is extremely old and dates back to the Paleozoic period. Most of it was worn down by erosion in Mesozoic times; later it was tilted again to form new valleys. These valleys were gradually eroded by rivers, such as the Connecticut, where fertile ground was accumulated. In more recent times an ice sheet spread over New England, thick enough to cover large hills, and in the successive advances and retreats of the glaciers the ice left the country filled with glacial remnants, with debris, lakes, streams, boulders, terminal moraines, eskers, drumlins and scraped rocks. New England thus became a typical example of a glaciated region, widely covered with a thin soil filled with rocks and stones.

The rivers which break this rocky region flow in the main north-south, starting as wild mountain streams and ending in broad estuaries. The most majestic are the Penobscot, the Androscoggin and and the Kennebec in Maine, the Merrimack in Massachusetts, and the Thames, the Connecticut River and the Housatonic in Connecticut. Longest of all is the Connecticut River, 400 miles long from its source in the Canadian border wilderness to its mouth in Long Island Sound. Most of its course it winds through long, narrow stretches of meadowland dotted with elm trees. These rivers are navigable only to a limited distance away from the mouth, where rapids make further passage of larger boats impossible. The rapids were favorite fishing places for the native Indians, and in the last century provided water power for the new industries. The Indians have left their mark in the names of the larger rivers and in some other geographical names.

5

In this glaciated country agriculture was from the beginning a back-breaking business, and the early settlers needed all their courage, training and Calvinistic faith to hold body and spirit together. Through the hard and continuous labor of several generations they reached a moderate level of prosperity. Despite the skill of the farmer and his desire to improve himself and his family, agriculture never rose far above subsistence level, with some exceptions, notably in the Connecticut Valley. The main reason for this was the lack of capital investment in agriculture which alone could have made it profitable. When large fortunes were made in New England, they were in commerce. The possibility of considerable returns on investments in trade, commerce, and also land speculation, discouraged the interest of wealthy men in the improvement of farming. At a time when agriculture in the mother country advanced technically, Yankee farming remained backward and poorly paying. The men and women of the countryside had to make the best of a difficult occupation by frugal living and a considerable display of ingenuity. Since the seacoast towns, even Boston, remained small throughout the whole colonial period, the backbone of New England remained fixed in its solid yeomanry.

Yankee farming was thus extensive in character and to a considerable extent self-sufficient. Capital and labor were thinly applied over a large extent of land, and the produce was mainly for home consumption. Poor transportation facilities discouraged production for the market, and made the farm a poor market for city industries. The farms ranged normally from 100 to 200 acres, but were often smaller near the larger centers of population, and increased in size in newly settled regions. There was plenty of work for all members of the family. The woman washed and cooked and sewed, spun yarn and churned butter. The man worked in the field, sowed and harvested his crops, and kept his tools in working condition. Hired labor was scarce, the poorer members of the community often trying their luck in fishing or in shipping. The farmer in coastal districts might also be a sailor or a fisherman during a part of the year or a part of his life.

The homestead produced food, clothing, farm implements, furnishings, all of a simple kind. It was dependent on the outside world only for the products of some local craft, a blacksmith, a miller or perhaps a cobbler or a fuller. The farmer had no occasion to forget whatever skill his ancestors had brought with them from the old country. Over several generations he developed into a many-sided artisan, a veritable jack-of-all-trades. He could, in the words of Chancellor Livingston, with equal address mend his plow, erect his walls, thresh his corn, handle his ax, his hoe, his scythe, his saw, break a colt or drive a team. Forced from early life to rely upon himself, he acquired skill in many different arts, a skill and a versatility uncommon in countries with greater division of labor, a more generous nature or a feudal land tenure.

6

This was an ideal background for the development of manufactures, the display of inventive skill and even the pursuit of the less speculative sciences, the more so since the Puritan religion sanctified the self-reliant man and his right to a profit. Francis Bacon's philosophy and the Puritan's exodus sprang from a common soil, and it might be expected that New England would become a testing ground of Baconian ideas, perhaps a "New Atlantis," Bacon's utopian community across the seas. The speed with which the howling wilderness was transformed into a civilized community showed that the many crafts learned in the Elizabethan homeland were put to excellent use. "Wonder working Providence of Zion's Saviour," exclaimed one of the founders of Woburn in 1654. The settlers were acquainted with the use of water and wind power. Grist, saw and fulling mills were built along the many fast brooks and streams, with an undershot or overshot wheel built to measure by the local carpenter. The rhythmic ticking of the wheel amid the rushing water of the mill creek—a sound still familiar to European ears but alien to modern America—not only was a sign of civilized life and its conveniences, but set many a Yankee lad dreaming of the use to which water power might some day be put.

Occasionally the mill creek had to be dug first. Then the new settlers turned canal diggers, if only on a small scale. At Dedham the Neponset and Charles Rivers were connected by a short canal and at Watertown a ditch was dug at the head of tidewater on the Charles River. Both canals were in existence before 1640. Around the same time a millpond with mill creek and causeway was constructed in Boston's North End, on which several mills were built, driven by tidewater. Cattle were introduced as early as 1624 and in increasing amount, so that by the time emigration from Europe ceased the stock was well diffused over the country. This promoted leather and shoe making. It is reported that the first settlers of Lynn excelled in stock husbandry and had large numbers of horned cattle, sheep and goats, so that they were the first to think of tanning and cobbling. At any rate, Lynn and the surrounding towns soon became leather centers, but many other towns also acquired tanneries at an early date. And in this way the many crafts of England were spread throughout the country, including the noble art of printing. Daye's press, the first in the colonies, was set up in 1638 at Cambridge near Boston, where in 1639 he printed, significantly, a *Freeman's Oath* and an *Almanack*.

At a very early date discovery was made of the existence of iron ore in numerous peat bogs and ponds in eastern Massachusetts.[1] In 1641 the younger Winthrop formed in London a "Company of Undertakers for the Iron-Works," which established an iron furnace with forge at Saugus, near Lynn, in 1643. Several ironworkers came on from Hammersmith near London, among them Joseph Jencks, metalworker, die cutter and handy man in general. He could work in brass as well as in iron, and provided Boston with iron house utensils, silver specie, and even in 1659 with "an Ingine to carry water in case of fire." This early mechanic, who has been called the Tubal-Cain of New England, was the father of the founder of Pawtucket, and for several generations there were Jenckses who pioneered in Rhode Island industries. This bog iron foundry was never very successful as a financial venture, hammering out, it is said, more contention and lawsuits than bars of iron, but it produced skilled ironworkers who established forges and "bloomeries" in other parts of the country. By 1670 the ironworks were abandoned. They have recently been rebuilt as an open-air museum, where we can again see the blast furnace, the forge, the rolling and slitting mill at work, driven by huge waterwheels. Here we see also, restored, the old ironmaster's house, with its steep gables, diamond-paned casements and huge central chimney, a delightful example of early colonial architecture.

The Saugus venture was followed by other foundries in different parts of New England, near places where bogs or

[1] This "bog ore," a form of limonite, consists of rusty brown lumps occurring in the surface of certain swamps fed by springs of water carrying iron in solution, and was "mined" by simply scrapping up.

rocks contained iron ore. Charcoal could easily be obtained from timber, at that time abundant in the colonies even near the Eastern shore. The most important iron mines were those of the so-called Salisbury district, which extended from southern Vermont through the mountainous country of western Massachusetts and western Connecticut along the Housatonic River to the eastern banks of the Hudson. The iron ore in this district was much denser than bog ore. Many ruins of forges and furnaces, subterranean shafts and local names such as Ore Hill, Red Mountain or Mine Mountain remind the traveler today of the importance of this region to the colonial iron industry. The district was named after the town of Salisbury in Connecticut, where, beginning in 1732, some of the most successful mining was done. Workmen were brought from as far as Switzerland and Russia. We know of forty-five important mines in this district, some of which are still being worked at the present day. There Ethan Allen, the later hero of Ticonderoga and deist philosopher, helped to establish in 1762 an iron forge and blast furnace, which—though under other owners—cast a large proportion of the colonists' cannon for the Revolutionary War. Alexander Lyman Holley, one of the leaders of the nineteenth-century iron industry, was the scion of a family which hailed from Salisbury.

Two processes were used, one to produce cast iron, another to produce wrought iron, from which colonial blacksmiths produced nails and other household articles, as well as weathervanes, candlesticks, firebacks and gates. Some steel was produced in small quantities from wrought iron by a trial-and-error process, the oldest attempt going back to the middle of the seventeenth century. Samuel Higby, a blacksmith of Simsbury, Connecticut, claimed in 1728, however, that he was the very first in America to convert iron into steel, which "curious art" he had found out and obtained "with great pains and cost." The legislature granted him and an associate a ten-year patent on his invention. In 1716 an act was passed giving another firm the right to erect a slitting mill at Stony Brook in Connecticut. Invention was at work in the iron industry and the product improved to such an extent that the British iron manufacturers became worried.

Among the experimenters we find the Reverend John Eliot, of Killingworth, also in Connecticut, who had iron smelted from black magnetic sand, a process he described in the *Transactions* of the Royal Society in London of 1762, and for which he received a gold medal from the Society for the Encouragement of Arts, Manufacturers and Commerce in London. Without the many furnaces the American Revolution would have been even more sorely in need of arms than it was.

There were also copper mines in the New England colonies, mainly in Connecticut. The mine at Simsbury, with which Samuel Higby was connected and from which he struck bootleg copper coins of remarkable perfection—now a choice collector's item—had two shafts up to 80 feet and vast caverns in the hills. The imperfect knowledge of mining, especially the difficulty of draining the water out of the shafts, never made it a very profitable enterprise, and on top of this came the prohibition of smelting in the colonies by Parliament. In 1773 the works were converted into a prison for felons, who had to do the mining. This was the horrible Newgate Prison at East Granby, which served during the Revolution as a place of confinement for Tories, and from 1790 to 1827 as the Connecticut State Prison. Mining continued in this place interruptedly until the middle of the nineteenth century, but even the use of modern mining methods could not make it profitable. The place is now preserved as a museum.

7

Yet, with all these possibilities, Yankee industries never advanced beyond this initial stage. The reasons can be found partly in the concentration upon shipbuilding and navigation, from which big fortunes were made, partly in the overwhelming amount of work necessary to clear land and raise harvests in the agricultural districts. However, the mercantilist policy of the British Crown was the main reason of the frustration of Yankee inventiveness. Cromwell, in 1651, prohibited the ships of foreign nations from trading with the colonists. This navigation law was not in every respect unfavorable to the colonists, and actually promoted shipbuilding in New England and elsewhere in America, but it was followed by a whole series of measures created to establish a protective system beneficial to English manufactures and detrimental to the industries of the colonies. The more these industries grew, the more they worried the manufacturers of the home country. Laws were passed by Parliament aimed increasingly at the full suppression of colonial industry and keeping the colonies only as providers of raw material for British factories. The expanding iron industry furnishes one of the best illustrations. Not only New England but also other colonies, especially Pennsylvania, began to threaten English iron manufacturers by the abundance and excellence of their product. Inquiries were made in 1731-1732 by the royal governors, in which it was shown that New England possessed six furnaces, nineteen forges, one slitting mill and one nail factory. Though they did not supply iron enough for the country's use, British iron interests became increasingly frightened. They realized, however, that America, with its unlimited resources of wood, could supply the British market with cheaper bar and pig iron than the home country, which was fast becoming deforested. The result was the Act of Parliament of 1750 which encouraged the importation from the colonies into the port of London of bar iron, and into any port of Great Britain of pig iron, to be exchanged for woolen and other manufactures of Great Britain, and which prohibited the erection of any "mill or other engine for splitting or rolling of iron, or any plating forge to work with a tilt hammer, or any furnace for making steel," or their continued operation, if already erected. The importation of raw materials into England was thus encouraged, and the export of finished products. Even the establishment of any manufacture for distant sale inside the colonies was discouraged, as the export of hats, wools, and woolen goods from one province to another by water, by land or upon horseback. Permitted were household manufactures or manufactures producing only for the immediate neighborhood, but products such as those of steel furnaces and slitting mills had to be purchased from British merchants and manufacturers.

A lucid description of this side of British colonial policy can be found in Adam Smith's *Wealth of Nations,* which was published in 1776. Smith had to agree that the British mercantilist system was more liberal than that of other countries, notably that of Spain or Portugal. Nevertheless, his criticism was severe:

> To prohibit a great people, however, from making all that they can of every part of their own produce, or from employing their stock and industry in the way that they judge most advantageous to themselves, is a manifest violation of the most sacred rights of mankind.

The system was complemented by the restrictive power exercised by the royal governors, who were instructed to veto all legislation tending to assist the development of manufactures. Limitations were imposed upon British capital, if it intended to invest in the colonies. Such investment was allowed only in land or in land speculation—in the South, also in tobacco—and the opportunity for such speculation was widely used. One of the most notorious of these speculators was Governor Benning Wentworth of New Hampshire, whose nephew John, also governor, was one of the men who spoiled young Benjamin Thompson, the later Count Rumford. No such gambles were possible in the colonial manufactures. The result of all this was that in the very period in which England was rapidly converted from an agricultural to an industrial country, and the factory system was spreading over large sections of the homeland, the colonies remained entirely behind in industrial development. Neither did they profit from technical changes in British agriculture, except in a few isolated instances.

It is remarkable that the mercantilist system did not enter more prominently into the demands made to the Crown by the rebellious colonies. The rebellion did not consciously center so much around the question of the repressive economic laws as around the political principle of taxation without representation. One reason is that the navigation laws were never rigorously enforced, notably in overseas commerce. Even in the industrial field there were ways and means to evade them. Bishop reports that copper smelting and refining at Simsbury were quietly conducted by German workmen notwithstanding the prohibition. Another reason was that, as long as the colonists had so many other economic possibilities, the prohibition of industries was not generally felt as a great injustice. The principles of the mercantilist system were taken more for granted than we might think at present. Even William Pitt, the great friend of the colonies in Parliament, while thundering against taxation without representation, maintained the right of Parliament "to bind, to restrain America," and claimed that its legislative power over the colonies was sovereign and supreme. Benjamin Franklin, in England during the period after the conquest of Canada, maintained that Canada should be kept because it meant cheap land, so that America could remain a nonmanufacturing country. The original form of Jefferson's agrarian democracy was based on a distrust of manufactures. The Declaration of Independence enumerates in detail the grievances of the colonies, but the mercantilist system is never directly attacked. Jefferson became convinced of the value of domestic manufactures only after the painful experiences during the long period of Napoleonic Wars, in which the supply of goods from Europe was curtailed. It is only when we compare the lopsided scientific activity and the restricted inventive productivity of colonial days with the steady growth of science and the outburst of inventive genius after the Revolution that we understand the deadly pressure of the mercantilist system on the colonies. The situation in India under British rule invites comparison.

8

In its restricted field, however, Yankee craftsmanship was certainly in full evidence. The simple and beautiful products of the colonial artisan are now highly valued as Americana—chairs, tables, clocks, glassware, pottery, silver, pewter and hand-woven textiles. Clockmaking, which after the Revolution became such a highly developed industry in Connecticut, had its roots in colonial craftsmanship. The products of Thomas Harland and of Thomas Danforth are still well preserved, as are those of the Willards of Grafton, Massachusetts. The Huguenot Paul Revere, who settled in 1723 in Boston as a silversmith, was the father of a more famous and more versatile namesake, who succeeded his father in his shop and became known as an engraver, a pewter worker, a gunpowder maker, a bell founder and a printer. Longfellow's poem celebrates him as a horseman and a patriot. Among these New England clockmakers were men with a considerable interest in science. Among them we find Joseph Pope of Boston, who made a planetarium (orrery, as they used to say), which can still be admired at Harvard University, John Fitch, the later steamboat maker, and Walter Folger, the Nantucket astronomer and technical jack-of-all-trades. Several of these craftsmen, despite their seemingly humble birth, were men of considerable social status in their community. Many leaders of the Revolution were able craftsmen themselves. The very fact that the mercantilist system frustrated the development of industries may well have had a beneficial effect on the growth of this often exquisite craftsmanship. When the Revolution had removed the mercantilist restrictions the more inventive of these artisans turned their minds to new and more ambitious schemes, and many early republican inventors and shipbuilders sprang from the ranks of these colonial carpenters and metalworkers.

9

Shipping and shipbuilding suffered least from the mercantilist system, and in many ways actually profited by it, either legitimately or by illegitimate means. This led a vast portion of New England's energy into fisheries and into navigation. Along the Atlantic Coast with its sheltered harbors in bays and river estuaries many fishing and trading centers arose, of which several grew into towns of some importance, though not into big cities. Shipyards flourished in or near the trading centers on the coast and along the navigable parts of the Connecticut, the Merrimack and the Kennebec Rivers. Here local carpenters displayed remarkable skill in ship construction. The Chebacco boats of Essex had a special fame and there is a legend which connects the invention of the schooner with Gloucester. Ship models seem to have been used as early as the beginning of the eighteenth century. These ships, mainly sloops and schooners, were used not only for the carrying trade, but also for the fisheries, and, in increasing numbers, for whaling and sealing. New Bedford and Nantucket sent out large fleets of whalers long before the Revolution, and provided the country with oil for illumination and lubrication.

Shipbuilding stimulated lumbering. The large primeval forests yielded a constant supply of fine pine, oak and maple wood. Famous were the towering white pines, of a size now nearly extinct. The botanist Michaux, studying logging operations on the Kennebec in 1806, reported trunks up to 180 feet long and 6 feet in diameter. President Dwight, on his travels through New England in the same period, was told that the trees went up as high as 250 feet. They were excellent material for masts on the large sailing vessels of the later colonial period. In the beginning of the eighteenth century foresters began to mark the loftiest of these pines by an arrow, to indicate that they were reserved for His Majesty's navy, a practice discontinued only by the Revolution. But not only the pines were excellent material for shipbuilders. The heavy primeval oak was excellent timber for hulks. There seemed such an inexhaustible supply of pines and oaks that the application of any scientific principles of forestry was neglected. The result was a horrible waste, which was encouraged rather than discouraged during the republican period, and for which the present generation is still paying.

10

The largest urban center of New England was Boston, with an estimated population of 12,000 in 1722 and 15,000 in 1760. This small provincial capital was flanked by some minor commercial centers, notably Salem, Marblehead and Beverly. Other relatively important coastal towns were Gloucester, New Haven, Stonington, Providence and, surpassing them all in wealth and elegance, Newport. Inland, on the Connecticut River below the rapids, were Hartford and Middletown. The majority of the townspeople were artisans, craftsmen and mechanics, such as bakers, chandlers, blacksmiths, silversmiths, shoemakers, wheelwrights, carpenters, carriage makers, calkers, masons and printers. These men and their families have been described as the rabble, but they were mostly skilled and God-fearing people, many of whom had a considerable social and intellectual standing in the community. From their ranks came Benjamin Franklin and Paul Revere, as well as Sam Adams and his followers, who constituted the backbone of the resistance movement in the towns. Then there was a small but highly influential group of merchants, grown rich in overseas commerce, especially in the well-known triangular trade, in which West Indian sugar, Negro slaves and New England rum changed hands. To this class of men belonged Faneuil, Hancock uncle and nephew, Boylston and Bowdoin in Boston; Redwood, Lopez, Channing and Gibbs in Newport; and the Brown family in Providence. Closely allied with this group was the colonial intelligentsia, consisting in the main of ministers, lawyers and physicians, often graduates of the two New England colleges, Harvard and Yale. A few of these men carried a European degree, normally from an English or Scottish university, but occasionally also from Leyden in the Netherlands; Scotland and the Netherlands being the countries whose theological leanings appealed most to the New England descendants of the Puritans. Among them were members of merchant families, sometimes merchants themselves, like James Bowdoin, but also men of farming stock, like John Adams, who by hard work had gained acceptance into the ranks of the provincial aristocracy. The Mathers of Boston were one of those families in which the tradition of intellectual proficiency passed from father to son during several generations.

Boston, from the beginning of the settlement, was the intellectual capital of New England. With the neighboring towns of Cambridge, Roxbury, Dorchester, Salem, Marblehead and Beverly it formed a center of culture, only surpassed in the British colonies by Philadelphia, and perhaps New York. Only one town in New England could be compared with Boston, if not in population, certainly in wealth and culture. This was Newport, in Rhode Island. Though only eighty miles away as the crow flies, Boston and Newport breathed an entirely different atmosphere. Boston was the stronghold of traditional Puritanism and all it conveyed. Newport, historically bound to the Rhode Island tradition, was a town of religious tolerance, where Quakers, Episcopalians and Jews were placed in leading positions. Newport's close connection with West India planters made also for greater proficiency in the slave traffic and greater use of Negro slaves. During the eighteenth century merchants settled in Newport with tastes more akin to those of Southern planters than to those of Yankee traders. In the spacious mansions of Abraham Redwood, John Channing and many others interest was displayed in music, architecture, belles-lettres, and even in the theater, a horrible thing in the eyes of the Lord according to the Bostonian point of view. To Newport, in 1729, came George Berkeley, the Dean of Derry, later Bishop of Cloyne, on a journey to establish a missionary college in the Bermudas: "Westward the course of empire takes its way."

The Dean bought a farm which is still standing, and stayed there for two years and a half with his large family, writing and, presumably, experimenting with his favorite panacea, tar water. He had already written some of his widely read philosophical books, which have done so much to give extreme idealism a respectable place in the college curriculum, but which at the time of their writing were also felt as an attack on the Puritanism of the traditional church. The Dean seems to have liked Newport's maritime and mercantile atmosphere; he left to Yale College his library of 880 volumes as well as his Rhode Island farm—hoping, it is said, that the college would turn Episcopalian. He helped to establish a cultural tradition in Newport, and during his stay Abraham Redwood established a philosophical society, one of the first in the colonies. Later Redwood gave £500 to purchase books in London, to be housed in a building completed in 1750, which is still known as the Redwood Library. In 1726, James Franklin, brother of Benjamin, came to Newport to print a newspaper and an almanac. All this made Newport in the later colonial period, during its "golden age," a center of culture in the Southern style, where men and women entertained each other gaily with scarlet coats and brocade, lace ruffles and powdered hair, high-heeled shoes and gold buckles, delicate fans and jeweled swords, all against the grim background of Negro trade and Negro slavery. It did not produce any scientific or technical work of outstanding value except perhaps Abraham Redwood's botanical garden, but it did create an atmosphere in which respect for science could grow. We shall meet the Newport names of Waterhouse, Gibbs and Channing in American science of a later period.

Boston's position as a colonial center of scientific interests can be better understood if it is compared to Philadelphia, the most advanced colonial town, considerably ahead of Boston in population, in wealth, in fashion, in learning and in science. The Quaker city made a deep impression upon visiting Bostonians, who, as John Adams testified, felt thoroughly provincial. Less dependent than the New England seaport towns on navigation alone, it had stronger ties with the agricultural hinterland, where, to the horror of the British manufacturers, industrial establishments were springing up with disconcerting rapidity. There was in Philadelphia less provincialism, a greater amount of religious tolerance and a closer cultural connection not only with England, but with freethinking France, than existed in New England. All these factors account for the great variety of learned minds concentrated in the Quaker City. There Benjamin Franklin, son of a Boston chandler and a Nantucket girl, was the center of the intellectual and political life of the city and the best scientific thinker of the colonies. There were, furthermore, David Rittenhouse, the astronomer, Benjamin Rush and Thomas Bond, the physicians, and many other gentlemen interested in the natural sciences. Philadelphia was the seat of the American Philosophical Society, founded in 1743, and the only scientific society in the colonies except for the little group at Newport and some other small circles of friends of science. It also had a hospital and a medical college, the first in the British Colonies, where Benjamin Rush taught chemistry. Near by on the Schuylkill River was a botanical garden, established by John Bartram, the greatest natural botanist in the world, as Linnaeus called him. To Philadelphia came industrialists and politicians, visiting scholars and roaming travelers. Lying between New England and New York to the north, and Baltimore and Virginia to the south, it was the natural meeting ground for all the varied elements which composed the colonies.

Compared to Philadelphia and the South, New England

in the eighteenth century appeared somewhat left behind in cultural achievement, especially in the field of natural history, after a good start by Paul Dudley and Cotton Mather. That charming "natural history circle," as it has been called, which embraced European and American students of plants and of nature in general in pleasant correspondence and exchange of specimen for herbaria and gardens, had few representatives in New England. Linnaeus's correspondents lived in Philadelphia or in the South. Scientific travelers like the Swede Peter Kalm avoided New England, and went from New York to Canada via Lake Champlain because of the horrible wilderness in northern New England. Most of the colonial botanists and chemists of some fame worked in Philadelphia or in the South. Jared Eliot, minister at Killingworth, Connecticut, was fairly isolated. New England had no man with the vision of Thomas Jefferson, the genius of Benjamin Franklin or the versatility of Benjamin Rush or David Rittenhouse.

11

All this notwithstanding, New England had its own scientific merits, though in a restricted field. Almost all scientific activity in colonial New England centered in Boston and some other coastal towns, and was the privilege of the merchant class and its allies in the liberal professions. The impetus came not primarily from scientific curiosity, nor did it come from a general thirst for knowledge. The approach to science of these colonial Americans was more direct and practical, more Baconian. The pursuit of science needs stimulation from technology, manufacture, agriculture, navigation or some other vital social activity. It is the result of the search for a better understanding of nature to promote the welfare of a group, a class, or of mankind in general. Under colonial conditions it could only grow when a certain reservoir of wealth and leisure had been established, either by direct accumulation of private fortunes or through endowment by a public or semiprivate body of citizens. The struggle for mastery of a new continent and mastery of the seas would naturally lead to scientific interest in the problems of agriculture and of navigation, with occasional study in the scientific approach to manufactures.

The attention of the merchant class was directed, in the first place, towards problems related to navigation. Sailing the seas was a risky enterprise, and required knowledge of the position and course of the heavenly bodies, of weather conditions and of methods of computation. Scientific interest thus concentrated on astronomy, and the related fields of mathematics, surveying and meteorology. Many a New England minister, lawyer, physician or teacher could compute an eclipse, work with a sextant, or even find a longitude. Some of the best early American weather records came from eighteenth-century New England coastal towns and notably from Salem. These colonial scientific observers also paid attention to the flora and fauna of their country, with an eye on the improvement of crops, the fur trade and the fisheries. There was considerable interest in medicine, stimulated by the needs of an expanding town population easily subjected to contagious diseases brought from overseas. Beyond the coastal towns there was little more than traditional agriculture, often on a low technical level and in a pioneering stage. This condition was not conducive to scientific work. The important work in scientific agriculture done in England during the eighteenth century remained for a long time unknown in New England except in a few select circles.

Interest in science came naturally to a people who had brought with them from their English homeland a general respect for learning. New England was the country of the town meeting and the school-house. Almost everybody could read and write and carry on the home political tasks. Elementary education was widely distributed and formed a foundation for more advanced studies, permitting money and leisure. During the eighteenth century some academies or seminaries were founded, such as the Dummer Academy at Newbury, of 1762; the Windham Academy in New Hampshire, of 1768, and the free school in the Berkshires founded by Colonel Williams and soon called after him. There were the colleges at Cambridge and at New Haven, both typical provincial institutions, but with some eminent talent on the staff. These schools were later joined by Dartmouth College, at first a school for Indian children at Lebanon, Connecticut, and after 1770 in the wilderness of Hanover, New Hampshire.

Rhode Island College, established in Warren, Rhode Island, and removed to Providence, later took the name of Brown University.

The first rank in the hierarchy of knowledge was, of course, reserved for theology. The "College in Cambridge" was founded in 1636, only a few years after the first pioneers had established themselves on the peninsula of Boston, at a time when the surrounding country was barely emerging from the primeval wilderness. Originally the main task of the college was the education of ministers, since the "ends of coming into these western paths were . . . to establish the Lord Jesus on his Kingly Throne, as much as in us lies, here in his churches," as one of the early Puritan statements declares. The attempts of the New Haven colony to found a school were long frustrated, but in 1701 resulted in what was soon to call itself Yale College, which was similarly directed towards the education of ministers as a primary object. Dartmouth College grew out of a mission school for Indians. College presidents, and the early principal teachers, were all clergymen, and the Calvinist church kept a strict eye on the colleges. Preoccupation with theology, however, never excluded the study of more worldly forms of science and industries. The broad possibilities of the new homeland gave new impetus and even a new meaning to this traditional interest. It was not without vision that the first university town in the colonies was named after Cambridge, where later in the seventeenth century the towering figure of Isaac Newton was destined to dominate British science for more than a century. There was always a good deal of Baconian philosophy in the outlook of the New England men of influence. And so it happened that from earliest colonial days through the whole period of British control, both within and without the colleges there were a number of able men of science, interested in the practical as well as the theoretical arts of colonial life. Almost all of these men were amateurs, conducting their study and research during hours of leisure. John Winthrop, Jr., son of the first governor of Massachusetts Bay Colony, and himself governor of Connecticut, was the founder of the Saugus ironworks, a chemist, a metallurgist, a physician and a collector of scientific books. At least two other

Winthrops entered the records of colonial science. One of them taught mathematics, astronomy and electricity at Harvard College for many years till his death in 1779. The Reverend Cotton Mather, of witchcraft fame, was not only one of the first scholars in the colonies to catch the spirit of Newton's work—his "Christian Philosopher" of 1720 is some kind of general survey of the sciences of his day—but was also a keen observer of the flora and fauna of New England, an early student of the hybridization of plants, notably of corn and squash, and enough interested in medicine to suggest the introduction of variolous inoculation against smallpox as early as 1721. Among the men connected with Yale College there were also some notable scientists. The most illustrious was Ezra Stiles, first a clergyman in fashionable Newport and after 1777 president of Yale, who found time to experiment with silkworms and electricity, and was a linguist of some distinction. In his Newport days he was for some time the Redwood librarian. All along the coast there were physicians, clergymen, lawyers, merchants and sea captains studying mathematics, observing eclipses, solving nautical problems or compiling meteorological data. They influenced the course of scientific studies at Yale and Harvard, where the scientific diet consisted largely of Newtonian mathematics and astronomy. The study of herbs, fruit trees and animals remained pretty well outside the colleges, and was indulged in by a few gentlemen interested in their farms and gardens.

None of the colleges really prepared for a scientific career, though a young man might pick up a good deal of useful information at the feet of Ezra Stiles at Yale, or, even better, of John Winthrop at Harvard. After graduation he might go to his home town or to some other place as a clergyman, a lawyer or a physician, and still keep his love of science, usually of mathematics or of astronomy. Such a man was Judge Andrew Oliver of Salem, whose *Essay on Comets,* published in 1772, was translated into French. Joseph Willard, a clergyman in Beverly, and Samuel Williams, a pastor in Bradford, became teachers of mathematics at Harvard. Even when going back to husbandry, a college graduate might well continue his study of the stars or of his book on "conick sections," as did Deacon Elizur Wright of South Canaan, Connecticut, Yale 1781, who imparted his passion for "fluxions" to his son, the future commissioner of insurance in Massachusetts.

12

The motherland recognized the scientific merits of the gentlemen of New England, as we can learn from the records of the Royal Society, which had several members from the New England colonies. This society, the most august center of learning in the English-speaking world, was incorporated by King Charles II in 1662, though its roots go back to the period of the English Revolution and the strong intellectual life it fanned. The society's first charter stated that it was Charles's purpose "to promote the welfare of the arts and sciences." Bacon's spirit ruled the early members; his bust is reproduced on the frontispiece of the earliest history of the Society, published in 1667. The most influential of the original members was the chemist Robert Boyle, one of the most faithful Baconians among the scientists. Other early members were the astronomer Halley and the physicist Hooke. Most prominent was Isaac Newton, whose spirit soon rose to dominate the history of the Society for at least a century. The *Transactions* published the most important contributions of the experimental sciences in Great Britain.

This Baconianism and Newtonianism greatly influenced the scientific gentlemen of New England. No less than three colonial Winthrops, all named John, were members of the Royal Society. The first of the three were the governor of Connecticut, a charter member of the Royal Society, who became, after his return to the colonies, its western correspondent. The second John became a member in 1734 and was a collector of minerals, fossils and stones. The fortieth volume of the *Transactions* of the Royal Society was dedicated to him. In its preface we find the interesting statement that the Royal Society might have been established in Connecticut

> had not the Civil Wars happily ended as they did, Mr. Boyle & Dr. Wilkins with several other learned men, would have left England, and, out of esteem for the most

> excellent and valuable Governor, John Winthrop the Younger, would have retired to his new-born Colony, and there have established that Society for Promoting Natural Knowledge, which these gentlemen had formed, as it were, in embryo among themselves; but which afterwards receiving the protection of King Charles II obtained the style of Royal. . . .

The third John Winthrop, who became a member, was the Harvard professor of astronomy, elected to membership in 1766. We shall meet him again as the most distinguished of New England colonial scientists. Other members of the Society were the Reverend William Brattle, of a Boston merchant family, John Leverett, a Harvard president, Cotton Mather, Thomas Robie, a Harvard tutor interested in astronomy, and Chief Justice Paul Dudley. In 1680 Thomas Brattle, brother of William, contributed comet observations, which Newton used in his *Principia.* Dudley's papers to the *Transactions* are typical examples of the way in which colonial scientists obtained their problems from practice. They deal with the making of sugar from maple trees and molasses from the apple tree, with vegetables, plants and animals of commercial importance, and even include a study of whales. In 1724 he pioneered in describing the spontaneous crossing of different varieties of New England corn. Again we have a glimpse of possibilities which somehow never materialized under colonial conditions, in this case the scientific side of the agricultural revolution, so well developed in contemporary England.

13

The teaching of mathematics, astronomy and surveying came to Harvard officially in 1728, when Isaac Greenwood became the first incumbent of the chair of mathematics and natural philosophy established a year before by Thomas Hollis, a Baptist London merchant. Greenwood was fired for intemperance, leaving some writings, including an English arithmetic, the first written by a British American. His grandson became an instrument maker and a dentist, whom we shall meet again. John Winthrop succeeded him in the Hollis chair, and taught the natural sciences in a way at least comparable to the very best that was offered in England or Scotland. His career in the Hollis chair, which lasted from 1738 till his death in 1779, illustrates the possibilities and the limitations of a New England professional scientist of the period. His tasks, like those of his predecessor, had been broad, but well defined. They included instruction in

> Natural Philosophy, and a course of Experimental in which to be comprehended, Pneumaticks, Hydrostatics, Mechanicks, Staticks, Opticks, etc. in the Elements of Geometry together with the doctrine of Proportions, the Principles of Algebra, Conic Sections, plain and Spherical Trigonometry with the general principles of Mensuration, Plain and Solids, in the Principles of Astronomy, and Geometry, viz. the Doctrine of the Spheres, the use of the Globe, the motions of the Heavenly Bodies according to the different Hypotheses of Ptolemy, Tycho Brahe and Copernicus with the general Principles of Dialling, the division of the world into various kingdoms, with the use of the maps, and sea charts; and the arts of Navigation and Surveying.

This is a good tableau of eighteenth-century exact science, especially of science in Newton's tradition. Winthrop has been reported as the first to teach Newton's fluxions in America. Interest in theory being small, most of Winthrop's work was practical and experimental. During his incumbency the scientific collections at Harvard increased considerably.

Thomas Hollis presented the college, with many books and "philosophical apparatus", and this magnificence was continued by his nephew and namesake. The collection grew, but was destroyed by fire in 1764, and then again replaced and even improved through donations by a third Thomas Hollis, Benjamin Franklin, John Hancock, Edward Holyoke and other gentlemen of scientific taste. By the time of the Revolution Harvard had one of the best collections in the New World, including telescopes, compasses, microscopes, thermometers, barometers, and electrical apparatus.

Winthrop's first preserved work is on sunspots, which he watched in 1739 on Boston Common. Later papers deal with transits of Mercury and Venus, eclipses and meteorological observations, many of which were published in the *Philosophical Transactions* of the Royal Society, to which his own merits and the interest of his friend Franklin gave him access.

This was the period during which European scientists were experimenting with static electricity. Glass was rubbed with silk and ebonite with fur, and the two different types of charges studied. The difference was found between conductors and insulators, and charges of considerable strength were produced by electrical machines and stored in condensers. The Leyden jar was constructed in 1745 and could produce powerful sparks. These discoveries had good entertainment value and American audiences flocked to see performances with electrical instruments.

A Dr. Spencer came from Scotland to Boston to perform an early barnstorming trip of electrical experiments. Benjamin Franklin witnessed one of his performances during the summer of 1746. He became so deeply interested that, upon his return to Philadelphia, he purchased "philosophical apparatus" and began his own investigations. These led to his theory of positive and negative electricity as well as to the kite experiment and the invention of the lightning rod—one of the first practical applications of the new theory. John Winthrop, who delivered lectures on electricity as early as 1746, followed Franklin's experiments with interest. In 1758 Franklin sent him electrical apparatus from London. We may safely assume that he explained the kite experiment, performed in 1752 in Philadelphia, to his Harvard audience, shocking part of the orthodox clergy by explaining away the supernatural character of lightning as a token of divine wrath.

To Winthrop's lectures, one summer day in 1771, walked Loammi Baldwin and Benjamin Thompson, lads from North Woburn, eight miles away from Harvard Hall. Walking "by shady roads and green fields, and easy hills and pleasant ponds," they seem to have profited by the instruction. Thompson, that warped genius who became Count Rumford of the Holy Roman Empire and an outstanding physicist, later described Winthrop as an excellent and happy teacher, and we still have an entry in his memorandum book from the same period, in which he gives "an account of what Work I have done towards Getting an Electrical Machine." Instrument makers were rare in those days. We shall meet both Rumford and Baldwin again, the last one as a revolutionary soldier and an early amateur in civil engineering. He remained interested in electricity, and later described how he repeated in Woburn Franklin's kite experiment, not without danger to his person.

In 1740 Winthrop observed a transit of Mercury and a lunar eclipse, which he reported to the Royal Society. Thirteen more papers followed. From 1742 to 1763 he recorded meteorological observations in Cambridge. The Lisbon earthquake of 1755, as well as the return of Halley's comet in 1759—that brilliant verification of Newton's theory of the solar system—gave him a chance to explain in some excellent public lectures the natural origin of phenomena which were generally feared as awesome tokens of the Lord's dissatisfaction with man. He observed in 1761 the important transit of Venus across the sun's disk in Newfoundland, where he was sorely tried by insects. In 1766 he became a Fellow of the Royal Society. In this, as well as in the honorary degree which the University of Edinburgh bestowed upon him in 1771, we may see the guiding hand of his friend Benjamin Franklin. Oxford refused a degree, rejecting Winthrop as a dissenter.

John Winthrop's scientific activities show clearly that colonial science was not exclusively utilitarian. Science, seriously undertaken, can never be purely practical, not only because practical questions need theoretical understanding for their solution, but also because the scientific mind simply does not function in such an exclusively practical way. Once the inquiring mind is thoroughly aroused, it does not stop at the solution of utilitarian questions, but begins to ask for pure knowledge. The reward lies in the discovery of a deeper truth than appears at first, and in eventual control of more powerful forces of nature. Franklin's study of electricity led as an immediate result to improved fire protection, but his studies, as well as those of Winthrop, Stiles, Rittenhouse and so many others, led eventually to the mastery of an entirely new force of nature.

Yale, in the eighteenth century, followed in its scientific instruction a pattern similar to that of Harvard. There the first systematic teaching of natural philosophy and astronomy seems to have been due to Thomas Clap, who from 1739 to 1766 was president of the college, and who wrote a pamphlet on meteors. Later came the talented Ezra Stiles, a lawyer, fashionable clergyman, linguist and experimenter with silkworms, the culture of which was encouraged by the British Crown. In addition to his predilection in astronomy and mathematics, he also experimented with electrical apparatus and corresponded with Franklin. From 1778 till his death in 1795 he was president of Yale, where he did much to improve the taste for science. However, his "latitudinarian" leanings, derived from his Newtonian outlook on the world, caused frowns to appear on many a Congregationalist brow in orthodox Connecticut. Neither he nor Clap was a professional scientist; both were clergymen of broad information and tastes. As such they were more typical of the colonial intelligentsia than Winthrop, who was one of the very few professional scientists of the period.

The transit of Venus was an interesting example of the amount of scientific organization possible in the American colonies. Transits of Venus are a rare occurrence, happening when the planet passes between the sun and the earth. They always occur in pairs, the two in a pair being eight years apart, the next pair not coming until more than a century later, and are useful to find the sun's parallax. The 1761 transit was poorly observed everywhere, but for the 1769 transit considerable preparations were taken. It was central on the Western Hemisphere. On the advice of the Royal Astronomer Maskelyne a number of observatories were established in the colonies, of which four were in New England, well equipped with timepieces and telescopes. Among the others were some as far apart as lower California, Philadelphia and Hudson's Bay. The observations as a whole were a success; Rittenhouse found by means of them a parallax of

8.805 seconds, a value which even modern calculations have hardly refined. Ezra Stiles, John Winthrop, the Rev. Samuel Williams, and Benjamin West of Providence participated.

14

The scientific interest displayed by Stiles, Winthrop and many other gentlemen of New England showed the existing possibilities for study and even occasional research. The other side of the medal, however, should not be omitted. All this work was pure observation, repeating of experiments, reading and some speculation, mostly on a rather low level when compared with first-class work in Europe. Colonial conditions account amply for this state of affairs. Too many energies of the people went into the endless pioneering struggle for the very essentials of food, clothing and shelter. The few existing colleges were not wealthy, their curriculum was not directed towards a career of science or invention. The schools of lower standing gave even less scientific training. The mercantile system frustrated many channels of possible technological and scientific endeavor. And even if the struggling scientist was able to overcome these handicaps and absorb the culture of the mother country, he found himself working in the scientific atmosphere of Georgian England, which was inferior in many respects to that of the Continent. During the major part of the eighteenth century England's cultural development was very lopsided, and though literature and some of the arts flourished, science was in a state of near stagnation. Routine and lethargy reigned at Oxford and even at Cambridge, which, though a little less conservative and clerical than Oxford, rested contented in a slavish cult of Newton. Teaching was equally poor. Fox-hunting squires and port-drinking clergymen had more influence on the colleges than was desirable for their healthy growth.

Even so, science in England had its devotees, and its group of brilliant pioneers, who blazed new trails in botany, chemistry and medicine. Among them were John and William Hunter, anatomist and surgeon, Stephen Hales, Joseph Black and Joseph Priestley, chemists. Several of them were Scotchmen and nonconformists, and in their general outlook on life closer to the colonial scientists than to the Church of England of their own country. They had considerable influence on American science, but mainly after the Revolution. Yet, despite all barriers, some influence of continental science upon colonial methods of thought was in evidence, partly through direct study of continental publications and partly through the medium of English scientists. Authors like Boerhaave and Musschenbroek in Holland, Stahl and Haller in Germany, and Linnaeus in Sweden were studied in the colonies. But the enormous development of continental mathematics and its influence on astronomy and mechanics, due to Leibniz, the Bernoullis, Euler and Lagrange and their followers, remained almost entirely unknown in the colonies, mainly because of the pigheadedness of English mathematicians, who remained contented with the study of Newton.

We get a glimpse of the background of a colonial scientist in studying the titles of those books in John Winthrop's library which have been preserved. It was large for its time and included more than a hundred volumes on the exact sciences. But only a few books deal with the new scientific works of French and other continental scientists; leading authors, the Bernoullis, Euler, Maupertuis, d'Alembert, Clairaut, are absent. Though some copies of the publications of the French Academy came to the colonies, we have the impression that Winthrop's information was almost exclusively based on classical and English sources.

The results can be seen in his work and in that of less fortunate scientists, not only of his generation, but even of a later date. They concentrated almost exclusively on routine observations, watching eclipses or transits, sunspots or the weather, experimenting with a Leyden jar or an electrical machine, or at most computing a lunar position or an eclipse from conventional prescriptions. When a genius like Franklin broke the spell, he remained an isolated phenomenon with nobody in the country to continue his work where he left it. Many years later it was said that if Lagrange, the great French mathematician, had ever come from France to America, he would have been able to make a living only as a surveyor.

All these factors contributed to the rather primitive condition of colonial science, despite the ardent work done by many amateurs and some professionals, and despite the relative prominence of colonial science in the whole framework of British science. This backwardness of colonial science was occasionally taken in continental Europe as a means to depreciate the intellectual possibilities of the colonies in general, as can be seen in the controversy between Jefferson and the Abbé Guillaume Raynal. France's admiration for Franklin counteracted this tendency to a certain extent.

The almost exclusive study of English courses was due not only to the traditional ties with the motherland; it was also a result of real economic interests. New Englanders were, after all, naturally interested in navigation and astronomy, and the path to these sciences went through Newton's *Principia.* This period in American science has justly been called "the Newtonian period," and New England science was even more Newtonian than that of Philadelphia or the South.

15

Interest in Newtonian science was not based exclusively on practice. The new mechanics was the first example of a modern science in its full development, equipped with a convincing set of axioms, a logical method, a developed technique and the power of forecasting events. Even those who could not master the details could admire and follow its general approach. This admiration, amounting often to awe, of Newton's theory of gravitation and its applications to the solar system affected the theological and religious outlook on life of many persons of culture, even in Calvinistic and theocratic New England.

These men sought confirmation of their religious views not so much from revelation or from strict acceptance of the Bible on faith as in the teachings of the deists, who stressed the rational basis of religion. It was possible to understand the truth of the biblical teachings by reason, and Newton's theory was a powerful means to this end. The universe, as Newton had explained it, was governed by laws, and the result of laws; the only task of the Deity was to provide the First Cause, after which natural law could take its course. The whole rationalist movement stressed natural religion, to be learned from the study of science and the dictates of conscience.

This eighteenth-century rationalism came to New England mainly through English writers. Locke, Shaftesbury, Bolingbroke and Hume were read. Locke and Hume were studied also for their teachings on government. This philosophy was cast into sonorous poetry by Alexander Pope, whose claim that "Nature and Nature's laws lay hid in night: God said, Let Newton be! and all was light" may have prepared many a New England reader for the consequences of modern rationalist philosophy. Since many British scientists were nonconformist, they were spiritual brethren of the men who settled New England, and searchers for harmony between the tenets of their creed and the results of science. Their influence was felt on those American gentlemen who indulged in the study of nature and in mathematics. The influence of the French philosophers on New England thought was relatively small, since many a good colonial Yankee was duly horrified by materialistic France, its radical philosophy and its colonial aspirations in Canada and the West.

The strict traditional Calvinism of the New England clergy and countryside was only a partial antidote to the influx of Cartesian rationalism, of Deism and of skepticism in its milder English form. The more tolerant doctrines found considerable support among the respectable merchant families of New England. Newport, the most extravagant and least puritanical of the mercantile towns, was a veritable center of Deism. From Newport came the library bequeathed to Yale by Bishop Berkeley. It contained many books written by religious dissenters and deistic philosophers, which had their influence on the Puritan mind of the younger generation. Although the merchant families held the purse strings of the colleges, they did not interfere with the unorthodox teachings of a Stiles at Yale or a Winthrop at Harvard. Many shared John Adam's horror of "the frightful engines of ecclesiastical councils." The clergy itself was split in its attitude towards science; many a minister encouraged the gradual extension of scientific teaching and research, even where it seemed to establish a conflict with the ancient interpretation of the Scriptures. Still, many believers frowned at the audacities of astronomy, and took the attitude that any prying into the mysteries of nature was fraught with danger to the soul. They saw in the colleges places where wordly learning might destroy the spirit. This attitude was to a large extent a reflection of the distrust the country folk felt toward the weathy and easygoing residents of the town. Till deep in the nineteenth century there were families who objected for that reason to their children going to college. It deprived Eli Whitney of a Yale education in his early youth. Even as late as 1860, Deacon Morse of Portland, the father of the Harvard conchologist Edward Sylvester Morse, objected on religious grounds to his son's collecting of shells and studying them under the microscope.

The picture had also its other side. Cotton Mather, Jonathan Edwards, Thomas Clap, Ezra Stiles and other scientifically inclined men were clergymen and great admirers of the new astronomy. "Newton," wrote Cotton Mather to the Royal Society, "is the perpetual dictator of the learned world." We have already mentioned how John Winthrop's lectures on comets and earthquakes gave comfort as well as discomfort to religious men and women. The opposition to his interpretation of earthquakes in 1775 was headed by Thomas Prince, pastor of the Brattle Street Church in Boston. Prince ascribed the Lisbon earthquake of that year to man's rebellion against God's laws as exemplified in Franklin's invention of the lightning rod.[1] Other clergymen, however, supported Winthrop's interpretation.

These lectures of Professor Winthrop indicated the limit to which opponents of the old-fashioned orthodoxy in academic circles were willing to go. French science, on the contrary, was part and parcel of French anticlerical philosophy, and therefore could find small favor in the eyes of New England respectability. Jefferson, with his admiration for France, grew

[1] Prince said after the earthquake of 1755: "In Boston there are more iron points invented by the sagacious Mr. Franklin than anywhere else in New England, and Boston seems to be more dreadfully shaken. Oh! there is no getting out of the mighty hand of God!" Winthrop answered: "It is as much our duty to secure ourselves against the effects of lightning as against those of rain, snow, and wind by the means God has put into our hands."

up in Virginia, not in New England. Winthrop himself was never a deist, but according to Ezra Stiles, remained "a Firm Friend to Revelation." Typical of New England was also John Adams, who did read Montesquieu and other French philosophers, but read Rousseau and Condorcet with scorn and indignation. We can get an idea of his position by taking his copy of Condorcet's *Outline of the Progress of the Human Mind,* which is preserved in the Boston Public Library, and reading his marginal protests against the author's "nonsensical Notions of Liberty, Equality and Fraternity." John Adams was a fair example of New England's ruling class, not only as it was during his lifetime, but also during later generations.

16

In the seventeenth century, the sick and ailing were usually comforted by the minister, who knew how to help the body as well as the soul. This was, in the words of Cotton Mather, an "angelical conjunction," for "ever since the days of Luke the Evangelist, Skill in Physick has been frequently professed and practised by Persons whose more declared Business was the Study of Divinity." Some regular graduates of medicine at English colleges practiced in New England, but most of the early physicians were professional men of other fields with some "Skill in Physick and Pharmacy." Medicine as an independent profession began to develop in the eighteenth century. One of the first regularly graduated physicians in Boston was Dr. William Douglass, who came in 1718 with a Paris and Leyden education, and practiced medicine in Boston till his death in 1752. As a rule the colonial physician was not academically trained, nor were there general hospitals or medical schools in the country where he could obtain such training. Shortly before the Revolution such institutes were opened in Philadelphia and in New York, but there were none in New England. Unless he could go abroad to study, usually to Edinburgh or London, a young man had to become apprentice to a practicing physician, if he liked to study medicine. One of the few Boston physicians of late colonial days with an M.D. from abroad was John Jeffries, whom we shall meet as a Tory and an early student of aeronautics. The majority of colonial physicians had to be satisfied with a local training. They began by doing all sorts of chores around the house of an older doctor, and prepared his medicines, pounded his drugs and accompanied him in his rounds to the patients, learning while assisting. This method might not provide the best scientific training, but it did turn out a group of general practitioners of considerable professional skill, who were often outstanding members of the community. One of the most famous native-bred physicians was the Salem practitioner Edward Augustus Holyoke, a Harvard graduate and son of a Harvard president, who began his practice in 1749. This noble citizen and doctor continued his daily rounds for seventy-nine years, never traveling, it is said, more than fifty miles beyond the city limits. He was the father of twelve children and lived to give a toast at the public dinner in honor of his hundredth birthday, in 1828. In his long career he set his stamp on much of Salem's public and intellectual life of the period.

The physician was likely to be the rationalist par excellence. Nobody had a better opportunity to watch the determination of the soul by the body, the dependence of mind on matter, or the constitution of the human body as a complicated piece of mechanism. In Europe rationalistic, and often extremely mechanistic, philosophy was promoted by physicians or by men to whom medicine was familiar. Lamettrie, the daring author of *L'homme machine,* was a physician. In England, and even more so in New England, the physician seldom was an open materialist, but his reasoning might be quite offensive to the old-time Calvinist. A classical example of the conflict between progressive medical thought and the old-time prejudices occurred in Boston during the fight for variolous inoculation against smallpox, in which, despite all colonial handicaps, the small provincial town actually placed itself in a leading medical position. Curiously enough, progress in this fight was represented by an orthodox clergyman, the formidable Cotton Mather. The medical profession was divided, the most consistent defender of the conservative point of view, however, being one of the ablest physicians in town, William Douglass. In the *Transactions* of the Royal Society of 1714 and 1716 Mather had read an account of inoculations in Constantinople. In his own home he saw an example in his slave, Onesimus, who had been inoculated in his native Africa, where inoculation was common practice among some tribes. Mather made the new method known in Boston, and shortly afterwards, in 1721, hardly later than in advanced parts of Europe, Dr. Zabdiel Boylston inoculated his son and two Negro patients. This was followed by the inoculation of seven more patients against the violent opposition of some of his Boston townsmen. A war of pamphlets followed, in which William Douglass excelled in attacking the new practice. For a while Mather's and Boylston's persons were in danger. Eventually, the successes of inoculation made the method more popular and provided considerable relief from the recurrent smallpox epidemics. It can be argued in defense of Douglass that Mather and Boylston's early campaign was based more on enthusiasm than on case study; when the practice showed results Douglas recognized his errors, though with little grace. During the smallpox epidemic of 1764 two inoculation hospitals were opened in Boston Harbor, one at Castle Williams, the other at Point Shirley, the first hospitals of New England. Feeling against inoculation, however, remained so strong that a hospital opened near Marblehead on Cat Island in 1773 was burned by a mob a few months after its opening. Four men suspected of carrying the infection were tarred and feathered, and drummed out of town. "The exquisitely droll and grotesque appearance of the four tarred and feathered Objects of Derision exhibited a very laughable and truly comic Scene," wrote a Salem newspaper. There was more trouble in Marblehead and it seemed one moment that a march of armed Salemites (pro-inoculation) against Marblehead (anti-inoculation) was in the making. In some fashionable circles the reaction was different and some inoculation hospitals became for a while the center of gay social activity. Daniel Lee's establishment at Ram Island on the Connecticut coast was established in 1788; it became a popular resort, inviting ladies and gentlemen who were afraid of the smallpox to have the disease in a safe and mild form, while at the same time enjoying the best of fishing, hunting and boating.

This inoculation was not the cowpox vaccination, which came into use after 1799. The eighteenth-century method attempted to prevent smallpox by the use of a serum from human patients who carried the infection in a mild degree. Despite primitive and often careless treatment it had considerable success in reducing the number of victims of this dreaded disease, which took periodically hundreds of lives. A description of the practice and its results was given in the *Transactions* of the Royal Society of 1766 by Benjamin Gale, successor to Jared Eliot in Killingworth. This is one of the first papers written in America in which a statistical study is made. It contains, in particular, a record of the population and mortality rates of Boston during epidemic years.

Another field in which physicians had to cross arms with popular prejudice, sometimes fanned by orthodox conservatism, was the dissection of human bodies for the study of anatomy. It was hard for physicians to get bodies, and even harder to get permission to dissect them. Public feeling ran so strong that in 1788 serious rioting occurred in New York when a mob detected evidence of dissection. At about this same time, Dr. Jeffries of Boston began to give a course of lectures on anatomy. A mob broke into his anatomical room on the evening of the second lecture and bore away the object, the body of an executed felon presented to the doctor by the governor. The course was never resumed. In Massachusetts dissection was not legalized till 1831; England followed in 1832. In many states of the Union dissection was not legalized till well into Civil War days. "Body snatching" was not uncommon, and led to unsavory incidents. However, unorthodox views concerning dissection crept in with the advance of anatomy and surgery. Some Harvard students of the classes of 1770-1772 formed their own anatomical society which met for secret dissection, if only of dogs and cats. To this society belonged John Warren, the brother of the physician who fell at Bunker Hill and the founder of the Warren family of Massachusetts physicians. The idea of such a society was kept alive, and during the Revolution it served as an inspiration to the founders of the Massachusetts Medical Society.

17

Thus a visitor to New England in late colonial days, meeting all classes of the population, would have found evidence of a considerable amount of interest in science, of scientific activity, and of technological skill and inventiveness. In the coastal towns he could have met with professors, physicians, lawyers and ministers who were following the general trend of eighteenth-century science in England, although as a rule on a slightly lower level. They were closely related to the merchant groups, among whom there were patrons of the arts and sciences, with occasionally a man who dabbled in science himself. Scientific interest was stimulated mainly by its usefulness in navigation, geography, mechanical invention, agriculture or public health, and less in the continental way as a means of rational understanding of the universe, although Cotton Mather had tried to use it as a means to strengthen a weakening orthodoxy. Attempts at organizing scientific work in the form of academies and societies, or in the form of hospitals, botanical gardens and expeditions were only in their initial stage. If our visitor went into the homes of less exalted sections of the population, penetrated into the smaller towns of the hinterland or into the countryside, he might occasionally have discovered a farmer or mechanic interested in astronomy and mathematics, but far more often he would be found spending his spare time technically improving some mechanism or craft. At the time that Winthrop observed the planets and Warren dissected dogs there were farmhouses where young Thompson was tinkering with electrical apparatus, Eli Whitney with nails and John Fitch with clocks. During and after the Revolution such men began to grasp the new opportunity and to utilize their inventions. Usually they were unable to share the sheltered existences of the professional men, but had to taste the ups and downs of the inventor in the struggle for recognition. The professional men also began, at a slower pace, to use the wider opportunities for scientific research presented by the disappearance of the British colonial masters. This period of expansion begins with the first shots fired at Lexington and Concord in the spring of 1775.

PART II

THE FEDERALIST PERIOD

Chapter 2

The Revolutionary Years

> All the arts may be transplanted from *Europe* and *Asia,* and flourish in *America* with an augmented lustre . . . the rough sonorous diction of the English language may here take its Athenian polish, and receive its Attick urbanity; as it will probably become the vernacular tongue of more numerous millions, than ever spake one language on earth.
>
> —EZRA STILES, 1783

1

THE AMERICAN REVOLUTION was a vast popular movement. Its success was due to the efforts and sacrifices of large masses of the colonial population. Reaching into every town and every homestead, it demanded its share from rich and from poor, from merchant, mechanic and farmer. It changed customs, it uprooted traditions, it installed into the victorious rebels an exalted feeling of patriotism. In Lenin's words, it was "one of those great, really liberating, really revolutionary wars of which there have been so few among the large numbers of wars of conquest."

With the disappearance of the British masters, their repressive taxation policy and their antiquated navigation and mercantilist laws, an unlimited amount of energy was set free. It could be turned into the channels of commerce, of manufacture, of technology and invention. New immediate possibilities and even wider opportunities for the future were opened. There were difficulties, but there was self-reliance and self-confidence to overcome them. They were needed, since the difficulties were manifold.

Before the conflict there were hardly any manufactures of finished products and few mechanics with the understanding of those machines that were already used to a large extent in England. Engineers were as yet unknown and though there were scientists, most of them were amateurs. There were no scientific organizations in New England, and few in America. Even in navigation, a field in which New England was outstanding, important shortcomings in training of personnel and in material existed, especially those necessary for warfare. The colonies had no navy to begin with. The manufacture of arms was primitive, as was the production of gunpowder and of other tools of war. There were no adequate means of taking care of the wounded. Agriculture was equally primitive and transportation was poor, except here and there on inland water roads or along the coasts. The American rebels, as the true progressives of their period, were actually able to overcome many of these difficulties in the course of the war, but often only in a tentative way and at the cost of endless sacrifices.

2

It was typical of the revolutionary leadership that so many ranking rebels either were interested in science and technology or were even themselves productive in some of these fields. Franklin was a scientist of world fame, a theoretician, experimenter and inventor, who, as Turgot declared, wrested the fire from heaven and the scepter from the tyrants:

eripuit coelo fulmen sceptrumque tyrannis.

He was in this respect the greatest of all, but he was no exception. Jefferson was a writer on many fields of science and technology and an able craftsman, Tom Paine an engineer of great vision, Benjamin Rush a physician and a chemist, Washington a surveyor and agronomist. In New England James Bowdoin, John Hancock and John Adams were patrons of learning and invention—Bowdoin was even the proud possessor of his own theory of light; Nathaniel Greene was a manufacturer and inventor, Josiah Bartlett a medical authority and Manasseh Cutler an astronomer and botanist. Under Tory Governor Hutchinson, Harvard College had the reputation of being heart and soul for the cause of the colonies against the arrogance of the Crown, especially because of John Winthrop's outspoken patriotism. Franklin had, among his correspondents, Winthrop, Bowdoin and Samuel Cooper, who were all Harvard men and with whom he exchanged opinions on scientific subjects as well as on more subversive topics. These leaders believed that to the rights of man belonged the pursuit of happiness, and that this end could be effectively promoted by the cultivation of the arts and the sciences.

Shackled for several generations by an antiquated regime, the new life blossomed forth almost immediately after the first liberating blows had been struck. Like the Hollanders of 1575, who founded Leyden University while the Spanish were still overrunning their country, like the later French of 1795 who founded the Ecole Polytechnique when the republic was menaced from all sides, like the Russians of 1920 who began electrification in the midst of foreign invasion, these Americans of 1780 began the organization of science and the establishing of manufactures "for the happiness of mankind" at a time when the enemy was burning their towns and ravaging the countryside. They were confident that they were building a new and better world.

A Woburn man by the name of Samuel Blodget has told a story which reveals that Washington, as early as 1775, thought in terms of a new great national institution of learning. He relates that, in October 1775, camped at Cambridge, he went to General Washington, accompanied by General Greene, to complain of the ruinous state into which the conduct of the quartered militia had brought the college buildings. Expressing the hope that after the war "we should erect a noble national university, at which the youth of all the world might be proud to receive instructions," they received from Washington the reply, "Young man, you are a prophet! inspired to speak what I am confident will one day be realized!" This idea of a national university as a carrier of the ideals of the Revolution remained for many years a program point of American thinkers in science and education. Washington remained faithful to it until his end. Joel Barlow, the poet and democrat from Connecticut, struggled hard to realize it. These men were thinking of the United States as a nation, and as a nation leading the world in enlightenment.

This budding nation, however, was far from homogeneous. The American Revolution was a popular upheaval, but like so many mass movements it was composed of very different elements. Many fashionable circles in the larger centers of population were leaning towards the Tories, and had family relations with the refugees in Halifax and in London. Among them were many families of landed aristocrats and merchants who used to live on Tory Row in Cambridge, now renamed Brattle Street after one of these families. They took wealth and culture with them, which the country could ill afford to lose. The revolutionary classes maintained a kind of unstable truce for the duration of the Revolution; they clashed bitterly as soon as independence was obtained. The backbone of the Revolution was formed by the farmers in alliance with the artisans and mechanics from the towns, the smiths, printers, wheelwrights, carpenters, masons, chandlers and brewers, who went into the army and bled and died on Long Island, near White Plains and at Valley Forge. With them were independent merchants, such as Hancock and Bowdoin in Boston, and Southern landowners such as Washington and Jefferson. That section of the merchant class of New England, which sided with the Revolution, a group of men wealthy, proud, interlocked by trade and marriage, and deeply class-conscious, tried to maintain its control throughout the Revolution and to keep it afterwards. It struggled to have its ideas adopted as the expression of the popular movement, and when this proved too ambitious an undertaking it strove to interpret or to rephrase the popular slogans in accordance with its own interests. Progressive in its struggle against colonial oppression and at least nominally the exponent of the abstract ideals of man's inalienable rights to life, liberty and the pursuit of happiness, the New England merchant class was at the same time conservative in its attitude toward the common people, whose democratic political ambitions it feared and detested. The ideal of the average New England merchant was a "gentleman's revolution," to establish the right of the individual to "quietly enjoy and have the sole disposal of his own property," as James Otis, one of the great spokesmen of his class, explained. We must therefore always be aware of the fact that the leading members of the New England merchant class, while often expressing the truly revolutionary aims of the American people, were bound by their own interests to restrict the interpretation of these ideals, and under some circumstances to dilute and distort them. Jefferson later accused many of them of being monarchists at heart. The same vigor which these gentlemen employed in combating usurpations of the British Crown was, only a few years afterwards, displayed in the struggle against the demands of the impoverished farmers of New England, who under the veteran soldier Daniel Shays sought redress of grievances and the reduction of their debts. After the Revolution all the hopes and fears of these New England merchants were embodied in New England Federalism. It also determined their position on science and technology.

One of the most characteristic of the revolutionary leaders of the New England merchant class was John Adams, himself a farmer's son, but with a Harvard education. For Adams not only economic and political action, but also the arts and sciences were the means to the advancement of his class, in which he saw the realization of greater human welfare and thus the vanguard of civilization. Starting as a young schoolmaster in Worcester, he had found solace from his "school of affliction" by burying himself in books, but not so deeply as to miss the fact that his "little school, like the great world, is made up of kings, politicians, divines, L.D.'s, fops, buffoons, fiddlers, sycophants, fools, coxcombs, chimney sweepers, and every other character drawn in history, or seen in the world." He became better acquainted with the French writers than most of his American contemporaries, though disagreeing thoroughly with their antireligious and democratic optimism. Adam's interests were quite universal. His library contained a considerable number of classical authors on natural science and mathematics, such as Archimedes, Euclid, Newton, Halley, Buffon, Linnaeus, as well as the classics of French enlightenment. There is an interesting parallel between Adams's and Jefferson's interest in science and philosophy, and though both men differed greatly in their conclusions, they agreed in the fundamental usefulness of scientific endeavor as a means of promoting human welfare.

Adams was interested in scientific collections. He was particularly impressed by a collection of birds and insects made by a Mr. Arnold, an innkeeper in Norwalk, which was sold to London, where Adams saw it again. In France, on his mission of friendship during the Ameircan Revolution, he visited the splendid collections there. All this, he wrote in his diaries, "increased my wishes that nature might be examined and studied in my own country, as it was in others." He was often entertained in France with inquiries concerning the Philosophical Society of Philadelphia, Franklin's creation, and this also stimulated Adams's wish to do something toward the organization of scientific work in his native Boston.

Adams's wishes were shared by others. We have already seen how the desire for co-operation was growing in medical circles; and this had been stimulated by the work in army hospitals during the war. Dr. Douglass had a group of physicians oranized as early as 1735, and Cotton Tufts, a practitioner in Weymouth and a friend of Adams, had planned a Medical Society in 1765. His work had received little encouragement and nothing came of it. In the same period the "American Society Held at Philadelphia for Promoting and Propagating Useful Knowledge" was established, but it lasted only a short time. In 1769 it merged with the American Philosophical Society, giving it the additional strength necessary to start the publications of its *Transactions* in 1771.

In 1779, the year of Winthrop's death, John Adams returned from France to Boston. At a public dinner at Harvard College in honor of the French Ambassador, he entertained Dr. Cooper of Boston with an account of his visits to scientific collections and of the compliments he had heard in France upon the Philosophical Society in Philadelphia. To quote his diaries:

> I concluded with proposing that the future legislature of Massachusetts should institute an Academy of Arts and Sciences. The Doctor at first hesitated, thought it would be difficult to find members who would attend it; but the

principal objection was that it would injure Harvard College by setting up a rival to it.

To this I answered,—first, that there were certainly men of learning enough that might compose a society sufficiently numerous; and secondly, that instead of being a rival to the University, it would be an honor and an advantage to it. That the President and principal professors would, no doubt, be always members of it; and the meetings might be ordered, wholly or in part, at the College and in that room [the Philosophy Chamber]. The Doctor at length appeared better satisfied.

This Dr. Cooper was Samuel Cooper, Harvard 1743, pastor of the Brattle Street Church in Boston, a classical scholar, a patriot, and correspondent of Franklin. James Bowdoin was easily won for the plan. The war did not deter these enterprising men, and in 1780 the charter of incorporation of the American Academy of Arts and Sciences was granted by the General Court of Massachusetts. James Bowdoin took the initiative for the first meeting, which was held in the "Philosophy Chamber" of Harvard Hall, and was found willing to become the first president. This was a fortunate choice, since Bowdoin was fitted for the chair as a patron of learning, a political leader and a man of means whose patriotism was mellowed with the right type of gentlemanly conservatism. Samuel Cooper became the first vice-president.

The aims of the Academy, as expressed in the charter, express the revolutionary eigtheenth-century conception of science, cultivated in the first place not for its own sake or for the sake of truth, but above all for the honor and dignity of the citizen as well as the ultimate material benefit of mankind:

As the Arts and Sciences are the foundation and support of agriculture, manufacture and commerce; as they are necessary to the wealth, peace, independence, and happiness of the people; as they essentially promote the honor and dignity of the government which patronizes them; and as they are most effectually cultivated and diffused through a State by the forming and incorporating of men of genius and learning into public societies. . . .

The Academy started with sixty-two constituent members. The prominence of the clergy was manifest. There were fifteen members with the title "reverend," and some more signed themselves D.D. There were political leaders such as John Adams, James Bowdoin and John Hancock, philosophizing ministers such as Samuel Cooper, Edward Wigglesworth and Joseph Willard, and physicians such as Edward Holyoke and Charles Jervis. The names of Francis Dana, John Lowell and John Pickering attest even more strongly to the domination of the Academy by the merchant aristocracy. The Academy soon grew, like Harvard College, into an intellectual stronghold of Boston Federalists.

During the next years the Academy elected prominent men in the country to its membership: in 1781 Franklin, Washington, John Warren, Ezra Stiles and Benjamin West, the Providence mathematician, in 1782 Loammi Baldwin and David Rittenhouse. Foreign scientists were added as honorary members. Thomas Jefferson was elected a member in 1792, before the deadly feud between him and the Federalists.

Manasseh Cutler, the versatile minister of Ipswich Hamlet, north of Boston, who was elected to membership in 1781, has preserved in his diary some facts concerning the early years of the Academy. On May 29, 1781, he went to the meeting at the Courthouse in Boston, together with Joseph Willard, and found twenty-two members present. He "communicated" a meteorological journal from July 1780, together with a list of the diseases most prevalent in his neighborhood. "I also presented the Society a sample of sheet-lint, from Dr. Spofford, who has contrived a machine for scraping it with great dexterity. It was much admired." And further:

Aug. 22, 1781., Wed. The Academy met at the Philosophy Hall. Several communications. Some members chosen. A gentleman from England recommended by the General Court to be employed under our direction as a mechanic.

May 28, 1782., Tues. Went to Boston. Attended the meeting of the Academy. Dined at Mr. Bowdoin's. Communicated my observations of the lunar and solar eclipses [these were observed March 28 and April 12. Cutler, who maintained a boarding school at his home, had all his "school lads counted clock"].

There is some importance in Cutler's note that a mechanic had to be imported from England. Though New England had a considerable number of craftsmen, specialized merchanics were rare. The Academy again met with this difficulty of finding skilled mechanics when it decided to publish a scientific journal.

This journal appeared in 1785 under the name of *Memoirs*, and was a stately tome of over 600 pages, the first scientific journal published in New England. The name *Memoirs* indicates that the Revolution had promoted French influence in New England. Cutler, in a letter of October 30, 1786, to the British Dr. Stokes, explained it somewhat apologetically as follows:

> You object to the title as savoring too much of the air of France, and as improper when applied to a Society. You will recollect that the American Academy was instituted at a time (1780), when Britain was viewed in this country as an inveterate enemy, and France as a generous patron. Although philosophers ought to divest themselves of all those prejudices which national contentions and combinations naturally excite, yet I doubt not it was the intention of those concerned in establishing the institution to give it the air of France, rather than that of England, and wished to be considered as following the Royal Academy, rather than the Royal Society. But, however this might be, it was of importance that the title should clearly and concisely distinguish this from a similar institution at Philadelphia, whose title was professedly copied from the Royal Society.

The same letter gives particulars about the technical

difficulties which the publishers of this first scientific periodical in New England met during the preparation.

> No book of equal magnitude in size, or numbers, or in any respect similar, had ever been published in this part of America—the Academy had no fund, few of its members conversant with the publications of such Societies, printers men of small capital, no aid to be obtained from men of fortune, those concerned in directing and those in executing the work much unacquainted with such an undertaking; in short, the publication has been a mere experiment.

There were more difficulties, all the result of lack of experience.

When the volume appeared, it breathed the practial and humanitarian spirit of the period. An apology was made that few, if any, of the papers "contain deep speculation . . . but they are chiefly of the practical kind." The physical papers were defended by words which in their eloquence contain more than ephemeral wisdom:

> It is the part of the patriot philosopher to pursue every hint—to cultivate every enquiry, which may eventually tend to the security and welfare of his fellow citizens, the extension of their commerce and the improvement of those arts which adorn and embellish life.

Joseph Willard and Samuel Williams reported astronomical observations, and James Bowdoin contributed his opening address of 1780, titled "A Philosophical Discourse," which is a lengthly refutation of Franklin's ondulatory theory of light. A surgeon major of a French fleet at the time stationed in Boston, J. Feron, described experiments concerning the composition of the water in Boston. This paper is the earliest attempt at the chemical examination of water in this country, and, almost a century later, was praised by Benjamin Silliman, Jr., the Yale chemist, as a work which "would not be esteemed an unworthy contribution today." Many articles dealt with improvements in agriculture, a significant item which later will be further discussed. John Prince, a Salem minister, described an air pump which for a while became quite famous, even abroad.

4

One of the longest and most interesting papers in the *Memoirs* was a botanical paper contributed by Manasseh Cutler. Cutler's life gives some idea of the type of professional man who contributed to science during and after the Revolution in New England, and to whom specialization was an alien conception. He started his career as a Yale graduate and storekeeper on Martha's Vineyard. Subsequently he studied for both the law and the ministry, and was ordained in 1771 as a minister to the parish which he served until his death in 1823. After a term of service as chaplain in the Revolutionary Army he returned to his flock, whom he served not only with spiritual council according to the Calvinist faith of their fathers, but also with medical advice. He combined the study of medicine with that of astronomy, and recorded several transits and eclipses; with the study of astronomy he also combined the study of botany. But he was not destined for a life of scholarly solitude. In 1787, he and General Putnam led a group of emigrants from New England over 750 miles into the Ohio wilderness, where Cutler became one of the founders of Marietta. A large covered wagon was used on this expedition, which one of Cutler's biographers has placed historically side by side with the *Mayflower*. On it was painted in white letters on a black background, "Ohio, for Marietta on the Muskinggum." The town was named after Marie Antoinette, Queen of France. The fate of this queen must have disturbed Cutler, for he became a great enemy of Jacobins and Jeffersonians. As a stanch Federalist he defended, from 1800 to 1804, his principles in Congress as a representative of his district. The Covered Wagon House in Hamilton, Massachusetts, preserves the memory of this active man, typical of the first generation of republican scientists. In Marietta he is remembered as one of the Founding Fathers of Ohio.

Cutler's paper in the *Memoirs* deserves some special attention as the beginning of systematic botany in New England. In the preface the author remarks that so far very little had been done in New England to promote natural history, although Canada and the Southern states had been visited by eminent botanists from Europe. Botany had never been taught in any New England college, owing to a mistaken opinion of its uselessness. It was time to change this point of view. And so, leaning on some English texts, on the magnificent description of Carolina and other southern plants by Mark Catesby (1731, 1743), and on his Linnaeus, the industrious divine described a few hundred plants, usually characterizing them only by their Latin genus, the local English name and the place of growth. The medicinal value of the plants is brought out wherever possible.

These beginnings of New England's botany have not the glamour and the charm of the pioneering efforts of America's first botanists who traveled through the wilderness of the Alleghenies and the Savannah. No man like Bartram, as far as we know, spent weeks and months tramping through New England swamps and woods, for away from human beings, to bring home the marvelous herbs and shrubs which gave fragrance to the wilderness. Trees and shrubs of New England seldom went to the botanical gardens of Europe. Between 1785 and 1796 the elder Michaux traveled through the United States, from Florida to Canada, collecting and classifying, but, except for a brief visit to New Haven to see an old scout, he avoided New England. The Lake Champlain route bypassed New England because of the horrible wilderness between southern Maine and the Canadian settlements.

And yet there is a charm in the efforts of these eighteenth-century herbalists like Cutler, who went through the woods around their houses and tried to identify their finds with the plants described in their foreign textbooks, with crude means and often in doubt as to whether their discoveries were already classified or were new additions to botany. We can follow Cutler along the Parker River in Newbury, to fields in Cambridge, to Lynn and to Chelsea. When he finds the ginseng, he tells us what he knows about this "famous panacea of the Chinese," which after its discovery in the Canadian woods by Father Lafitau early in the eighteenth-century had brought for a while a fortune to New France, and would soon again bring one to New England. Black currants, Cutler tells us, are frequent near the Kennebec, and Pyrola, the "rheumatism-weed," abounds near the White Mountains, as everyone can verify today. It was all very primitive and amateurish, but as a start it was as stimulating as Winthrop's experiments with a Leyden jar.

This piece of information concerning Pyrola may well have been inserted during the printing, since Cutler made his trip to the White Mountains in July 1784. This tour, made in a party of seven, all men of some distinction, was the first scientific expedition into the mountainous region of New Hampshire. At least three members of this party have left a description of the trip—Jeremy Belknap, Daniel Little and Manasseh Cutler, all scientifically inclined ministers. They approached the summit, the "Sugar Loaf," from Conway through what is now Pinkham Notch, and probably gave it the name Mount Washington. Cutler had great difficulties in getting his instrument up the mountain through the howling wilderness of what is now called Tuckerman's Ravine, damaged his barometer, and found 9000 feet for the height of the summit though the actual altitude is 6288. Cutler's River, in Pinkham Notch, carries the memory of this first scientific explorer of Mount Washington. The trip is described in several books on the White Mountains.

Another remarkable expedition related to the early years of the Academy is the journey undertaken by Samuel Williams, Winthrop's successor at Harvard, to Long Island in Penobscot Bay. He braved the wilderness of the coast of Maine in order to observe the solar eclipse of October 27, 1780. Harvard College and the Academy took the necessary steps and made arrangements with the General Assembly. This body was responsive and gave orders "to fit out the State Galley with proper stores and accommodations" to serve the scientific party. The war was on and the Penobscot region was in British hands, but the Assembly and Professor Williams hoped that the British commander would share their gentlemanly feelings concerning the conduct of the war and permit observations. Their expectations were justified; the British commander allowed the scientific rebels to stay for eleven days on Long Island. They observed the eclipse, but just missed the region of absolute totality because of their inaccurate maps. What they lost in one respect they gained in another, though they did not realize it too well at the time. They saw a fine thread of light remaining on one side of the sun, but this thread was broken into drops. This later became known as Baily's beads, after an Englishman's observations of 1838, but the Penobscot expedition seems to be the first on record in which these beads were consciously observed.

5

The Massachusetts Medical Society was founded in 1781, a result of the attempts of Cotton Tufts, John Warren and some of their colleagues to organize medical work. Their aspirations received encouragement during the Revolution through experience in army hospitals. The Revolution had shown the necessity of a stricter examination of physicians, and the Second Provincial Congress, in 1775, created a committee to examine surgeons for the army, which practice was continued in later years. The act of incorporation contained a rule that president and fellows could "examine all candidates for the practice of physic and surgery," after which those who passed would get "letters testimonial to such examination." The Society, in this way, set itself up as an examination board with monopolistic tendencies soon to be challenged by the foundation of the Harvard Medical School. The Medical Society began its career with thirty-one members, eight of whom were also members of the Academy, almost all from coastal towns, and again closely connected with the merchant aristocracy. The first president was Edward Holyoke, the Salem practitioner.

The foundation of the Harvard Medical School came soon after the Massachusetts Medical Society was established. One of the main driving forces behind both organizations was Boston's leading surgeon, John Warren. He was a farmer's son from Roxbury, then a rural community south of Boston. An admirer of his twelve-year-older brother Joseph, who combined an extensive practice as a physician in Boston with a leading position in the patriotic movement, he felt a calling to be a physician as well as a patriot, and began to work under his brother's direction. Tradition has it that he was one of the Indians of the Boston Tea Party. He studied under Dr. Holyoke in Salem, and after the outbreak of the war joined the Revolutionary Army as a physician. The only hospitals known in New England at that time were some inoculation hospitals. Under the necessity of the war army hospitals were erected. Warren, who had lost his brother Joseph at Bunker Hill, where a monument preserves the memory of the gallant physician, became a senior surgeon in the Cambridge army hospital, and later a surgeon at a hospital on Long Island. He served two years in the field sharing the hardships of the soldiers, and returned to Boston in 1777, where he set up a private practice and continued his surgical work in the army hospital.

Warren soon became the leading surgeon of the city, and a leading citizen as well. In the army hospitals the medical profession had reached a maturity and corps consciousness which it had previously lacked. New ideas and methods began to be appreciated. Warren's zeal for anatomy and surgery led him to the dissection of the bodies of some soldiers who had died unclaimed by relatives. Considerable popular prejudice against dissection was prevalent, and the private course of anatomical demonstrations which he delivered in the winter of 1779–1780 at the Boston military hospital had to be held with great secrecy, attended by only a few intimates. The prejudice began to wane and the second course, endorsed by the newly founded Medical Society, was held with less circumspection. The third series was attended by the whole senior class at Harvard. Among the audience at the second series were several Harvard men, namely Joseph Willard, the pastor of Beverly, near Salem, an amateur astronomer and a member of the American Academy, who became president of Harvard College in the same year 1781. These men became convinced that Harvard should at last emulate Philadelphia and Columbia and establish a Medical School. Money was found in an ancient legacy to Harvard from a Hingham physician, Ezekiel Hersey, to which some other funds were added, and the school was founded in 1782. John Warren became professor of anatomy and surgery, and two other members were added to the staff: Benjamin Waterhouse to teach theory and practice of "physick," and Aaron Dexter to teach chemistry and materia medica.

There was at first some uneasiness between the Medical School and the Medical Society because of the society's monopolistic tendencies regarding the licensing of practitioners. A committee was appointed to look into the situation. It concluded that the school did not interfere with the society, and this settled the matter. It was a help that the men prominent in the foundation of the school were also important in the society, although this did not remove the monopolistic tendencies of the group as a whole. Sometime later one of them, Dr. Waterhouse, was to complain of it bitterly.

John Warren remained the leading spirit of the Harvard Medical School until his death in 1815. He was one of the best surgeons of his day, pioneering in abdominal operations and in amputations at the shoulder joint. His duties at the school did not prevent him from publishing several monographs, from carrying an extensive general practice and maintaining a large family. The distinguished Warren family of physicians traces its ancestry back to John and Abigail Warren of Massachusetts. The oldest of their seventeen children was John Collins Warren, who succeeded his father at the Medical School; the youngest, Edward, became his father's biographer. A plaque in the Massachusetts General Hospital keeps their memory alive.

New Hampshire followed with a Medical Society in 1791, followed by a Medical College at Dartmouth in 1798. The entire faculty was the able Dr. Nathan Smith, who in 1812 went to Yale to organize the Medical School, not before he convinced the Yale authorities that he was not an "infidel." Here Smith taught and organized until his death in 1829. A son became the guiding medical man at the University of Vermont.

6

The Revolution stimulated not only the organization of natural science and medicine, but also the growth of manufactures and the development of technological improvements. There was, as yet, little connection between the two currents. Manufacture and technology were advanced primarily because the war necessitated the production of guns, cannons, gunpowder and saltpeter to make needed arms and supplies; and this in a country where the free unfolding of industrial and inventive genius had been suppressed. The existing forges and furnaces were used to the utmost, and new ones were erected. They were necessarily small, since there was little capital available and skilled craftsmen were rare. Even gunpowder was lacking. In 1775 Congress issued instructions for the manufacture of saltpeter, signed by John Hancock, and as a result several saltpeter workshops were established. John Peck, whom we shall meet as a shipbuilder and naval architect, had such a shop in Watertown. Paul Revere established a powder mill at Canton, south of Boston. A member of the Jencks family produced ammunition and firearms in his metalshop and forge at Pawtucket. Eliphalet Leonard at Eaton, north of Pawtucket, established one of the first steelworks in Massachusetts, providing some of the badly needed steel. Beginning around 1775 he constructed several furnaces, from which he supplied the neighborhood with firearms. Some of these enterprises, such as those of Jencks and Leonard, remained in existence after the war and even expanded. For many of them, however, peace meant the end of their activity because of the sudden influx of imported commodities against which the country was not yet able to successfully compete. England strengthened its navigation laws during and after the Revolution, which originally worked to the detriment of the United States. In the long run, however, it acted as a stimulating factor. The appetite for economic enterprise was greatly whetted, and many first attempts at invention during revolutionary days yielded remarkable fruits in the days to come. The development of the textile industry began almost immediately after the Revolution and was followed by the use of mass-production methods in other industries.

There was a Connecticut Yankee by the name of David Bushnell, whose ingenuity had already led to spectacular results during the Revolutionary War. He was a farmer's son, and graduated from Yale in 1775. Bushnell played with the idea of a submarine, an idea which had occurred to others before him, notably to that curious genius, Cornelis Drebbel, a Dutchman at the court of James I. Bushnell was interested in a machine that would blow up vessels from under water; shortly after his graduation he actually constructed one to be used against the British. It was composed of several pieces of large oak timber, scooped out and fitted together in the shape of a round clam, so that it became known as "Bushnell's Turtle." It was bound together with iron bands, the seams were calked and the whole was tarred to make it watertight. It was large enough to contain one engineer, and was fitted out at the top with glass, so that on a clear day and with clear water the man could read at three fathoms depth. Seven hundred pounds of lead at the bottom kept it upright, 200 pounds more were used to increase the buoyance, which could be discharged at any moment. The machine could move vertically in the water by means of two pumps which pressed water out at the bottom, and horizontally by means of paddles. There were 130 pounds of gunpowder stored in the back of the machine, which could be discharged at vessels provided the machine could be brought near enough to them. Between August 1776 and December 1777 Bushnell tried his invention out on several British ships in New York Harbor and in the Delaware, but he was never really successful because of the difficulty he encountered in navigating the unusual contraption. In a last attempt he prepared a number of machines in kegs to be floated by the tide upon the British vessels in the Delaware near Philadelphia, but he blew up only a small boat. Though the explosions of the kegs greatly alarmed the British, the attempts were so ridiculed among Bushnell's countrymen that he gave up invention to become a captain in the Revolutionary Army. Later he went to France, was for a long time lost to view, and ended his long career in 1826 as a physician in Georgia. Francis Hopkinson, the Philadelphia poet, wit and patriot, described the "Battle of the Kegs" in a poem to the tune of "Moggy Lawder," of which the following verses may give a sample:

> Gallants attend, and hear a friend
> Trill forth harmonious ditty;
> Strange things I'll tell, which late befell
> In Philadelphia city.

. . .

The Kegs, 'tis said, though strongly made
Of rebel staves and hoops
Could not oppose their pow'rful foes
The conquering British troops.

• • •

Such feats did they perform that day
Upon those wicked kegs
That years to come, if they get home
They'll make their boast and brags.

It is reported that this ditty gave many a weary patriot a laugh during that hard winter of 1777-1778.

7

This is perhaps the best place to deal with two scientists born in New England, whose main scientific activity took place on foreign soil. They left their native country during the Revolution as Tories, one of them, the better scientist of the two, never to return. We mean John Jeffries and Benjamin Thompson.

John Jeffries was a Boston physician with an M.D. from the University of Aberdeen in Scotland. He accompanied the British troops to Halifax after the evacuation of Boston, and served for a while as a surgeon major of the British forces in America. Later he became a successful physician in London and was one of the first to become "air-minded," after the brothers Montgolfier in France had shown the possibility of sending up a balloon by heating the air within it, or by filling it with the newly separated hydrogen gas. In 1782 this new field of "aerostation," as it became known, received a spectacular stimulation when first Pilâtre de Rozier, soon followed by other men, actually began to navigate the air in a *montgolfière*. New England was interested, and both the *Salem Gazette* and the *Boston Magazine* gave large accounts of the experiments, including one in Philadelphia. Jeffries, in London, was also greatly interested, mainly because of his New England interest in meteorology. He correctly foresaw in such balloon ventures the possibility of obtaining new information on temperature and air currents at different levels, and even hoped "to throw some light on the theory of winds in general."

Jeffries established contact with Francois Blanchard, a French "aeronaut," and the two ascended in 1785 from Dover Cliffs in a balloon. They crossed the Channel and landed in the forest of Guines near Calais. It was the first attempt to cross the Channel by air, and its success made the two adventurers the heroes of the moment. They were fêted for two months by ladies and gentlemen of French polite society, received attention from learned societies and were decorated by the King. Back in London, Jeffries read an account of his experiments before the Royal Society. He never repeated his adventures with balloons, but returned to his practice. His companion, Blanchard, continued to navigate as an aeronaut and also performed in America. He is credited with the invention of the parachute (1785).

Jeffries returned to Boston in 1790, where he practiced as a successful physician until his death in 1819. We have already mentioned how his first public lecture on anatomy was broken up by a mob. He seems to have been forgiven for his Toryism. Oliver Wendell Holmes, who, as a boy of eight, had a chance to see the doctor, remarked in later days that "it was something to have seen the man who first crossed the British Channel, not by water but by air; who walked among the dead of Bunker's Hill battle and pointed out the body of Joseph Warren to those who were searching for it in the heaps of slain; and who helped to deprovincialize the medical science in Boston." Jeffries's son, whom Holmes knew as "the old doctor," died in 1876.

8

Benjamin Thompson, or Count Rumford as he later called himself, is one of the greatest scientists America has ever produced, but all his scientific activities were carried on outside of the United States and independent of his native land —unless we see in his eminently practical approach to even his most theoretical results the spirit of the Yankee farmers from which he came. He was born in 1753 in Woburn, and led the simple country life of a boy in a small colonial town.

He became a clerk in a Salem store, and later a teacher in the town of Rumford, New Hampshire, now the capital, Concord. Already in Woburn he had begun to study optics, electricity and mathematics as much as his circumstances allowed. He established a friendship with Loammi Baldwin, thirteen years his senior, which even his disaffection from the American cause could not destroy. Young Ben Thompson was able to get some instruction not only from Baldwin, but also from John Winthrop at Harvard and from Samuel Williams, the astronomer-clergyman of Bradford near Haverhill, whom we have met as Winthrop's successor. He took good care not only of his formal instruction, but also of his own person, and liked to move in the best social circles possible. While he was teaching school in Rumford, his dashing ways earned him the affection of Mrs. Rolfe, the richest widow in town. The resulting marriage introduced Thompson to the circles of the provincial elite, led by New Hampshire's Tory governor, Wentworth. His fellow Yankees began to distrust him, and as early as 1774 he had to answer complaints of "being unfriendly to the cause of liberty."

During the siege of Boston, Thompson, suspected of Toryism, was actually arrested, and it has become increasingly clear that he was in actual communication with the enemy. Thompson now became open Tory. He went to London, served as informer to the English and was taken into the household of the Colonial Secretary of State. He returned once more to America, as a lieutenant colonel of the King's American Dragoons, and helped devastate Long Island. After his return to Europe he set out on one of the most remarkable careers in the history of science. On his way to Vienna he watched a parade in Strasbourg where his handsome appearance attracted the attention of Prince Maximilian, son of the Elector and King of Bavaria. He entered into the service of the Elector and remained for many years at his court in Munich, in a variety of high advisory functions. George III of England conferred a knighthood upon him; the King of Poland, the Order of St. Stanislaus; and the Elector of Bavaria, in 1792, the title of Count of the Holy Roman Empire. Thus the Woburn lad, who walked eight miles with Baldwin to hear Professor Winthrop lecture, became Count

Rumford—a name taken after the town which he entered as a poor schoolmaster and left as a Tory friend of the governor of New Hampshire. In Bavaria he laid out the English gardens at Munich, improved the breed of horses and cattle, established a gun foundry and was active in attempts at industrialization. For this last purpose he organized man hunts to drive the poor into workhouses, and used his studies in nutrition to prepare cheap diets for the workers. This achievement earned for him the praise of the Elector of Bavaria and of the majority of his biographers. Karl Marx was less enthusiastic and called him an "American humbug." "With the advance of capitalistic production," continues Marx, "the adulteration of food rendered Thompson's ideal superfluous."[1]

Later on Thompson tried to establish himself in England as the Bavarian ambassador, but in this he failed. After the death of the Elector, Rumford gave up his citizenship in the electorate and finally settled in Paris. Here he married in 1804, as his second wife, the widow of the chemist Antoine Lavoisier, and settled with her in the villa in Auteuil near Paris where Lavoisier had lived. The marriage was not a success, and the last ten years of the life of the Yankee Count were spent in rather embittered retirement. His daughter Sarah, born in America of his first wife, was with him for a while, but he made her far from happy. After Rumford's death in 1814 the "Countess Rumford" returned to Concord,

[1] Lest we find Marx's judgment too sharp, let us consider the judgment of the Swiss botanist A. P. de Candolle, who met Rumford in Paris:

> "We found him a dry, methodical man, who spoke of benevolence as a discipline, and of the poor as we should not have dared to speak of vagabonds. It is necessary, said he, to punish those who give alms; the poor must be forced to work, etc., etc. Great was our astonishment at hearing such maxims: however we did our utmost to profit by his advice in practical matters. . . ."

De Candolle had the greatest admiration for Rumford's genius, but found him "cold, imperturbable, obstinate, egotistical, prodigiously occupied with the material part of life, and in inventions of the smallest matters. He was engrossed with chimneys, lamps, coffee-pots, and windows made after a peculiar fashion."

where for many years she lived as a quaint and unhappy spinster.

Rumford's merits lie in his contributions to physics, primarily in his discovery of the nature of heat as a form of motion. In his work of modernizing the Bavarian Army he had a factory built at Munich, where guns were constructed according to his designs. Observing the large amount of heat continuously generated by friction in the boring of cannon (then a novel process) he came to criticize the prevalent theory of heat, which claimed that heat was a fluid substance, called caloric, held in the pores of bodies, which could be squeezed in and out like water from a sponge. If, he reasoned, caloric is squeezed out of metal by boring, then the powder produced in the process must have less heat in it than the original solid metal. It should accordingly require more heat to raise it to a given temperature than the metal. Tests showed that the specific heat of the borings and of a piece of gun metal were the same. Then Rumford decided to find out how much heat was produced by a certain amount of friction. His experiment, which has often been described, gave as the mechanical equivalent of heat 847 foot pounds—only 10 per cent from the now acepted value of 779. "It appears to me to be extremely difficult, if not quite impossible, to form any distinct idea of anything capable of being excited and communicated in the manner the Heat was excited and communicated in these experiments, except it be *motion*," is Rumford's conclusion in his report to the Royal Society. This result is the more astonishing when we realize not only that most of Rumford's contemporaries believed heat to be a fluid—though there were notable exceptions, even before Rumford—but that this misconception remained popular until 1845, when with Joule's experiment the truth of Rumford's principle was generally recognized. Rumford's paper went even beyond the statement of the equivalence of heat and mechanical motion. He connected heat, light, chemical action and mechanical movement together as capable of being converted from one into the other, and extended the principle to organic and even animal life.

The many papers published by Rumford during his life and collected by the American Academy deal with a variety of subjects, from improved fireplaces and "Rumford" stoves to scientific studies of cooking and the economic management of heat. They led him to experiments on conductivity and convection of heat, and to the invention of what in principle amounts to the "Thermos" bottle. The style of his papers is very clear and today many of them can stand the test of rereading. They can make us forget the curiously uneven character of the man who wrote them, a man who always remained somewhat of an outcast in every circle he penetrated, but who was able to penetrate socially most exalted circles. He struggled to overcome his uneasy feelings by hard scientific work, by personal snobbery and by large donations to scientific organizations. He founded the Royal Institution in London, perhaps the most permanent monument to his memory, where Davy and later Faraday worked. He gave $5000 to the Royal Society in London, and a like sum to the American Academy of Arts and Sciences—*"dulces moriens reminiscitur Argos"*—the interest to be given every two years as a premium to the person who made the most important discovery or useful improvement in heat or light "as shall tend most to promote the good of mankind." The "Rumford Medal" of the Royal Society has been regularly awarded to distinguished scientists, beginning in 1802 with Rumford himself. The donation to the American Academy came somewhat prematurely, and at first the board did not know what to do with it. No award was given for many years; eventually the money was turned into a fund for research in heat and light. Not until 1839 did the Academy begin to award the Rumford Medal, but again a gap appeared which lasted until 1862. The Rumford Medal is now regularly awarded. Rumford also left a sum of money to Harvard for a professorship "of the application of science to the art of living," of which Jacob Bigelow, in 1816, was the first incumbent.

It must be said that if Rumford had stayed in America he might have become a better citizen and a happier man, but certainly his science would have suffered. For its full development, his work needed the stimulation and the admiration of an older civilization. Americans appreciated this even during his lifetime, and have preferred to think of Rumford as a scientist rather than as a Tory. His birthplace is now a museum, and a statue adorns a central spot in Woburn—a copy of the statue in the English gardens at Munich which Rumford laid out. It presents a serious Yankee figure draped in pompous habiliment, and reminds posterity of the conflicting traits in the make-up of Woburn's famous son. The Rolfe and Rumford home, in Concord, where Thompson married Mrs. Rolfe and where his daughter spent her later days, has been converted into a home for orphan girls.

Chapter 3

The Practical Navigators

Nil mortalibus arduum est
Audax Japeti genus
Rien d'impénétrable aux mortels,
À la race audacieuse de Japet.

Si ces vers peuvent s'appliquer à quelque peuple, c'est bien aux Américains libres. Aucun danger, aucune distance, aucun obstacle ne les arrête. Qu'ont-ils à craindre? Tous les peuples sont leurs frères; ils veulent la paix avec tous.

—J. P. BRISSOT, 1791

1

SHIPBUILDING AND SEAFARING had always given character to the towns along the coast, from the St. Lawrence River to New York and farther south, especially along the Delaware and the Chesapeake. Navigation laws restricted the cargoes of the shops and prescribed places of destination, but did not interfere with shipping as such, and even stimulated it insofar as they protected British-built vessels. The magnificent forests of the colonies were an excellent source of timber for ships at a time when England was already suffering from continued deforestation; they provided oaks for the hulls and pines for the masts of ships built in the colonies and in Great Britain. Colonial trade was mainly conducted with European countries, West Africa and the West Indies. It kept many men busy as sailors, coopers, carpenters, calkers, chandlers and storekeepers; it stimulated craftsmanship of many forms, and it built the fortunes of many a colonial aristocrat. In a typical year, 1769, the colonies built and launched no less than 389 vessels, of which 276 were sloops and schooners.

They were small, these ships, many under a hundred tons, but, once at sea, they performed remarkably well. Yankee sailors were known in many ports of the world for their boldness and efficiency in handling their craft, and the youth of their skippers. The personal element, the skill and daring of the men, counted at least as much as design and rigging, since the best shipbuilder never progressed beyond the trial-and-error method. A sailing ship, it has been said, is an exceedingly complex, sensitive and capricious creation, quite as much so as a human being; no one has ever lived who could predict with accuracy the results due to particular design, construction and rigging. There is no doubt, however, that improvements were gradually made, and that an important factor in this development of American shipbuilding was the experience gained in the years of revolutionary warfare.

2

During the Revolution the all-important fact in the coastal towns was the British blockade. The Royal Navy was now an enemy instead of a protector. It tried to strangle oversea trade, and it severely handicapped shipbuilding. Yet ships were launched, but for war use—above all for privateering. The skippers and shipbuilders of these towns on the oceans and the estuaries were old hands at this game. Another way to fight the blockade, but a more difficult one, was to build men-of-war able to combat His Majesty's huge frigates and ships of the line. There was, unfortunately, no experience at all in this type of shipbuilding. Although the European powers had been building these castles of the seas for centuries, the colonies had never been allowed to participate. In the America of the 1770's shipwrights scarcely knew what a frigate was, and much less thought of building one. The best hope of the Yankee skippers lay, therefore, in building craft which could outrun the towering, many-gunned British monsters, and which could at least hold their own against the smaller ships of the enemy. This meant concentrating on speed, even if on the whole traditional patterns were to be followed.

Hence with the Revolution a change in shipbuilding set in. In the many small shipyards hidden away in the coves of the Atlantic Coast and Long Island Sound small craft were built, constructed entirely for speed and weatherliness, efficiently armed privateers as well as unarmed cargo vessels. The small privateers, built for short ranges and not equipped for defense against large vessels, were able to capture many supply ships and to keep armed cutters at bay. Their crews were highly trained for this type of warfare; later they manned the merchant marine of the United States in its rapid and astonishing expansion. During the War of 1812 new experience was gained in outwitting the enemy by speed and bold seamanship. Blockade-running skippers, it was said, could "smell their way from Hell Gate to Providence with their eyes shut."

The attempts to meet the impressive British seapower by a display of American seapower met with less success. In December 1775, the new Continental Congress ordered the building of thirteen large warships; an order for six more followed later. Several of these vessels were constructed in New England shipyards. The frigate *Raleigh* was launched at Portsmouth, the *Hancock* at Newburyport and the *Warren* at Providence. All these vessels carried thirty-two guns. Other frigates were built at Norwich, Salisbury Point, and at Middletown on the Connecticut River below Hartford.

Despite the best efforts of America's shipbuilders, it proved impossible to produce a navy able to break the blockade. The ships that were completed were too lightly built; those which would have been heavy enough were never finished. They were constructed of insufficiently seasoned timber; workmen and sailors were new to their tasks. Despite the seamanship and heroism of the crew, all ships of the American navy which participated in the war were either captured or destroyed, often after some initial successes. The only American-built frigate left in the Continental service at the end of the war was the *Alliance,* built in 1777 at Salisbury Point, on the Merrimack.

The *Alliance* was built by the Hacketts, who were among the best of the New England ship designers of the period. William Hackett was born and reared in Salisbury, across the Merrimack from Newburyport, where his family had been building ships for many years. As a lad of twelve he worked in the family shipyard instead of going to school, and during his whole lifetime he thought almost exclusively of ships. When his father died he entered into partnership with his uncle, and also collaborated with his cousin James at Portsmouth in New Hampshire. In the shipyard conducted by the uncle and nephew the *Alliance* was built. It is also possible that William Hackett designed the frigate *Hancock*. After some gallant action, the *Hancock* was captured by the British, who measured and described it, thus preserving information concerning these early American attempts at the building of warships. The British seemed impressed. They praised the *Hancock* as "the finest and fastest frigate in the world"—a sign not only that its builder knew his craft, but also that the Yankee emphasis on speed had been carried over into the construction of the new frigates.

Another center of shipbuilding lay in the North River section of Massachusetts, mainly in the towns of Pembroke and Marshfield, north of Plymouth. Like the Salisbury center it is now entirely gone; a region distinguished by decaying wharves, ancient homes with ells and secret passages and an occasional memorial tablet. Yet here, in the Brick Kiln Shipyard alone, now marked by a plaque, more than 120 vessels were built between 1730 and 1848, among them one of the ships whose cargo provoked the Boston Tea Party. Sites of old furnaces and forges preserve the memory of the former bog iron industry, which provided anchors and fittings for the ships as well as cannon balls.

From this region hailed Josiah Barker, one of the most characteristic among the early shipbuilders. He was the son of a blacksmith, and as a boy served on the frigate *Hague* in the Revolutionary War. A veteran at nineteen years of age, he returned to Pembroke in 1782 and became a shipbuilder. After some travel he settled at Charlestown near Boston in 1795, where he built ships for twenty years. During the War of 1812 he was master carpenter of the *Independence*, the first ship of line in the United States Navy; later he became naval constructor at the new Charlestown Navy Yard, where, in 1834, he rebuilt the frigate *Constitution* in the new naval dry dock. Though not a creative naval architect, he built vessels throughout his long life along traditional lines; among them several ships of the line and frigates for the United States Navy. His imposing figure, six feet tall, his "dignity, urbanity and hospitality," made him for many years a prominent figure in Charlestown. He kept his government position until 1843, when, an octogenarian builder of frigates and other wooden ships then somewhat out of date, he was transferred to Portsmouth. He died in 1846.

Most ship constructors of the early days of the republic were men like Barker, practical, highly skilled craftsmen, who built according to the demands made upon them by private merchants or by the government. We may consider them the first known American naval engineers, though we should bear in mind that most of them were also shipbuilders whose achievements were the result of practice rather than theory.

Somewhat more of an intellectual was John Peck, the father of the first Harvard entomologist, who started as a successful Boston merchant with ship designing as a hobby. During the Revolution he was made inspector of saltpeter in Watertown. He designed many ships, but since he did not build them himself during the better part of his life, he may be primarily considered an early naval architect. It is believed that he designed the Salem privateer *Rattlesnake*; it is certain that he designed the famous *Empress of China,* built at Boston in 1783 and the first American ship to make the voyage to China. Shortly afterwards he retired to Kittery near Portsmouth, where he operated a small shipyard. His son's exquisite entomological drawings may be the product of a skill he inherited from his father, the naval architect.

3

The large toll of ships lost in the Revolutionary War or captured by the British showed that American sailors had much to learn in the handling of the newer crafts, but they learned fast. So did the designers and builders of the ships in their endeavor to satisfy the demands of their customers; they profited by the various experiences of the crews and had a good look at the foreign vessels which came to the American shore. There is a tradition that American shipbuilders were influenced by the French men-of-war which visited American harbors during and after the war with England. There is really no reason to disbelieve reports that the design and behavior of such vessels were discussed, since French warships were reputedly the best constructed vessels of the eighteenth century. It was well known that they were the result of experience and technical skill as well as of the theoretical work of the continental mathematicians and engineers. From the *Memoirs* of the American Academy of Arts and Sciences we can see that Bostonians were perfectly willing to listen to a visiting Frenchman, when he talked chemistry. There is, however, little evidence that Americans actually imitated Frenchmen when they discussed naval architecture. Our experts see little or no French influence in American shipbuilding. When, after the Revolution, shipbuilding in the United States reached a new state of expansion, it was marked by a singular independence of development.

Freed from the protection offered by the British to their erstwhile compatriots, American skippers were also freed from the British navigation laws. This cost them some of the trade they had enjoyed before the Revolution, but it stimulated them to find new routes of trade. Emboldened by the experience gained in the war, they entered on new and world-wide cruises to India, China, and to the Pacific. As early as 1783–1784 Peck's *Empress of China* sailed from New York and blazed the trail to the old land of Cathay. The first Massachusetts ship to arrive at Canton was the *Grand Turk,* which sailed from Salem and returned in 1787 with comfortable profits for its owner, Elias Hasket Derby. From then on, the Canton trade became more and more popular, and Boston and Salem became the centers of this new traffic. Boston, already a provincial center of considerable importance, obtained a touch of metropolitan flavor. Salem changed from a fishing port into one of the leading seaports on the Atlantic Coast. The "romantic" days of New England shipping had begun. They bore fruit for American art, technology and science.

Salem's skippers sailed to China via the Cape of Good Hope, India, Java and the Philippines. In a division of the sea routes, Boston's captains selected a western course through the Strait of Magellan and by America's northwest coast. Pepper from Sumatra, coffee from Arabia, tin from Banka, cottons from India, crockery, silk and tea from China, furs from the northwestern shores, found their way into the coastal towns of Massachusetts. Nantucket's and New

Bedford's whaling, and Stonington's sealing, received new impetus. Fortunes were made, later to be invested not only in transportation, but also in the new mass industries.

This activity was conducted against a background of considerable danger to American shipping. England, France and Holland were at war during a greater part of the period ending with the War of 1812, and men-of-war of all the three countries searched, detained or seized American ships for contraband. The embargo of 1808 was a heavy blow to the Atlantic towns of New England. Piracy increased all over the world. To counteract the danger to shipping, Congress had already authorized the building of a new American navy. The program was accepted in 1794. Six warships were to be built. One of them, the *Congress,* which saw little service, was built under the supervision of James Hackett at Portsmouth. The *Constitution* was built at Boston, under the supervision of George Claghorne, and was launched in 1797; its career has become a legend of American naval glory. This "Queen of the Navy," which fought in the war with Tripoli and the War of 1812, now lies anchored at the Boston Navy Yard as a national monument. The place at Constitution Wharf in Boston where "Old Ironsides" was launched is marked by a tablet in bas-relief.

During the War of 1812 new ships were added to the American navy and from this time on there were regular additions and improvements. American shipbuilding was a flourishing business.

4

Many a savant or inventor hailed from these salty towns, where knowledge and mechanical skill were widely appreciated. Interest in meteorology, astronomy and theoretical navigation came naturally in places where fortunes were made and lost in shipping and where lives were risked on every voyage. Under such circumstances, understanding of the weather, knowledge of stars and tides, and correct place determination on the seas were both an economic necessity and an affair of deep personal concern. The observation of eclipses and transits, the study of the orbits of moon and planets were more than a gentlemanly pastime. Almanacs and books on navigation were added to the Bible and a book of sermons in the literature of the home. In the atmosphere of shipbuilding and ship maintenance mechanical inventiveness, already present everywhere in New England, was sharpened.

A great number of technical improvements made in those days have never been specifically connected with any name, and may become the subject of interesting historical research. It is not certain, however, that sufficient material exists, since many master builders built without elaborate drawings, and found their way by strictly empirical means. Some improvements have been definitely connected with some place along the New England shore, though often tradition is not strongly substantiated by facts. We have already mentioned that the invention of the schooner is ascribed to colonial Gloucester, but the story lacks authenticity. Another tradition connects Orlando Merrill, a Newburyport shipbuilder, with the invention of the so-called lift models of ships, built to improve hull design. The lifts represent longitudinal sections of the ship along the water lines, which were cut to the desired shape and joined to float in water. The claim for Merrill is refuted by the fact that such models were constructed in New England long before Merrill's time, but he certainly made wide use of them. These lift models became widely used and were of paramount importance in hull construction, especially in the years of the clipper ships.

Once aroused, inventiveness may lead into many directions. Many a good townsman hoped to make a little extra money from some tinkering of his own. For better or worse, one idea gave birth to another. A clever craftsman might begin to experiment with the mass production of some household commodity, such as nails or tacks; he might end with a wild scheme for steamboats or new textile machinery. Rumors and stories about inventions and technical improvements in foreign countries abounded along the water-fronts. One man had seen canals in England or Holland, another had met a foreign savant or visited an observatory, a third had witnessed the luxury of the Oriental aristocracy. Foreign lore and foreign influence lost much of their strangeness and terror along the wharves and quays; foreign languages were freely spoken and studied.

Salem had several such linguists, some even of considerable distinction. There was William Bentley, Democrat and Unitarian clergyman, who spoke most European languages and had a reading knowledge of many others, including Arabic and Persian. John Pickering, son of the Federalist Colonel Timothy Pickering, was an authority on Hebrew and on Indian languages, and prepared the first American lexicon of Greek and English. Cosmopolitan Salem seemed to have been an excellent training ground for such men. Nathaniel Bowditch, the mathematician, studied languages during his entire lifetime, using a New Testament and a dictionary as textbooks. Eventually he numbered among the books of his library twenty-five New Testaments in different translations. Nor was Salem the only town with linguistic leanings. In Boston, Cambridge, Providence and New Haven were men not afraid of grammars in foreign tongues, if only French, German or Spanish. On the island of Nantucket Walter Folger, a colleague of Bowditch, studied languages in a similar way, with the aid of New Testaments.

Astronomy was one of the primary scientific activities in these coastal towns. Behind much of the serious work in this field lay the search for exact longitude determination at sea. The problem was an ancient one, connected with the names of Galileo, Huygens, Newton and other great scientists and inventors of a former age. While latitudes can easily be estimated by observing the positions of the true North near the polar star, longitudes depend on comparison of local time and Greenwich time, and can be exactly determined only when good timepieces are available. In the middle of the eighteenth century reasonably satisfactory timepieces had been invented by the Harrisons in England, but the "chronometers" were expensive and did not come into general use until 1820 and later. The determination of a ship's place in mid-ocean remained a difficult problem.

Methods of finding longitudes at sea had been developed for many years, long before accurate chronometers were constructed. Scientists and inventors were spurred by rewards offered by governments and public bodies. These methods merit more study than they have so far received because of the large extent to which they influenced seventeenth- and eighteenth-century mathematics, astronomy, invention and, through these, philosophy and religion. A method suggested by Galileo in the early seventeenth century was based on the observation of the eclipses of the moons of Jupiter, which could be predicted well in advance and tabulated. More practical was the method based on lunar distances, which are the angular distances from the moon to some suitable fixed stars. On board ship lunar distances can be measured by means of a sextant; comparison with tables in Greenwich time can then give the local time. The advantage of using the moon as a reference body lies in the fact that the moon has, of all the bodies of the solar system, by far the most rapid apparent motion. This motion, however, became only better known in the middle of the eighteenth century, when Nevil Maskelyne, the Astronomer Royal, was able to publish appropriate lunar tables in his new enterprise, the *Nautical Almanac* (1767). After this, the method of "lunars" came into more general use. They were never very satisfactory, because they required good observers with mathematical skill, and an error of only one minute of arc produced an error of about thirty miles of longitude.

For this reason, and above all because of tradition and the expense involved, New England captains continued to rely upon dead reckoning, supplemented by uncanny seamanship. It was not unusual, however, for a ship to grope toward her destination almost as wildly as a bat in a cave. On a voyage in 1790 the *Massachusetts* from Boston missed the famous landmark called Java Head and lost three weeks of time wandering through the Indian seas. In an age where a good clock on board ship was an exception, the chronometer had to establish its value by a tedious process of demonstration and technical perfection extending over a period of years. It was not generally introduced till, in a period of increased prosperity, some of the ancient Yankee notions of thrift had lost their stringency. As late at 1823, the captain of a ship owned by Bryant and Sturgis of Boston, who had purchased a chronometer for $250, was informed in well-chosen words that he had to foot the bill himself:

> . . . could we have anticipated that our injunction re-

specting economy would have been so totally disregarded, we would have sett fire to the Ship rather than have sent her to sea.

One of the first makers of chronometers in New England was William Cranch Bond, who in 1839 became the first director of Harvard Observatory. But even Bowditch, at any rate in his seafaring days, recommended chronometers only "in a short run." "In a long voyage," he wrote, "implicit confidence cannot be placed in an instrument of such delicate construction, and liable to so many accidents." In later days, ships used to take three chronometers in order to eliminate individual shortcomings. At present, radio time signals are used to supplement the chronometer information.

5

Boston, Salem and Newburyport, closely knit together, formed an important intellectual center. Other such centers were New Haven, Providence and Newport. Interest in science and technique was widely spread, in the whaling ports of New Bedford and Nantucket, along the Connecticut coast in places like New London, Stonington and Norwalk, and along the northern coast in Portsmouth, Bath, and in Hallowell, that interesting settlement along the Kennebec founded by the Gardiners and the Vaughans. At Brunswick, Bowdoin College was founded in 1794 by a grant from James Bowdoin in Boston, the son of the governor of Massachusetts, who emulated his father as a patron of science. There were colleges in Cambridge, Providence, and in New Haven, where some nautical science was taught by landlubbers who could not escape the influence of the sea. There were men who indulged in the cultivation of astronomy, mathematics or meteorology, since they were useful for navigation, and occasionally studied new inventions which might well offer a profitable method of investment. Medicine, with its importance to public health, remained a topic of interest in many circles. Architecture began to prosper with the growing wealth, and books appeared to teach some of the arts and sciences. Much care was bestowed on the collection of meteorological data, especially temperature and rainfall. They were kept by professors at Yale, Harvard and Bowdoin, but those of Dr. Holyoke of Salem, which began in 1754 and ended only with his death, belong among the most complete records of this period. Most of all this scientific and technical activity was in close imitation of England, with which the new republic had first of all to catch up, but by the early decades of the republic an observer can detect the beginning of some intellectual independence.

Newburyport had its highly intellectual merchant families such as the Jacksons and the Lowells, from which came, among many other luminaries, an excellent Boston physician as well as the American inventor of the power loom. Here worked, in a goldsmith's shop, the versatile Jacob Perkins, who invented methods of plating shoe buckles and a machine for cutting and heading nails and tacks in one operation. He also made the dies for the copper coins which the Commonwealth issued in 1787. These old Massachusetts cents, with their Indian and Eagle, were coined only until the Federal Government stepped in, and are now a prized collector's item. To Perkins they meant the beginning of an interesting career, which led him in 1805 into the engraving of banknotes, a contribution to that constant fight against counterfeiters which was typical of the period. Early paper money banking in the United States has a history of insecurity. Perkins's method was adopted by most banks in New England and elsewhere in the United States. He substituted steel engraving for the customary engraving of copper, but it was not an easy process at that period. Not finding enough appreciation in the United States, Perkins went, in 1818, to England where he founded the firm of Perkins, Bacon and Company, Ltd., which printed the first British penny postage stamps. In the next forty years it produced over 22,000,000 British postage stamps by a process which Perkins had invented. The firm is still in business in London, and still engraves banknotes and stamps.

In 1823, at fifty-seven years of age, Perkins started an entirely new type of activity, new to him and almost entirely new to science. This was his experimental work with steam of high pressure. He embodied his results in many inventions, on which he continued to work until his death in 1849. His son Angier and his grandson Lofton distinguished themselves as engineers working in the same field of high-pressure steam.

There were more remarkable men in the flourishing seaports on the Merrimack River. We shall meet Timothy Palmer as one of the first great bridge builders of this country, whose wooden bridge connecting the banks of the Schuylkill stood for many years as a model for other builders. Another example, if on a humbler level, was set by Nicholas Pike, schoolmaster and magistrate, who published, in 1788, the first American arithmetic text to attain wide popularity, and to which we return in Chapter 6. In near-by Bradford was the pastorate of Samuel Williams, the amateur astronomer, who succeeded John Winthrop in the Hollis chair at Harvard, and who led the expedition to Penobscot Bay organized by the American Academy of Arts and Sciences. A curious and influential figure also was Edmund March Blunt, who came to Newburyport from Portsmouth to establish a newspaper business and bookstore near the Wolfe Tavern. He became widely known when he began to publish a series of nautical works. His first big venture was Captain Furlong's *American Coast Pilot,* which appeared in 1796 and for the first time offered badly needed information on harbor and sailing conditions along the coast of North and part of South America. It described every port in the United States and gave considerable further information for safe sailing. A veritable classic in its own restricted field, it passed through nineteen editions in the United States and was translated into many foreign languages. Many persons have been saved from shipwreck because of the information contained in this pilot's handbook. The enterprising Blunt now set out on an undertaking which reached even wider fame. In 1799 appeared Bowditch's *Practical Navigator,* continued in constant new editions and republished till this very day. These works, together with a number of charts, made Blunt's Newburyport bookstore the center of American nautical publications. He later moved to New York, perhaps as a consequence of a silly quarrel, in which the temperamental Blunt threw a heavy skillet at his adversary, who avenged himself by having a caricature of the publisher engraved on household crockery. The business in New York continued to expand, first under

Blunt himself, later under his sons, whose work was basic in the organization of the present United States Hydrographic Office, which obtained the copyright of Bowditch and other Blunt publications.

6

Interest in science and the mechanical arts abounded in and around Salem. It was in 1790 the sixth largest town in the country, with 8000 inhabitants, not counting its neighbor Beverly and its rival Marblehead across the harbor. Here there were several learned ministers and lawyers who could interpret the tenets of the stars and the theories of Newton as well as the texts of their Bible or their Blackstone. Joseph Willard, while pastor at Beverly, observed eclipses and computed longitude distances, before he was called to the presidency of Harvard College in 1781. Near Salem lived Manasseh Cutler, the amateur botanist and astronomer. The Reverend John Prince, at Salem, who served the souls of his flock almost as long as his contemporary, Holyoke, tended to their health, had a solid reputation as an expert instrument maker, notably as the constructor of an air pump. William Bentley, the linguist, was so well known outside of Salem that Jefferson offered him the presidency of his new University of Virginia—which he declined. We have already mentioned the legal luminary Andrew Oliver, who not only had his own theory of comets, but also wrote on meteorology and on water spouts.

A different and more original figure was Nathan Read, who hailed from a farm in Worcester County and graduated from Harvard in 1781. After a period of tutoring at his college he came to Salem shortly after the Revolution, where he studied medicine and opened an apothecary's shop. Read was a learned gentleman—he specialized in Hebrew—and a skilled mechanic to boot, uniting the elements of his backwoods experience with those of his Harvard training in a rather unusual way. He took out patents on several inventions typical of the period right after the disappearance of the mercantilist oppression, such as a machine for cutting and heading nails in one operation. His principal claim to glory lies in his experimentation with steam engines for their use in transportation, especially in navigation. The multitubular boiler with 78 vertical tubes and the double-acting steam engine he invented and patented—probably without knowing much about Watt—were distinguished contributions which eventually helped in the invention of a practical steamboat. During 1789 he also experimented with paddlewheels, turned by a crank (by hand, not steam), on the Porter River between Danvers and Beverly, to test its possible application to steamboats, entertaining with this curious contraption Pastor Prince and other friends from Salem. Read, however, discontinued his work. "I was too early in my steamboat projects. The country was too poor," he confessed much later in an autobiography. In 1796 he was one of the organizers of an iron foundry near Salem and for eleven years engaged in the manufacture of iron cables, anchors and other iron materials for ships. Later he moved to Belfast, Maine, where he lived to a ripe old age, published a history of his life in 1834 and served for many years as Chief Justice of Hancock County. He died in 1849.

Through the influence of such men Salem acquired one of the best scientific libraries of those days, perhaps the best north of the Delaware. Already in 1760 a number of wealthy townspeople, led by young Dr. Holyoke, had brought money together to purchase a so-called "Social Library." Then, during the Revolution, a privateer operating in the Irish Channel captured a British vessel. Among the booty, sold at Beverly in 1781, was the library of one of the best-known scientists of the day, Richard Kirwan, the Irish chemist and naturalist, now perhaps best remembered for his defense of the phlogiston theory against Lavoisier. Some leading Salem persons, among them Holyoke, Willard and Cutler, effected the purchase of these books by a subscription. This became the nucleus of the so-called "Philosophical Library." It included the best of European science, the *Philosophical Transactions* of the Royal Society, the *Histoire de l'Académie Royale* of Paris, the Berlin *Miscellania,* the works of the Bernoulli brothers, of Boyle, Newton, Maclaurin and the *Encyclopaedia Britannica.* This was a library not only of English, but also of continental books, a great rarity in the United States. Kirwan later refused an offer of repayment, "expressing his satisfaction that his valuable library had found so useful a destination." Joseph Willard became the first librarian. In 1810 the Social and Philosophical Libraries were united into the Salem Athenaeum, a flourishing institution where the ancient books can still be consulted.

7

The most famous of early New England scientists, Nathaniel Bowditch, hailed from Salem. His father, a sailor and a cooper, had never been able to provide a halfway comfortable existence for himself and his family of seven children. Nathaniel was born in 1773, had some elementary education, was apprenticed to a ship chandler and drew the attention of the local men of education by his studious habits. His patrons granted him access to the Philosophical Library, where he studied the mathematical and astronomical books, learning for that purpose Latin and French—largely self-taught. By studying books in the French language Bowditch became, with Irish-born Robert Adrain in New York and New Jersey, the first in the United States to understand continental mathematics and thus to break the spell of the exclusively English tradition in this field.

From 1795 to 1803 Bowditch participated in the China and East Indian trade on five long voyages. He started as a clerk, became a supercargo and on his last journey was captain of his ship. During the long and uneventful days on board he studied mathematics, especially the elaborate French texts of Lacroix and Laplace, and checked up on the existing charts and tables of navigation. He was struck by the many errors in the texts, tables and charts. Carefully compiling his results, he obtained an enormous stock of information which he offered to Blunt in Newburyport. Blunt saw money in the idea, and in 1799 published an improvement of an English text by J. H. Moore, with the title of *The New Practical Navigator,* first published in London in 1772. The title page stated only that "a skillful mathematician and navigator" had revised and corrected the text and had added new tables, but did not mention Bowditch by name. However, some recognition of his labor was given the young mathematician by his election to the American Academy of Arts and Sciences, quite an honor for a poor cooper's son. Harvard conferred an honorary degree upon him in 1802, three months after the first edition under Bowditch's name appeared.

This was the third edition of the book, prepared on a voyage to Batavia and Manila. Bowditch had found more than eight thousand errors in Moore's text, some of them serious indeed. The year 1800 had been tabulated as a leap year, and this blunder had actually been responsible for shipwrecks. Bowditch added an introductory text, as well as a new method for working lunars, and had it published by Blunt in 1802. The book was also published in England at the same time. It was *The New American Practical Navigator,* by Nathaniel Bowditch, A.A.S., with 247 pages of instruction and 29 pages of tables, explanations of sea terms, marine insurance, bills of exchanges, and other such useful information for the sailor. The instruction dealt with geometry and trigonometry, Gunter's scale and sector, logarithms, the handling of log and glass, quadrant and sextant, and led up to the determination of altitudes, declinations, time and lunars. It was a complete handbook of navigation, the best written until then in the English language.

In this book Bowditch revealed himself not only as a conscientious computer, but also as an able teacher, who knew how to write a text which sailors could understand. His book became extremely popular in nautical circles and was adopted as a standard text by teachers of navigation and by marine societies. It passed through one edition after another. Bowditch kept on improving the ten editions which appeared during his lifetime; later editors have continued to modernize it, and even now it is issued regularly as a publication of the United States Hydrographic Office. At an early date a kind of folklore had already grown up around the book. Captain Robert Bennet Forbes, who became known as a writer and an inventor, and who was one of the first to believe in iron hulls, has related how, in 1817, as a lad of thirteen, he went to sea with a "Testament, a Bowditch, a quadrant, a chest of sea clothes, and a mother's blessing." The study of "Bowditch"—from the book or from the author himself—became the fashion on many an American vessel, a study in which master and mate, supercargo, and occasionally even the cook participated. We know that when the Salem brigantine *Cleopatra's Barge* stopped at Genoa in 1817 the well-known astronomer von Zach was amazed at the skill of its crew in theoretical navigation. Even the Negro cook could compute lunars, and knew the advantages of one method of computation over another. In its way the "Bowditch" became a seaman's bible, and even now it is a standard book of reference. Its fame is comparable to that of the dictionary compiled by Bowditch's contemporary, Noah Webster. The *Practical Navigator* is one of the outstanding landmarks of New England's seafaring past in the world of books, as are also Dana's *Two Years Before the Mast,* Melville's *Moby Dick* and the American edition of the *Nautical Almanac.*

In compiling his book, Bowditch was aided by a series of manuscript journals, compiled by the East India Marine Society. This society, organized in 1799, required that every member keep a log, to be examined upon his return. The logs were copied in large volumes for further use and deposited in the East India Museum at Salem. According to Bowditch this stock of information, thus made available, was unsurpassed in the world. The curious collections of the society can still be admired in Salem's Peabody Museum, which also contains a set of early navigating instruments. Some of these instruments belonged to Bowditch.

As captain of the 260-ton three-master *Putnam,* Bowditch ended his last voyage in a sensational way, by making port in a heavy fog. Many years afterwards sailors used to tell this story of marvelous seamanship, and this added to the Bowditch legend. But now, at thirty years of age, quite an advanced age for a New England captain of that period, he decided to remain on land. The Essex Fire and Marine Insurance Company was looking for a president. Insurance companies were still a novelty in New England, although in 1752 Benjamin Franklin had established his "Philadelphia Contributionship." The practice of underwriting of marine risks by individuals or an occasional marine society was slowly superseded by corporate underwriting, but distrust of corporate practices impeded the progress of insurance companies. By accepting the presidency of the Salem company

Bowditch gave the weight of his prestige to the new venture. Again it was a field which required the skill of a businessman as well as of a mathematician. With his usual good business judgment and discretion, Bowditch made a success of his office, which he held from 1804 to 1823. The company paid the stockholders an average annual dividend of 10 or 12 per cent during the whole twenty years of Bowditch's presidency and had a large surplus when he left. Even the bad years of the Embargo and the war with England did not affect the solidity of his reputation.

8

The duties of an insurance director did not entirely absorb Bowditch's interests. He raised a thriving family, bought a beautiful house and, with his methodical habits and indefatigable energy, became more and more a public figure. He could be seen on many public occasions, a small-built, very active man with prematurely gray hair. As a settled and respectable Salem burgher he took an interest in the various literary, scientific and charitable institutions, and shared faithfully the Federalist convictions and prejudices of Salem's higher middle class. Sometimes he grated on the nerves of fellow citizens: "The little Mr. Bowditch puffed up by the flattery of his mathematical studies and destitute of every degree of literature or manners, has attempted to sacrifice me to party," wrote the Democratic Reverend William Bentley in his journal during 1804. However, his mathematical studies in the small seafaring town gave Bowditch a kind of Olympic position which was difficult to touch. In his spare hours he also engaged in astronomy, theoretical navigation and chart construction. Between 1804 and 1806, in a period when good maps of the New England coast were still rare, he prepared a chart of the harbors of Salem, Marblehead, Beverly and Manchester, which was widely praised for its beauty and exactness. It had all the familiar landmarks mapped so well in their true position that the pilots claimed that Bowditch had discovered the professional secrets, charting points and bearings which they had thought known only to themselves. The merits of such work are better understood when we realize that at this time President Jefferson was just beginning the organization of the Coast Survey, which did not accomplish much for several decades to come.

Bowditch's best-known scientific work, his translation of four volumes of Laplace's *Mécanique Céleste* into English, also belongs to his Salem days. The books were published in Paris between 1799 and 1805, with parts of the fifth volume appearing between 1823 and 1825, and were the work of Pierre Simon Laplace, one of the greatest of French mathematicians and theoretical astronomers. They form a full exposition of Newton's theory of gravitation, applied to astronomy and enriched by a hundred years of continental research. Bowditch had studied Laplace during the long hours on board ship; he continued his study during his retirement from active sea life, after the claims of social and domestic duties had been answered. His decision to translate the enormous work and to enrich it with notes provided the English-speaking world at last with a full exposition of continental mathematics and theoretical astronomy, and at the same time brought the Newtonian period of American science to a climactic end. The translation was made between the years 1814 and 1817, at which time the four volumes of Laplace —all that had appeared up to that date—were ready for publication. No work showed better Bowditch's methodical habits and his indefatigable energy. In his desire to make it a full introduction to the long neglected study of continental mathematics he added to the text nearly an equal amount of notes. It was difficult to publish such an opus in America; it came out in print only after Bowditch had saved enough money to defray the expenses. This did not happen before he had moved to a more lucrative position in Boston.

To Bowditch's Salem period belong thirty-one papers, the more important of which were published in the *Memoirs* of the American Academy of Arts and Sciences. The Academy had accepted him as a member after his first book was published and soon counted him among its most prominent representatives. Two of the papers published in the *Memoirs* deserve special mention, one because of the singularity of its topic, the other because it shows Bowditch as an independent investigator. An original research man was still the rarest of species in the America of his days.

On an early December morning of the year 1807 an unusual phenomenon occurred, which astonished all Americans and excited alarm in many witnesses. A huge meteor swept over New England like a flaming ball and crashed to earth with a number of violent explosions in the Connecticut town of Weston. Meteorites had been noticed before, but this was the first time that the descent of such a heavenly body had actually been observed in North America. Not so long before, in 1794, the German physicist Ernst Chladni had proposed the theory that meteorites were of cosmic origin; but even French Academicians had shrugged their shoulders at such fantastic prattle. Here, in New England, was one of the proofs of Chladni's theory, and scientists were present to study the situation. Professor Silliman, the young chemist of Yale College, supported by Yale's librarian James Kingsley, went to near-by Weston to collect fragments of the meteorite, which had smashed into many pieces. The largest segment they could locate weighed about two hundred pounds; they were in all able to account for about three hundred pounds. Chemical analysis show 41 per cent silica, 30 per cent oxide of iron, and some magnesia, nickel and sulphur. Some of these fragments have been preserved at Yale.

Bowditch, on hearing of this striking celestial event, set out to study its astronomical aspects, publishing his conclusions in 1815. He compared the different observations which had been made of the meteorite from the first moment it had been observed, near Rutland, Vermont, until its explosion in Weston. A distance of 107 miles had been covered. He concluded that the meteorite had been moving at an altitude of 18 miles at 3.5 miles per second with respect to the earth, and determined direction and magnitude of the meteor. Its original bulk, he claimed, must have been about six millions of tons, which others have estimated to be the contents of the pyramid of Cheops. Bowditch's contribution was well received in Europe, where it added substantial material to Chladni's catalogue. Since that time only a few more meteorites have fallen on New England.

The other paper of Bowditch's deserving special notice was published in the same number of the *Memoirs* as that discussed above. Professor James Dean, of the University of Vermont, was interested in the apparent motion of the earth as viewed from the moon. He had experimented with the motion of a pendulum suspended from two points instead of one. Not only was the heavy ball subjected to oscillations in a vertical plane about its point of suspension, but this point itself was subjected to oscillations in a perpendicular plane. Bowditch found it a case which could be analyzed by the methods of Laplace. The result was the systematic generation of the different figures which the heavy ball could trace, among them segments of a straight line, circles, ellipses and even more complicated figures. The importance of this paper lies in the fact that many years after Bowditch the French physicist Jules Lissajous, in a problem on vibrating strings, found these curves again, and they are now taught as part of the regular college curriculum.

Bowditch's Salem period came to an end in 1823, when he accepted the directorship of the Massachusetts Hospital Life Insurance Company in Boston, newly founded by members of that city's merchant aristocracy. The salary, $5000, was more than three times his salary at the Essex Fire Insurance Company. He now moved to Boston with his wife and family of six children—four sons and two daughters. The scientific and actuarial activities of his later days will be described in their Cambridge and Boston settings.

9

In moving to Boston, Bowditch did what many other prominent Essex County people were doing. They had originally established themselves in the Essex seaports, but saw larger rewards, economical as well as political, ahead of them in the provincial capital. Some may have had an understanding of the coming decay of these smaller towns. The Jacksons and Lowells from Newburyport, the Cabots, Lees and Higginsons of Salem, all moved to Boston, and in this way strengthened their financial position. At the same time they built their narrow political wing, first known as the Essex Junto, stronghold of Federalism and doomed to dismal failure after the Hartford Convention in 1814. It later became a dominant force within the Whig party. Their very narrowness and intolerance was at the same time a source of strength in the limited domain of New England politics and culture. Most of them tended toward Unitarianism in religion and toward Newtonianism in science. In politics they changed from free trade to protection as their interests changed from commerce to industry; this helped to change their interests in science from astronomy to chemistry and natural history.

Boston was not only a seafaring town but also a provincial capital, and though its streets smelled salty enough near the wharves, it did not, like Salem, depend completely on the sea. The amateur astronomers at Cambridge were less windbeaten than their colleagues near the shore, and as far as their professional interests were concerned, the learned men at the Medical School were rather indifferent to navigation. Life around 1800 was still simple enough, and even the local aristocracy had little show of wealth. Apart from a number of meetinghouses and such colonial mansions as we can still admire on Brattle Street in Cambridge, little art was originally displayed in building. With the growth of Boston as a port, new and beautiful mansions began to appear on Beacon Hill, built with money originally obtained from commercial enterprises, though gradually invested in the new industries. In 1794 there were some 80 wharves and quays, later followed by new and impressive ones, like India Wharf of 1805, with 32 stores built by Bulfinch; and Central Wharf of 1819, a quarter of a mile long and with 54 brick stores, a structure at that time perhaps unprecedented in the United States. These were undertakings compared to which even Salem's Derby Wharf looked insignificant. The history of Boston in the nineteenth century is that of a rapidly expanding city with an excellent harbor and an active industrial and intellectual life.

The foremost architect of the Boston merchants was Charles Bulfinch, a Harvard graduate but, like all technical men of the period, self-taught as an architect. Technical education could not be obtained in schools. An extended trip to England and the Continent stimulated his taste, and he returned with a heart full of ambition and a trunk full of books. His great talents were recognized after 1795, when he built the new State House on Beacon Hill. He attached his name to public and private buildings of many kinds, including the stores on India Wharf, the Massachusetts General Hospital, the Insane Asylum of South Boston and University Hall at Harvard. Many of these buildings still stand, and in their chaste neoclassicism exhibit a quiet and delicate charm. After a dozen years in Washington as the architect of the Capitol —half burned by the British in 1814—he returned to Boston to build again in New England, notably the State House in Augusta, Maine. In 1844 he died at the age of eighty. He was the first professional architect of New England and opened a period of remarkable activity. Part of his library can be seen at the Massachusetts Institute of Technology; it contains some books on the mathematics of perspective.

Bulfinch's influence was extended through the men he influenced. One of the best known of them was Asher Benjamin, who came to Boston in 1803 from Western New England. Through Benjamin's books Bulfinch's influence was carried into the village and the home. The first of these books, *The Country Builders Assistant,* appeared in 1797; the last of the group, *Elements of Architecture,* in 1843, only two years before Benjamin's death. All these books passed through many editions. There is hardly a town in New England which does not, to this very day, reflect his influence, whether in the lines of its church spires, or in the moulding profiles, trims or cornice details of its homes. The "colonial" revival after 1910 has again made us familiar with Benjamin's influence. It is significant that his last book was partly devoted to the study of strength and stiffness of material. This shows how, in the eighteen-thirties and forties, the very requirements of architectural practice began to demand an interest in more theoretical questions of applied mechanics. Among Benjamin's pupils we find Ithiel Town of New Haven, about whom later.

With the new prosperity, Boston found funds for works of public improvement, for bridges, buildings, causeways and harbor works which changed almost beyond recognition the aspect of the old revolutionary hub. The execution of these works also helped to transform the traditional mechanics and craftsmen of the older days into men of more modern engineering outlook. The leading engineer was Loammi Baldwin, Jr., whose career we shall later discuss in more detail. With him were other technical men, among them Alexander Parris and Solomon Willard, who were half contractor and businessman, half engineer. To this same group belong Uriah Cotting, who did a good deal of work in improving property for commercial purposes. He projected the Mill Dam, a causeway leading from Charles Street to Brookline through the Back Bay, and was its first engineer. A considerable number of technical and commercial enterprises sprang up in connection with this dam, among them a tide mill. Later the dead water on the inner side of the Mill Dam was filled in and became the Back Bay district of Boston; Mill Dam became Beacon Street. When David Stevenson, a British civil engineer, visited Boston in 1837, he was impressed by the many engineering projects which he saw in this city:

> The population of the town is about 80,000. Its situation is curious. Placed on a peninsula having deep water close inshore, and almost entirely surrounding it, it is connected with the adjoining country by means of a dam and seven wooden bridges, of which the most extensive is about a mile and a half in length. The dam consists of an embankment of earth 8000 feet in length, enclosed between two retaining-walls. It serves the double purpose of affording a means of communication, and also forming a large basin, in which the tide-water being collected, a water power is created for driving machinery.
>
> The quays of Boston are constructed in the same style, and of the same materials, as those of New York [wooden piles with diagonal braces and earth], but more attention has been paid by the builders to the durability of the work. Some of the wharfs extend about a quarter of a mile into the harbour, and are of sufficient breadth to have a row of warehouses built on them. In the suburb called Charlestown, which is connected with Boston by means of three

wooden bridges, is situated the navy-yard of the United States, and the graving-dock already mentioned.

Indeed, few cities in the world have changed their geography so much as Boston and in so short a period.

10

Scientific stimulus for all this engineering and nautical activity came only in small amount from the mildly dusty college in Cambridge, which was slow to attract new talent and new ideas. It recognized in the sciences the leadership of Bowditch, who spent the last decade of his life in Boston and showed deep interest in the college. After all, Bowditch worked in the traditional Newtonian line. With the Salemites moving into Boston came Bowditch's young friend and fellow mathematician Benjamin Pierce, one of the men who eventually stimulated the college to a more modern curriculum. Though scientific interests arose gradually in Boston and Cambridge, owing partly to these incoming merchant families, there remained for a long time a certain narrowness in Boston and its surroundings, long after its Federalist days were past.

The career of William Cranch Bond was closely connected with the sea. His father, an unsuccessful Portland (Falmouth) lumber exporter, had settled in Boston in 1793 as a struggling silversmith and clockmaker. The firm still exists under the old family name, overlooking Boston Common on Park Street. Young Bond learned his father's craft to perfection and around 1804 constructed a ship's chronometer. It was the period when sailors frowned on such expensive innovations. Yet chronometers were in demand and for several years Bond's chronometer was taken as the standard by which other chronometers of ships sailing out of Boston were measured. The young expert's interest in this helpful field of navigation stimulated again his interest in astronomy, but like Bowditch in Salem and Folger in Nantucket, he had to study without an expert teacher. He built a transit instrument in his Dorchester home—"a strip of brass nailed to the east end of the Champney house, with a hole in it to see a fixed star and note its transit," according to his brother. In 1811, with the aid of this primitive contraption, he was able to discover a new comet. The comet of 1811 was also discovered by others, but the fact that there was a young instrument maker in Dorchester who had found it independently attracted considerable attention. Bowditch, who spent much time in computing the elements of the orbit of the comet, was impressed, and so was Professor Farrar, the mathematician at Harvard. When, in 1815, Bowditch and Farrar were appointed to a committee of Harvard College "to consider the subject of an observatory," and they heard that Bond was about to make a trip to Europe, they were influential in procuring from the college a commission for Bond to visit Greenwich Observatory and to gather data concerning instruments. When Bond returned, he found that the college was not willing to establish an observatory. He did the best he could at his own expense, and had a model dome and telescope constructed in the parlor of his own house—a huge granite block in the middle of the room serving as the foundation. The ceiling was intersected by an opening in the meridian. During the day Bond was occupied by his Boston business; his spare time he spent in constantly gathering data on chronometers, performing astronomical observations, and later also observing terrestrial magnetism for the United States Government. In 1839 he moved to the Dana House in Cambridge as the "Astronomical Observer" of Harvard College, and when at last Harvard began to set up its own astronomical observatory he became its first director.

11

Deep in the ocean lay Nantucket Island, home of Quakers and whalers. Abiah Folger, second wife of the Boston chandler Josiah Franklin, and mother of Benjamin Franklin, was a Nantucket girl. Already during colonial days Nantucket had sent its tiny ships as far as the Davis Strait and the coasts of Brazil. In 1775 it possessed 150 whaling vessels, of which it lost 134 during the war. Throughout the Revolutionary War many Quakers, with their doctrine of nonresistance, tried to be neutral; the result was that the island was attacked by both sides and suffered terribly. After the war Nantucket tried to stage a comeback. It is said that the "rebellious stripes" of the new American republic were first displayed in the Thames flying from the masthead of the *Bedford,* owned by William Rotch, a wealthy merchant of Nantucket. Recovery was slow, and the War of 1812 was another blow. But the recovery gained a new momentum after 1814, and Nantucket headed towards an era of considerable prosperity.

The so-called "glorious age" of whaling in Nantucket and in its daughter town New Bedford has been the subject of considerable literature, as detailed as that of the "glorious days" of Salem. Out of it has come at least one literary masterpiece, Herman Melville's *Moby Dick.* In 1835 Nantucket was rated as the third richest municipality in Massachusetts, being outranked only by Boston and Salem. Like their colleagues of Salem, the Nantucket captains were interested in longitude determinations and astronomy in general. This interest produced a Bowditch in Salem; it produced a Folger and the three Mitchells on Nantucket. Since it was smaller than Salem and far more isolated, Nantucket never produced the stately array of cultured merchants or learned ministers and lawyers of which the former could boast, and had no library as did Salem. Nantucket's captains, however, ventured deeper into uncharted seas than those of Salem. They were whalers and sealers, and whaling and sealing vessels had to penetrate deeply into northern and southern oceans. Nantucket captains belong to the early explorers of the Antarctic. The best known geographical contribution of a Nantucket captain is Matthew Folger's rediscovery of Pitcairn Island in 1808. Here he found one survivor of the nine mutineers of the *Bounty,* and thus solved the mystery of their fate, which had been unknown since 1789. Many other islands were charted by these captains, and somewhere in the Pacific Ocean they baptised a rock with the nostalgic name of "New Nantucket."

The career of Abiah and Matthew Folger's kinsman, Walter Folger, invites comparison with that of Bowditch in Salem. He was born in 1765 and received his early education mainly through self-study and evening school. Though never apprenticed to any craftsman he became an expert clockmaker. His craftsmanship can still be admired on the island, where he constructed an astronomical clock which tells seconds, minutes, hours, days, months, years and centuries, not forgetting the leap years. The rising and setting of the sun and moon, their paths, indicated by a disk and a ball respectively, moving in correct astronomical time, as well as the changes of the moon and the time of high tide in Nantucket, were given for each day. This clock, now kept in the Historical Association Museum, was constructed in 1787–1788, and still keeps time. Folger made many more instruments—a telescope, a barometer and a thermometer. When he had invented something he gave it away, saying that money would do him more harm that good, and in view of the tribulations of Whitney, Fitch, Morse, Goodyear, Bell and Morton with their patent fights, we may well find Folger's homely ways the better part of wisdom.

Like Bowditch he was a devotee of the exact sciences, and studied mathematics, navigation, astronomy, even medicine and the new arts of the textile industry. He was probably less systematical than Bowditch, but his interests seem to have been more varied. He acted as surveyor and as engraver, calculated eclipses and like the sage of Salem used a New Testament to learn French. In 1783, during a sickness, he learned lunar observations from a book which one of his uncles had procured abroad. He also taught lunars, and his pupil, Captain Joseph Chase, whom he instructed in 1789, may have been the first American captain to find his longitude in this way. He computed almanacs, but because of religious scruples refused to go to college. During the War of 1812 he established his own cotton and woolen mill, which he operated successfully with his family, and which was one of the first to use a power loom. There is a quaint story about him with a Keplerian flavor, according to which he computed the form of casks to hold a maximum amount of whaling oil with the least surface "by a difficult fluxional process," and "had his casks made to his plans, contrary to the cooper's views." "Gradually the people adopted his ideas, and by this means a great amount of money has been saved on this island, as well as in other places where whaling has been carried on."

That such a man corresponded with Oliver, Prince, Bowditch and Jefferson, directed a bank on Nantucket, discovered (in 1812) a method of annealing wire without changing its color, and in his later days (1826 to 1832) presided over the Nantucket Philosophical Institute is not astonishing. When we hear that he became a councilor at law in 1807 and practiced this profession for twenty years, we are somewhat amazed. And when we are further informed that in his later life he followed a successful career as a state senator, a Congressman and a judge, we must conclude that Walter Folger was one of the most versatile men New England ever produced. This talent, concentrated on a smaller variety of subjects, might have produced a first-class scientist and inventor. There is, moreover, an extremely human touch to this Nantucket Leonardo. Staying close to the people, giving gratuitous legal advice, teaching without compensation, he remained a Jeffersonian Democrat all his life. He never strove to be a local Olympian, whether consciously or unconsciously, as did the Federalist Bowditch, but on the other hand he never achieved as much as his purposeful colleague. In his own way he performed some scientific work, observing the comets of 1807 and 1811 and the total eclipse of 1806; he wrote on aerolites and on the comet discovered by Encke which kept the astronomers of that period busy for many years. However, most of his work exists only in manuscript. He died in 1849. He was certainly a man who deserves more attention than he has so far received.

Folger was not the only Nantucket astronomer. William Mitchell, a generation younger than Folger, enjoyed a similar though perhaps less varied career. Like all the men and women of Nantucket who achieved some fame, he was entirely self-made. He became a cooper, teacher, state representative, and an insurance and bank director on the island, holding the last job for thirty years. At the same time he "inherited a love of astronomy and diffused a scientific atmosphere around his home and neighborhood," to use the words of Professor Lovering. In his house were several telescopes, with which he made observations of star positions, partly for the United States Coast Survey and partly for his own use in rating chronometers for the Nantucket whaling fleet. Where Folger reminds us of Bowditch, Mitchell reminds us of his friend and contemporary, Bond. His work lasted till the fifties and sixties of the century; in 1860 Harvard awarded him an honorary degree.

William Mitchell is probably best known as the father of Maria Mitchell, one of the outstanding woman scientists of nineteenth-century America—a lady as distinguished in her chosen field as was that other remarkable Nantucket woman, Lucretia Mott, in the field of abolition and woman's emancipation. Since her father was "not merely fond of mathematics, he was addicted to that dismal science"—to use the words of one of Maria's admirers—she and her brother obtained a taste for his work. At an early age they were already helping the older Mitchell in his astronomical work, and Maria spent long nights watching the stars from their little observatory. In 1847 Maria Mitchell discovered a comet, which brought her a medal from the King of Denmark and a "fellowship" from the American Academy of Arts and Sciences in Boston—a grave upset in that august body's notion of the proper use of English words. In 1865 she became professor of astronomy at the newly founded Vassar College in Poughkeepsie, and lived to enjoy a quiet old age in Lynn. Her brother, Henry entered the Coast Survey and became the leading hydrographer of the United States.

The little gray house in the side yard on Vestal Street in Nantucket, where Maria Mitchell was born, is now maintained as a memorial to her. Across the street, where her father taught school, there is a scientific library also named after Maria. In the Memorial House one may see the little 2⅞-inch telescope with which she discovered the comet. Adjoining this house an astronomical observatory was built in 1908.

We cannot leave the New England whalers without saying a word about Lewis Temple, the ingenious Negro blacksmith of New Bedford, who by 1836 or earlier operated his own whalecraft shop on Coffin's Wharf. Nearby was the place where Frederick Douglass lived for several years, and the men may well have known each other. Temple deserves notice as the inventor of an improved harpoon, called the toggle, which had a removable head so mounted on the shaft that the shaft could be withdrawn after the head had plunged into the whale. When the whale began to run, dragging the boat behind it by means of a rope looped through a hole in the harpoon's head, the whale would force this head, now imbedded in its body, to turn at right angles, that is, to "toggle"—thus forcing it firmly into the blubber. This harpoon, which Temple first manufactured in 1848, was soon widely used, and has been called, by proper authorities, the most important single invention in the history of whaling. Temple died in 1854 after tripping in the dark over a plank. Damage was ordered by the City Council, but never paid, and Temple's widow was left struggling with debt.

12

We must confine ourselves to only a few words about the coasts of Rhode Island and Connecticut, about Newport and Providence, Stonington, New Haven and New London. The general picture was similar to that of the northern shore. Intellectual life was perhaps somewhat less intense, as in southern New England there were no centers such as Boston and Salem. Still, south of Boston, along the coast of Rhode Island and Connecticut, there were many towns with enterprising merchants, skillful and daring skippers, and friends of the mathematical and nautical sciences. Although Newport never entirely recovered from the terrible damage it suffered during the Revolution, when nearly a thousand buildings were destroyed, it remained the home of many a cultured family. In 1805 George Gibbs, the "Colonel" as he was called, returned to Newport after extensive travel through Europe. He extended generous hospitality, especially to those guests who liked to inspect his valuable collection of minerals. We shall see how one of these guests, Professor Silliman, was eventually able to obtain the collection for Yale College. It was also at Newport that David Melville was, in 1806, the first in the country to use coal gas to light his house. From Newport came Benjamin Waterhouse, a pioneer in vaccination, and

Walter Channing, a pioneer in the application of ether to childbirth.

Providence was not so seriously damaged by the war, and counted among its residents some of the most aggressive and enterprising merchants of America. For a while the commercial classes of Providence, with their strong Quaker element, vied with New York in the number of their ships. The outstanding figures were the four Brown brothers, wealthy merchants, whalers, China traders and patrons of science, one brother, Joseph, being himself an amateur astronomer and physicist. Another brother, Moses, was an early abolitionist, who with his brother John was instrumental in moving Rhode Island College to Providence, where it received the name Brown University after Moses's nephew Nicholas. The Browns—"John and Jose, Nick and Mose"—a curiously brilliant family, half merchant-adventurers of the Elizabthan type, half humanitarians with one eye on their purse and another on their Bible, set their stamp on the entire mercantile development of early republican Rhode Island and on the emerging textile industry of the United States.

Here, in Providence, lived Benjamin West, an almanac maker, not to be confused with his contemporary and namesake, the painter. West came from a farm near Taunton, Massachusetts, and in 1753 settled in Providence, where he had a drygoods store and later a bookshop. From 1802 until his death in 1813 he was also postmaster of the town. As an almanac maker West was a representative of that socially important group of men, who, from the Renaissance until deep into the nineteenth century, enriched the yearly book market with their curious almanacs, sources of astronomical information and of ethical edification, often filled with astrological and meteorological fantasies as well as homespun wisdom. Early American almanac makers were the Franklin brothers, James at Newport (1728) and Benjamin at Philadelphia (1732), where his *Poor Richard* received a fame which lasts till the present day. Boston had its almanacs throughout the seventeenth century. Another almanac maker was Nathaniel Ames of Dedham, an innkeeper and country doctor, father of that Federalist oracle Fisher Ames. Ames's

"Almanack", first published in 1725, was a household necessity in New England homes for at least half a century. Its lines

All men are by Nature equal
But differ greatly in the sequel

express the type of grass-root philosophy typical of these almanacs. Some of their authors were skilled astronomers, as was Benjamin West, one of the observers of the Venus transit of 1769, who also taught at Rhode Island College and at Brown University. Such men computed many of the ephemerides for their almanacs themselves.

13

Among the early republican mathematicians and astronomers of the Connecticut coastal towns there was one man whose influence has been more than local. Jared Mansfield was the son of a New Haven sea captain and merchant, who sent his son to Yale during the revolutionary period. Young Mansfield seems to have followed Stiles's lessons in mathematics with considerable success, but also seems to have suffered from what a biographer has called "certain unfortunate differences with the faculty in matters of standards of behavior that impelled the said faculty in his senior year to withhold Jared's diploma." After several years there was somewhat of a change of heart, for in 1787, when Mansfield was twenty-eight years of age, the sheepskin was forthcoming. Until 1802 he was a schoolmaster in his native town. During these years Mansfield wrote his *Essays, mathematical and physical,* published together in 1801, which show some mastery of theoretical navigation, Newtonian mathematics, astronomy and ballistics. In a country so barren of scientific publications as the New England of those days, these *Essays* were an actual contribution, and placed Mansfield in the front rank of American scientists. His work drew the attention of President Jefferson, who in 1802 appointed him a captain of engineers and a member of the first faculty of the newly established military academy at West Point.

Mansfield stayed at West Point only a short time. In 1803

President Jefferson appointed him Surveyor General of the United States, doing work in the newly opened Northwest Territories, particularly in Ohio and Indiana. The city of Mansfield, Ohio, carries his name. In 1812 he returned to West Point, where he stayed till 1828 as professor of natural and experimental philosophy. Here he was responsible for the very comprehensive course in natural philosophy, which was considerably ahead of most other American college curricula of the period. Serving first under the autocratic Alden Partridge, later under the majestic Sylvanus Thayer, he was long remembered at West Point as a venerable figure and a strict but kindly teacher:

> Mansfield, gentle, nearsighted, although coldly silent when listening to a poor recital, was unsparing in praise for work well done. For one of his scholarly, absent-minded temperament, his previous thralldom under Partridge's régime must have been a veritable purgatory. He expanded under Sylvanus, but could never adapt himself to a purely military system. Thus we find him, in 1821, protesting against disciplinary punishment, for cadets. "It is a mistake to suppose that because the Academy is military . . . it is therefore under martial law . . . thereafter Professor Mansfield doubts whether he will ever recommend a cadet for dismissal."

Mansfield's son, Edward Deering Mansfield, became an author and editor in Cincinnati. His daughter married Professor Charles Davies, a Connecticut Yankee like his father-in-law, who taught mathematics at West Point for many years, and was the author of a large number of widely used textbooks.

14

Stonington was a strategic port on a rocky peninsula at the eastern end of Long Island Sound. It was, like Nantucket, a town of Quakers, whalers and sealers; hardy men who, in President Dwight's words, suffered "in religion from the nearness of Rhode Island.' Its strategic position made it a target for the British fleet during the Revolution and during the War of 1812. In August 1814, a British squadron bombarded the town throughout four days, but the defenders withstood the attack with uncommon brilliancy, inflicting severe casualties upon the enemy:

> It cost the king ten thousand pound
> To have a go at Stonington.

Running the British blockade in the narrow waters of Long Island Sound gave the Stonington skippers a reputation for daring and skill even among the Yankee sailors. In the period between and after the wars they ventured into unknown and uncharted Southern Seas in search of seals. The Fanning Islands, near the equator, 1200 miles south of Honolulu, are named after Captain Edmund Fanning, a Stonington skipper who made his maiden voyage to the Southern Seas in 1792 on a hunt for sealskins, and continued his voyages for many years. From 1797 to 1798 he sailed in the 93-ton *Betsey* around Cape Horn, hunted seals at Juan Fernández Island, rescued a missionary in the Marquesas, sailed to Canton where he exchanged his skins for tea and silks, and returned home by way of the Cape of Good Hope. On this voyage he discovered several new islands, including the group named after him. Sighting new islands was as common an experience among the Stonington sealers as among the Nantucket whalers, who charted them as well as they could—though usually not for public enlightenment. The secret knowledge of an island with extensive rookeries was a precious thing, not to be shared with others, or at most only with one's business associates. Such knowledge contributed to Captain Fanning's success, and he was eventually in a position to retire on a handsome fortune. The intensive hunt for new sealing grounds defeated its own ends, since it resulted in the ruthless extermination of the seals without any attempt at conservation. It drove the sealers deeper and deeper into the Antarctic. Edmund Fanning, even after his retirement to his captain's mansion at Stonington, encouraged the exploration of the South Seas by word and by investment, reading up on the voyages of ancient Dutch and other explorers to obtain geographical information of use in seal hunting.

Among the men whom Fanning influenced were his six brothers, several of whom have made names for themselves through their intrepid seamanship. His best-known "pupil" was Nathaniel Brown Palmer, thirty years his junior, who first went out sealing in 1819 on the *Hersilia,* 88 tons burden, in which Fanning had a financial interest, and on which one of his brothers was supercargo. The expedition was organized not only with an eye to financial profits, but also as a voyage of exploration. Early in 1820 the *Hersilia* went as far south as the South Shetland Islands, which had been discovered only a year earlier by an English sailor, and the existence of which was not yet known to Americans. After a wholesale slaughter of seals the expedition went home and reported to Fanning. The *Hersilia* was sent back again to the Antarctic in the same year, now accompanied by five other vessels commanded by Captain Benjamin Pendleton, the twenty-year-old Nathaniel Palmer being the captain of his own boat, the *Hero,* a sloop of 44 tons. Stonington skippers considered it quite natural to send a 44-ton sloop out on a 10,000-mile voyage into unknown and dangerous seas under a twenty-year-old captain. It was on this *Hero* that young Palmer made Antarctic history. Having arrived in the South Shetland Islands, he ventured even further and discovered new barren land with high and desolate mountains. In the words of Captain Fanning:

> [Pendleton] being on the lookout from an elevated station . . . during a very clear day had discovered mountains (one a volcano in operation) in the South; this was what is now known by the name of Palmer's land. . . . To examine this newly discovered land, Captain N. B. Palmer, in the sloop "Hero," a vessel but little rising forty tons, was despatched; he found it to be an extensive mountainous country, more sterile and dismal if possible, and more heavily loaded with ice and snow, than the South Shetlands; there were sea leopards on it's shore, but no fur seals; the main part of its coast was ice bound, although

it was in the midsummer of this hemisphere, and a landing consequently difficult.

Palmer lingered a few days to explore the coast, but finding no seals he cruised northward again to report his findings. His log shows that he discovered the new land at 4 A.M. on November 18, 1820, and recorded its position at 60° 10′ W.L. and at 63° 45′ S.L. The following season, in 1821–1822, Palmer returned, now in the sloop *James Munroe,* twice as large as the *Hero,* and explored the coast for 15°, from 64° to 49° W.L. Palmer found plenty of sea leopards and penguins, but no seals, and returned to Captain Pendleton and the South Shetlands.

The land discovered by Captain Palmer is now known as Palmer Land or as Graham Land, a name which the Englishman Biscoe gave it in 1832. The interest in this discovery lies in the fact that many authorities see in Graham Land a peninsula branching out from the Antarctic mainland. On the strength of this assumption, Palmer has been declared a second Columbus, the discoverer of the Antarctic continent. This claim, however, depends on the interpretation of our knowledge of the connections between Graham Land and the mainland. Sir Herbert Wilkins, flying over the region in December 1928, reported a passage between Graham Land and the continent, which he named Stefansson Strait. In accordance with this observation some maps now speak of the Graham Islands and of the Palmer Archipelago. Modern opinion seems to favor the name Palmer Peninsula.

Whatever the final result of Antarctic research may be, Palmer deserves to be honored as one of the explorers of Antarctica. He discovered not only Palmer Land, but also Deception Harbor in the South Shetlands and the volcanic crater in Deception Island, and he charted parts of the South Shetlands. His merits were recognized as early as 1822, when Captain George Powell, an English sealer, published a chart of the South Shetlands with "Palmer's Land" indicated in outline, together with a short description of Palmer's discoveries. Palmer was also mentioned, with many other American and British sealers, in the account of an official Russian expedition led by Fabian von Bellingshausen, who cruised in the Antarctic regions with two corvettes between 1819 and 1821. Bellingshausen spotted the *Hero* in 1820 near the South Shetlands, as can be verified from the log of the *Hero,* now in the Library of Congress. We are free to think that the young Stonington sealer impressed the Russian commander by his knowledge of the unknown shores, or, as an admirer of Captain Palmer writes:

> It must have been a little annoying to a famous explorer forty-two years of age, when a youth of twenty-one offered to pilot him along the intricacies of what he until a few moments before had thought to be with one exception the most inaccessible land on the map.

This Russian expedition was an example of the way in which European governments supported expeditions to unknown seas in behalf of their sealers and whalers. The Stonington skippers felt that the American government should follow this example. Their very rugged individualism was thwarted by superior and organized competition, and this made them ardent supporters of a naval expedition to be organized by the Federal Government. The result was a lively agitation, carried out in letters to newspapers, in lectures and in memorials to Congress. Captain Fanning was one of the most ardent agitators. In 1828 the House of Representatives was won to the plan, but a change in national administration intervened, and under President Jackson the plan came to nothing. Now the wealthy Stonington captains actually sent out their own expedition consisting of two sloops and a schooner, under command of Captains Pendleton and Palmer, which between the years 1829 and 1831 sailed around Cape Horn as far as Chile. On the way they made a cruise north and west of Palmer Land. Some scientists went on this expedition, among them also the lobbyist Jeremiah N. Reynolds, who spent many years of his life propagating the idea of an expedition into the Southern Seas. It was not a great success as a scientific expedition, partly because America did not yet have scientists adequate to such a task, but mainly because the crew insisted on hunting rather than on exploring. After all, the principle prevailed of "no furs, no pay." The English scientist W. H. R. Webster, who met Captain Palmer on this voyage, has left us a description of the Stonington Columbus:

> When he made his appearance on board the brig with Captain Foster, we took him for another Robinson Crusoe in the shape of some shipwrecked mariner. He was a kind and good-hearted man; and thinking that they would be a treat to us, had brought with him a basket of albatross' eggs, which were to us a most acceptable present. . . . It is the time and manner of making a present that gives it all its value.

The indifferent results of the exploration made the Stonington captains even more insistent upon a government-sponsored expedition. The agitation was continued, and when public interest was at its height in 1833, Captain Fanning published his *Voyages around the World,* which is one of our main sources of information concerning the contribution of Stonington to geographical exploration. Public support for science increased with the growth of the democratic movement in the thirties, and the result was the Act of 1836, which at last authorized the federal expedition.

This expedition took place between 1838 and 1842 and became known as the Wilkes expedition, after Captain Charles Wilkes, who had already been appointed astronomer to the voyage authorized, but never realized, in 1828. This expedition was a landmark in the history of exploration and natural science in America and its influence will later be discussed.

As to Nathaniel Palmer, he continued his colorful career as a captain for many more years. He took an active part in the Latin-American rebellion against Spain by ferrying arms and recruits for Bolivar, took part in the Stonington exploring expedition of 1829–1831, was captured by escaped convicts on Juan Fernández Island, and later became a captain on the new and speedy packet boats. In 1840 his packet, the *Sid-*

dons, sailed from Liverpool to New York in the record time of fifteen days. This made Captain Palmer a natural leader in the coming decade of clipper ships, and in the middle forties we find him in New York as a builder of some of the finest clippers in the American merchant marine. He lived until 1877; in 1914 the American Geographical Society unveiled a bronze plaque on his New York house.

Chapter 4

Turnpike and Towpath

> Among the causes which gave the impetus to the great improvements by which this nineteenth century has been distinguished, the principal has been, in my judgment, the American Revolution. The common mind of the time was set free to think, particularly in the United States, where the mind was not hampered by the prejudices and unwieldy habits of former ages. . . . An open field and fair competition have been the causes of the singular success in improving the useful arts which has distinguished the period.
>
> —JOSIAH QUINCY, 1855

I

IN THE TOWNS and on the homesteads all over New England men and boys were experimenting with simple tools, seeking to improve some operation or invent some new contraption. They were the people who had brought the Revolution to victory; one of the best equipped and trained armies in the world had been defeated by their blood, sweat and tears. They were confident and self-reliant men. Theirs was a simple creed of freedom, the desire for an independent existence, in which they could work out their own destiny, in either their shops or their homesteads. Now that with political independence the restrictions of the navigation laws had been removed, they tackled their tasks with greater hope of success.

The merchant class shared this new feeling of self-reliance, but with better chances for rapid success. The small artisan and the man on the farm had to start from scratch. The war had left most of them penniless or in debt; the land had been poorly tilled, trade was in decay, mechanics and artisans lacked raw materials and were able to supply only their immediate neighborhood with the product of their toil.

The merchant class, on the contrary, had not lost its money

in the war. Many a representative of this class had been able to make an honest—or dishonest—penny at the expense of the English or even his own countrymen. There was capital to start new enterprises. It was invested first of all in new mercantile ventures, in ships and in commodities to buy and sell. But the possibilities in the development of industries became more and more apparent, and the merchant class began, slowly at first, and then with increasing vigor, to participate in the industrial revolution.

The industrial revolution had just begun in the colonies. The effect of the British navigation laws had been strengthened by the nature of the colonial economic system itself. Modern industry can flourish only where its products can be marketed. This implies a relatively large population with purchasing power and good transportation facilities. Few such conditions existed in the new republic. Furthermore, under the conditions of the late eighteenth and early nineteenth century, it also needed the pauperization of a part of the population, to provide wage earners for the factories.

It is true that the war had impoverished a large number of men and women. But there were obstacles to the use of these people for the building of industries. It was often possible for an enterprising mechanic or farmer to emigrate to what was then the West. Many people trekked from the Eastern Coast to western Massachusetts, Vermont, western New York and the Ohio territories. With them they brought to the frontier and to the wilderness their Yankee ingenuity. A remarkable interest in science and in new inventions sprang up in the backwoods countries, not only in what we now call the Middle West, but also in western Massachusetts and in Vermont. Depopulation began in Connecticut at an early date and followed in western Massachusetts and Vermont only some decades later. From the point of view of the eastern entrepreneur, this emigration to western parts of the country meant serious shortages in the available supply of labor. A struggle ensued for the creation of an adequate supply of wage earners, a struggle lasting all through the nineteenth century. It led to the importation of foreign laborers, but also took many other forms, such as the opposition against the granting of cheap land in the West.

The Federal Government was weak, untried, and without funds; the state governments were as poorly equipped to help in the industrialization of the country. The popular mind was divided on the desirability of manufactures. And though the war itself had stimulated the industrial development of the country, notably the iron industry, many infant industries succumbed after the signing of the peace treaty.

In order to understand the trend of the inventive mind of the period and the related thinking in the fields of business and science we must constantly keep in mind the fact that conditions on the farms and in the cities were still comparatively primitive, even in places and among sections of the population which had not suffered from the war. Division of labor was only beginning in the cities, and was often nonexistent in the country. There was no mass production in manufacturing. Tools were made to order by hand. Able craftsmen, coopers, carpenters, blacksmiths, silversmiths, calkers and wheelwrights produced whatever was needed, and did it well, one object at a time. Most bakers, smiths and shoemakers worked individually in their shops, with at most a few apprentices. The farmer did almost all his work by himself. His wife made her own soap and candles, spun and dyed yarns, wove cloth and did the sewing and knitting. The plow was still the ancient wooden instrument used from times immemorial, and there were no agricultural instruments except handmade spades, hoes, sickles and flails. There were a few forges, iron foundries and potteries, an occasional paper mill. The musket and the rifle of the War of Independence were the products of the local gunsmith; the clothes, of the local tailor; the shoes, of the local cobbler. Some attempts at co-operation of labor forces were made in hand-loom weaving and shoe-making. Shipbuilding was perhaps most advanced in division of labor.

This system did not depend on extensive facilities for transportation, nor did it require widespread marketing possibilities. As long as transportation was poor and the British discouraged manufactures, there was no incentive for the development of mass production methods. The fame of the new British inventions spread to America, but with the royal governors in authority nothing much could be done to introduce them, even if there had been enough initiative in the colonies.

Neither the mineral nor the botanical and zoological resources of the country were tabulated or even partially explored, and few people were interested. Iron, and here and there some other metals, were found and worked, the iron in simple furnaces and forges. The woods provided timber, charcoal and potashes. Shipyards and fisheries were in relatively good shape and yielded considerable incomes.

Houses, even those of the wealthiest Americans, were heated by fires or at best by Franklin stoves, a native invention of colonial days, followed later by stoves of the Rumford type. Light was provided by candles; cooking was done over an open fire or in the ashes. Museums and old-fashioned private houses give us a good picture of this rather primitive life with its heavy demands on housewives and servants. The arts and crafts of colonial days could flourish only so long as general economic conditions did not radically change because of the growth of mass production methods. Windsor chairs, originally imported from England, and later from Philadelphia or New York, were first made in Boston about 1786. Bennington pottery began to appear after 1800. Clockmakers continued their work with increased ingenuity. Decorative glassware as well as pewter and artistic work in wrought iron began to be produced in greater quantities than before the Revolution. As long as there was a lag in industrial expansion these arts could prosper without manufactured products interfering.

2

The primary model of industrialization was, of course, the old mother country. In the past decades England had undergone a vast economic revolution, and all signs pointed to an even greater future expansion. England guarded the secrets of its revolution as carefully as possible, but the basic facts were very well known. Isolated centers in Great Britain had been connected by an extensive system of turnpike roads and canals, and were rapidly increasing both in industries and in population. Craftsmen were improving tools, tools were being changed into machines. By 1750 the Darbys of

Coalbrookdale had enormously improved the production of cast iron, using coke instead of charcoal; in 1783 Henry Cort extended the process to the production of wrought iron. Hargreaves's spinning jenny dated from 1764, Arkwright's water-driven spinning frame from 1768, and the combination of these two, known as Crompton's mule, from 1779. The Reverend Edmund Cartwright constructed his first power loom in 1785. There were a large number of Newcomen steam engines in use, mostly engaged in pumping water out of mines, which again made possible the mining of deeper lying layers. These clumsy engines were first patented in 1705; they were somewhat improved during the next decades, but it was not until James Watt revolutionized their use by the introduction of the condenser that the steam engine became widely useful as a source of power supply. Watt's first patent dated from 1769; in 1774 he entered his history-making partnership with Matthew Boulton. From that year on Boulton and Watt began to produce steam engines on a regular production scheme.

New types of mechanics now appeared on the scene, men who were trained not only in the use of machines, but also in the improvement of parts, in precision work and in an understanding of the strength and durability of materials. New professions, those of civil and mechanical engineer, emerged, typified by such men as Smeaton and Brindley. Despite the loss of the American colonies trade and navigation flourished as never before. Scientists began to disregard the Newtonian tradition and to intensify their study of chemistry, geology and applied mechanics. And most decisive development of all—two classes emerged on the world scene, the bourgeoisie and the proletariat.

How was all this accomplished? The technique of building roads, canals, machines and engines remained the secret possession of a few British gentlemen and craftsmen, whose talent was permitted to serve only members of British ruling classes. However, everyone who traveled in England was aware that the roads had become less wretched, that stagecoaches began to move more rapidly and frequently along the turnpike systems and that transportation in boats along the new canals had actually changed traveling from a grueling and dangerous experience into a pleasant and leisurely occupation. The factories of Arkwright and Strutt, the works of Boulton and Watt, the many other new mills were there for everybody to see, but seeing did not necessarily mean understanding, and even understanding was of little avail if there was no money to invest.

The improvement in transportation was due to the new roads and canals. British road making made considerable progress during the eighteenth century, although many of the main roads remained in a wretched condition until well into the nineteenth century. The turnpikes, chartered by private companies, came into fashion after the Jacobite rebellion of 1745, when the miserable state of the roads had given the nimble Highlanders, who traveled without baggage or wagons, an initial advantage over His Majesty's heavily armed regulars. In 1763 only one stagecoach ran between Edinburgh and London, the journey consuming two weeks. Arthur Young, the author and agricultural expert, traveled from London to Manchester in 1770 and as a result of his experience cautioned all travelers to avoid this route abounding with four-foot ruts floating with mud. But within the next half century great progress was made in the construction of roads. The industrial districts, especially those of Lancashire and Yorkshire, began to improve their roads against much local opposition. Between 1760 and 1774 no fewer than 452 acts were passed by Parliament for the making and repairing of highways. The roads were laid out by private companies, who obtained the right to levy tolls. This gave such roads the name of "turnpikes," a term which has remained in the language to denote a main rural thoroughfare, even when no charge is made for its use. Thus the professional road builder came into existence, a species typified by "Blind Jack of Knaresborough," a Yorkshire trader and fiddler by the name of John Metcalf, who, guided by his uncanny feeling of direction and sense for soil and material, laid out 180 miles of turnpikes along former horse tracks and mill roads, across brooks and bogs. His work was continued in later decades by Thomas Telford and John Loudon McAdam. The result was that in the early 1800's Great Britain possessed an excellent system of highways, interconnecting the larger centers of population. This opened the period of the Merry Old England of Dickens, celebrated nostalgically on our present Christmas cards, when stagecoaches drawn by two or four horses carried a contented gentry along the highways with little likelihood of paying for the pleasure with bruised limbs or broken legs.

These were also the days of leisurely canal traffic. The canal period in British history begins with the Duke of Bridgewater's canal, built by James Brindley in 1760 and 1761 to connect the Duke's coal fields with Manchester, a distance of ten miles. It was a waterway without locks, carried over the River Irwell in a lofty aqueduct 39 feet above the water, and which ended in a subterranean extension near the Duke's mines, so that boats could load at the pits. The financial success of the canal encouraged further canal building, and in the next years Brindley connected the Duke's canal with the Mersey, Severn and Trent Rivers. An outstanding feature of this canal system was the Harecastle Tunnel, 2280 yards in length and 70 yards below the surface. Brindley's work was continued by Telford and other engineers, so that by 1830 there were an estimated 2200 miles of navigable canals in England.

Practically all the early technical leaders of England's economic revolution acquired their technical skill by sheer practice, outside of the schools. There were no trade or vocational schools, and the universities had, as a whole, no understanding of engineering. Brindley was a self-made man, and so were Metcalf, Telford and Watt. One of the first university men to show appreciation of the scientific importance of technology was John Robison of the University of Edinburgh, whose articles on this subject in the third edition of the *Encyclopaedia Britannica* (1786–1797) did much to spread technical knowledge not only in England, but also in America.

3

This development in England showed that one of the first conditions for economic expansion in America had to be the improvement of inland transportation. Colonial roads were as poor as, and perhaps even worse than, those in early eighteenth-century Britain. Transportation depended upon horses, oxen, boats, or upon a sturdy pair of legs. In the early years of the republic there were few passable roads, so that people preferred water transportation, when available. Stagecoaches had been gradually introduced on the more frequented roads. In 1775 the stagecoach journey from Boston to New York lasted a week, and in 1800 four days. Quagmires were numerous and the passengers had to be ready to lend a hand in emergencies.

The turnpike era began in America in 1792 with the famous Lancaster Turnpike, which ran from Philadelphia to Lancaster, a distance of 66 miles. It was built by a private company working under a charter granted by the Pennsylvania Legislature. In the same year, 1792, the Mohegan Road was built between New London and Norwich, in the state of Connecticut. This road was operated by the counties through which the road passed, and was therefore not typical. The first legislature in New England to give a charter to a turnpike corporation was that of Rhode Island, which in 1794 incorporated the West Gloucester Turnpike, running west of Providence from Chepachet Bridge to the Connecticut line. From this date the turnpike corporations followed each other in rapid succession. The Legislature of Connecticut issued its first charter in 1795, those of Massachusetts, New Hampshire and Vermont in 1796. In the next eighteen years Massachusetts chartered 97 turnpike corporations, 16 in the banner year 1803 alone. Few new companies were chartered after 1814, though there was a moderate revival in 1826, when 6 appeared. The first Massachusetts turnpike was in Worcester County, from Warren via Palmer to Wilbraham. Most turnpikes never turned out to be a satisfactory investment, and when railroads appeared the business gradually came to an end. There were some exceptions, such as the Mount Washington Summit Road, opened in 1861, which still operates as a toll road, but most of the turnpike companies were dissolved, and the roads turned over to the counties or the municipalities for maintenance.

Although the turnpikes were not very profitable as an investment, their economic importance should not be underestimated. They rendered great service in the economic expansion of the country. They speeded up traffic and greatly reduced the hazards of travel; they opened up the backwoods country to markets, made farming more profitable and allowed the products of manufactures to reach a far larger section of the population. The Yankee trader peddling his wares from village to village, and from farm to farm, became a more and more familiar figure.

The lack of curves was an outstanding feature of most turnpikes. Most of the early Massachusetts charters directed that the turnpikes be built as straight as possible. "The shortest line is a straight one and can not be rivalled, and as such merits the first consideration," wrote a turnpike philosopher, S. W. Johnson, in his *Rural Economy* of 1806. We can still recognize the old turnpikes by this token as they go uphill, down into the valley, across swamps and lakes, with never a curve if a straight course is possible. The result was often unfortunate, since the grades were frequently too steep for traffic and the centers of population were too easily bypassed. The Newburyport Turnpike, from Boston to the north, which runs in a straight line for more than twenty miles, does not pass through any village, though there are several near by. It is now in its turn bypassed by a super highway, which avoids Newburyport itself.

4

Turnpike builders used no labor-saving devices. Their tools were shovels and picks, obtained from local blacksmiths, one at a time as they hammered them out. The construction of these roads stimulated the old iron industry south of Boston, where millwrights, nail makers and artificers in iron were quite numerous. It has been claimed that Robert Orr of Bridgewater, later master-armorer at the Springfield Arsenal, introduced the iron shovel industry into Massachusetts. In near-by Easton, Oliver Ames founded a shovel factory in 1804, where, using primitive methods of mass production, he made about a dozen shovels at a time, which he then sold in town to procure the materials for another dozen. This was the humble beginning of the large Ames shovel works of today. Orziel Wilkinson, blacksmith at Pawtucket, also participated in the production of tools for turnpike building. He had the contract for building thirteen miles of the Norfolk and Bristol Turnpike in 1805–1806, and had to set up a shop of his own in Pawtucket to manufacture the necessary shovels and picks. This was the same Wilkinson who, through his son David and his son-in-law Samuel Slater, became so prominently connected with the beginning of the American textile industry. The Wilkinsons play a leading role in American industrial pioneering; Orziel was also the first to make cold cut nails, about 1779.

The whole construction of these turnpikes reminds us of the stories concerning the building of the Burma Road. However, the dynamite or time fuse had not as yet been invented, so that the ancient method was used of laying a train of gunpowder to explode the blasting gunpowder charges. The course of the road was staked by means of a surveyor's compass like those in use today, or with a circumferentor, which was a compasslike instrument for measuring horizontal angles. Occasionally, when greater precision was required, better instruments were used, but they were few and expensive. Yet the "mearsmen," who operated these crude instruments, did a good job, as everyone can see who has traveled down the long straight stretches of the old turnpike roads.

The engineering ingenuity of these road builders was severely taxed when ponds or rivers had to be crossed. On the Boston–Salem Turnpike, chartered in 1802, a "floating" bridge was built across Collins Pond between Lynn and Salem, this pond being of great depth and with a soft, peaty bottom. Here a long raft was built of 511 feet length and 28 feet width, and laid across the pond; then upon it was placed a course of logs hewn on the upper side, and at right angles to it a course of timbers one foot square. This operation was repeated five times after which a top course of plank was laid, making the bridge about 5½ feet deep. The construction of this bridge in 1804 delayed the completion of the turnpike about a year. Repairs in later years usually consisted in adding new timbers until the bridge was over 15 feet thick. An accident later happened because a drove of cattle passing the bridge gathered on one side, so that the bridge began to list. Since the animals had thrust their heads under the railings their horns were caught, and several were drowned.

The advent of the railroads came as a heavy blow to the

turnpikes. Much traffic was diverted from them to the equally straight tracks of the new system of steam-driven cars. The long rural thoroughfares became quiet lanes. Parts of them eventually became city streets, such as Hampshire Street and Western Avenue in Cambridge. Parts of other turnpikes became overgrown forest trails. But the railroad met with a powerful competitor in the automobile, and the automobile brought new life to the ancient turnpikes. Many old turnpikes are now modern roads; some have become four-lane highways or even better. Their very straightness, which worked to their disadvantage in early days, now in the age of speed works to advantage. Modern bus and truck traffic along these widened roads is, in many respects, a return to the ancient days, when stagecoaches and covered wagons moved along the same thoroughfares. Deep in the country, however, there remain some of the ancient turnpikes, untouched by modern techniques. It is possible to walk over them mile after mile, uphill and downhill, without meeting a single person. They are long sandy tracks bounded by stone fences on both sides, which separate the roads from fields and woods, where cellar holes mark the sites of old homesteads. Here and there an ancient inn remains. Sometimes the road has entirely disappeared, and can only be traced through the woods by the ever-present stone fences. This change of the turnpikes into mere ruts in the wilderness on the one hand, on the other into modern highways, affords an instructive reflection on New England's industrial and agricultural history during the last hundred years.

5

Road construction leads of necessity to bridge building. The new republic had few bridges worth boasting about. Brooks and small rivers were bridged by simple single-stringer structures; larger streams had to be crossed by ford or by ferry. Sewall's bridge near Kittery, Maine, was one of the first pile bridges in America; it was built in 1757. The Pawcatuck River, in Westerly on the boundary between Rhode Island and Connecticut, was bridged as early as 1712 to facilitate the traffic on the Boston–New York Post Road, but there were few such ambitious undertakings.

As one of the few colonial bridges which were more than

single-stringer affairs the Westerly structure is of interest. The bridge of 1712 was a frail affair "hardly wide enough for an ox cart to pass over." It was probably constructed by means of wooden cribs with interlocking joints at their corners, bound together with wooden pins. Such cribs were placed at certain distances from each other in the river, after which they were filled with stone to keep them in position. Wooden girders were then placed on these stone-filled cribs, and the floor planking rested on these girders.

The maintenance of this bridge was a source of contention between the government of Rhode Island and Connecticut, and between the towns of Westerly and Stonington. During 1734 and 1735 the bridge was almost entirely rebuilt and widened to serve for many years the needs of the traffic. It stood until 1873, when it was replaced by a modern bridge.

Such a structure was quite exceptional in colonial days, even though the river it bridged was a small one. Large rivers such as the Connecticut or the Kennebec had to be crossed by ferry, which meant not only considerable loss of time, but in bad weather even personal danger. Yet bridge building was an old art, known to the ancients and even practiced with great success in the European Middle Ages. The continent of Europe, as well as England, had remarkable examples of solid bridges across large rivers, some dating back to Roman times. During the eighteenth century considerable attention was paid to bridge building, and engineers as well as mathematicians contributed to the old art by computing and constructing new and daring structures, usually of masonary. The old London Bridge, finished in 1209, for five centures and a half was the only structure across the Thames, but in 1750 Westminster Bridge, and in 1769 Blackfriars Bridge, were added. William Edwards, a Welsh stonemason, built a one-arch bridge of 140 feet across the River Taff near Cardiff in the form of a circular arc of a circle with a radius of 85 feet. This Pont-y-Pridd, or Rainbow Bridge, was finished in 1755, and was of larger span than any other bridge built in England until that date. It still graces the river at Newbridge. The english ironmaster Abraham Darby tried his hand at a cast-iron bridge, the first in the world. His Coalbrookdale Bridge, across the Severn near Shrewsbury, has five nearly semicircular ribs, each of three concentric arches and 106 feet long. It was built in 1779, and it also is still standing.

The young republic needed bridges as desperately as it needed roads. Among the first promoters of bridge construction was Tom Paine, the English mechanic who became one of the prophets of American democracy. During the Revolution he planned a bridge across the Harlem River near New York to be financed by Gouverneur Morris, but nothing came of it. Like Abraham Darby, with whose work he was probably familiar, Paine believed in iron as material for bridge building, perhaps as a result of his conversations with master ironworkers in Philadelphia. After the Revolution he built a model of a 400-span iron bridge over the Schuylkill River at Philadelphia. "Great scenes," he wrote in 1789, "inspire great ideas. The nature of America expands the mind, and it partakes of the greatness it contemplates." The Schuylkill Bridge remained another of Paine's dreams, but he had the model exhibited in Paris, where it received considerable attention.

Bridge building remained a great necessity, and after the Revolution men appeared willing to try it, even without adequate engineering training. Though led in the first place by economic motives, these men were also inspired by patriotic sentiment and the grandeur of the landscape. Thomas Pope, a New York shipbuilder and the author of a *Treatise on bridge architecture* which appeared in 1811, described such feelings in eloquent form.

> It is a notorious fact [he wrote] that there is no country of the world which is more in need of good and permanent Bridges than the United States of America. Extended along an immense line of coast on which abound rivers, creeks and swamps, it is impossible that any physical union of the country can really take place until the labours of the architect and mechanic shall have more perfectly done away the inconvenience arising from the intervention of the waters. Nature, ever provident for man, has, however, afforded us ample means of remedy.—Our forests teem with the choicest timber, and our floods can bear it on

their capacious bosoms to the requisite points. Public spirit is alone wanting to make us the greatest nation on earth; and there is nothing more essential to the establishment of that greatness than the building of Bridges, the digging of canals, and the making of sound turnpike roads. Necessity has already produced some handsome and extensive specimens of Bridge-building in the United States, agreeable to ancient models; and we shall proceed to mention a few of them.

At the time of Pope's writing, 1811, there were already some impressive bridges in the United States, but, with very few exceptions, they were made neither of stone nor of iron. They were constructed of wood, a material which America had in more abundance than Western and central Europe. Large wooden bridges existed in other parts of the world, but nowhere in great numbers. The best known were the Swiss bridges, built by the Grubenmann brothers at Schaffhausen over the Rhine and at Wettingen over the Limmat, built in 1756 and in 1764 respectively. The Schaffhausen Bridge, described by Pope, was composed of two wooden arches with spans of 193 and 172 feet, supported at their ends by abutments, and at their junction by a stone pier. The Wettingen Bridge had a span of 390 feet, and has never quite been equaled in wood. These bridges, and several others built by the Grubenmanns, were burned by the French in 1799.

The Grubenmanns were simple carpenters, without engineering training of any sort. What they had performed, Yankee carpenters could also accomplish when money and opportunity became available. The first New Englander to try his hand at spanning a large river by means of a wooden bridge was a man by the name of Enoch Hale from Rowley near Newburyport. Hale belonged to a group of Rowley men who in 1759–1760 laid out the wilderness village of Rindge, New Hampshire, in the Monadnock region. A trail from this region into Vermont crossed the Connecticut River at Bellows Falls. Crossing this majestic river had always been difficult in colonial days. It is no accident that the first attempt at bridge building on a larger scale than before occurred along the Connecticut, but it is remarkable that this first ambitious attempt was immediately successful. Hale—Colonel Hale—obtained shortly after the Revolution a grant from the legislature to build a toll bridge near Bellows Falls between Walpole, New Hampshire, and Rockingham, Vermont. The enterprising Colonel had to endure much criticism and ridicule, but he persevered and completed the bridge in 1785. Using white pine, which was plentiful, he selected a narrow part of the river with a rock reef in the middle. Dividing the distance into two spans, a central pier was built on the reef, and the bridge of 365 feet was laid about 50 feet above the water. The support consisted of four sets of braced stringers. It was the first bridge built in America with spans greater than those which could be negotiated with single-stringer sticks, and it grew quite famous in its day. People came from afar to visit it; newspapers heralded the great feat, and in his narrative of his trip through New England, President Dwight of Yale expressed his admiration for the bridge. To this bridge the turnpike builders led their roads. In 1799 both the Third New Hampshire Turnpike and the Green Mountain Turnpike were chartered, and connected at the Bellows Falls Bridge, thus opening the middle Vermont region to the coastal towns of New England. The bridge stood until 1840, when it was replaced by another wooden bridge, the Tucker Bridge. This bridge has now been replaced by the Vilas Bridge, a cement bridge dedicated in 1931.

Another early bridge was built in 1786–1787 across the Charles River, connecting Boston and Charlestown, and which, with its "forty elegant lamps," 75 piers, its 1503 foot length and 43 foot width, was the talk of the town. This time it was not a backwoods jack-of-all-trades who had taken the initiative, but a group of prominent and wealthy Bostonians which included Governor John Hancock. It was opened with much pomp and festivity, and was duly hailed in Morse's *Geography* as the greatest example of private enterprise of this kind in the U.S.A. Several other bridges followed after these first successes.

One of the most ingenious wooden bridge builders of this period was a Newburyport man, Timothy Palmer. We know little of the man himself, except that his family, like Hale's, came from Rowley and that he was a "gentleman," which may mean that he belonged to the merchant caste of Newburyport. He became famous for the Essex Merrimack Bridge over the Merrimack, three miles above Newburyport, which he built in 1792. The bridge consisted of two arched spans, one on each side of an island in the river; the southern arch was 160 feet in length and the northern arch 113 feet. The second Palmer Bridge was constructed in 1794 across Little Bay northwest of Portsmouth, N. H. It also was in two parts, and had an arch 244 feet in length with a rise of 27 feet, approached by pile trestles, making its total length more than half a mile. This "Piscataqua Bridge" was praised and admired far and wide for its impressive arch in the middle of the broad estuary and its long approaches through the water. It became the eastern terminus of the first turnpike in New Hampshire, connecting the capital, Concord, with Portsmouth and with Massachusetts. Here again the bridge antedated the turnpike, which was chartered in 1796. The bridge withstood the elements till 1855, when it was carried away by the ice. The beautiful General John Sullivan Toll Bridge, dedicated in 1934, stands about half a mile below the site of the old bridge.

Private corporations throughout the whole country now sought the services of Timothy Palmer. The Newburyport builder could adapt his methods with great ingenuity to the different problems which he had to solve. Like the Swiss wooden bridge builders he used the arch to a great, and for his period daring, extent, strengthening it by a structure of short timbers, built up in sections, and forming a rigid triangular or polygonal frame, a so-called truss. Little was known about the actual load which each member of the truss bore, so that the truss-arch bridges of Palmer and his contemporaries contained much waste material. They were, however, strong and durable structures. Many of them stood up very well under the moderate loads of early nineteenth-century traffic and the heavy attacks of water, ice, snow and wind. Palmer built such bridges across the Kennebec, the Delaware and the Potomac, the last one opposite Washington, the new Federal Capital. One of his best-known bridges was the so-called "Permanent Bridge" across the Schuylkill River at Philadelphia, built between 1801 and 1805 as the terminus of the Lancaster Turnpike. It was a truss-arch bridge with three arches. Many difficulties had to be overcome in the construction of the stone abutments and the two piers on which the bridge was built. Palmer could not have done it without the assistance of William Weston, the English engineer, whom we shall also meet as a canal builder.

The "Permanent Bridge" was not so permanent after all, since it was destroyed by fire after approximately fifty years of service. It was one of the first bridges protected against the weather by a roof and side planking. The structure was built on the spot where Tom Paine had dreamt that his iron bridge would stand, long before Timothy Palmer obtained his commission. A memorial tablet on Market Street, near the Thirtieth Street station, now marks the site of the Paine and Palmer projects.

Timothy Palmer died at Newburyport in 1821. One of the first bridge constructors to use the trussed-arch type of wooden bridge, he contributed greatly to its independent American development. He pointed out the advantage of covering the trusses to protect bridges against the weather, especially against the heavy snow of New England, and in this way increased their average life from ten to forty years. The period of the covered bridges with all its historical and nostalgic associations began with Timothy Palmer.

The full development of the truss without the arches—a method of bridge building which became typical of New England—came into its own with Ithiel Town, a New Haven architect, and a younger contemporary of Charles Bulfinch. He was the son of a farmer and house carpenter in Thompson, Connecticut, who died when his son was eight years of age. Young Ithiel was sent to live with his uncle in Cambridge, where he probably attended a local academy and studied architecture at the school of Asher Benjamin, a local architect. This was during the first decade of the nineteenth century, when Charles Bulfinch enjoyed his great wave of popularity, and his interpretation of Palladio's and Wren's Renaissance classicism was taught by Benjamin. When Town returned to New Haven, sometime before 1812, he built the graceful Center Church on New Haven Green, choosing the orthodox Christopher Wren style. Then, in 1814–1815, he built gloomy Trinity Church, next to Center Church, one of the first examples of Neo-Gothic in America. It was also a first example of that versatility in style for which Town and his firm of Town and Davis became famous throughout the United States in the next decades. For more than thirty years this firm was engaged in building state capitols, churches and other impressive buildings, first in Gothic and Greek, and later, especially after Town's death, in a motley variety of styles ranging from Tuscan, French and Swiss to Moorish and American Log Cabin. The American businessman who wanted to live in fashionable elegance could order his type of mansion from the catalogue of Town and Davis.

The first Town bridge on record was built in 1816 across the Connecticut at Springfield, replacing an older structure. It combined the arch with the simple king-post truss tied to stringpieces. Gaining experience in the construction of other bridges, he developed the pure lattice-truss bridge, on which he took out a patent in 1820. Town's design, published in a pamphlet of 1821, was a self-contained rigid structure of diagonal bars, which could be simply placed to rest on piers. With a vertical load it exercised an almost entirely vertical pressure on the supports, and did not have the horizontal thrust characteristic of the arch type of bridge. Its first application was the covered bridge near the Whitney armory works not far from New Haven. This bridge, built in 1823, had a 100-foot clear span and side bracings of 3-inch planks crossing each other at an 80-degree angle and spaced 4 feet center to center. These were securely pinned together and held top and bottom by stringers on each side. There were only pin connections, so that mortising of the timbers was unnecessary. The vertical symmetry was an outstanding characteristic of this type of bridge—it would have served its purpose equally well if turned upside down. In 1860 the Whitneyville Bridge was lifted bodily from its abutments and moved a quarter of a mile upstream. In 1891 it was replaced by a steel bridge.

The covered Town truss bridges were destined to become an intimate feature of the New England landscape. They were very well adapted to rural conditions in America, especially in the backwoods country. The truss bridge as well as another type with a double web, on which Town took out a patent in 1835, required no more material or ingenuity for its construction than was available to the average local carpenter. There was little or no ironwork; the construction called for ordinary planking and had no mortises and tenons. There is a case on record where such a prefabricated bridge of 200 feet span, 120 feet above a river bed, was erected by local labor in two weeks. Town himself was not a practicing bridge builder but sold the use of his patent at one dollar per foot span, which added up to a nice income. He spent his later days in comfort with an office in New York, an agency in Washington and a beautiful home in New Haven, which he built himself. He befriended young Samuel F. B. Morse and was one of the founders of the National Academy of Design (1826) and the American Institute of Architects (1836). His library on architecture and the fine arts was one of the largest and choicest in the country. Bewteen 1827 and 1831 he built the Connecticut State Capitol on New Haven Green, a replica of the Parthenon; this was unfortunately demolished in 1889. Town died in 1844; his house on Hillhouse Avenue near the Yale Campus has been preserved, though considerably altered for the worse, and is now part of the Sheffield Scientific School.

It is characteristic of the prominence of New England wooden bridge construction that some of the outstanding trusses are named after men who hailed from Boston's hinterland. The Burr patent, a widely used combination of arch and truss, was named after Theodore Burr, from Torrington, Connecticut, who spent most of his life in Pennsylvania. Very little except his work is known about Burr. He built the bridge at Springfield over the Connecticut, which spanned the river for 104 years. The Long truss was named after Stephen Harriman Long of Hopkinton, New Hampshire; it was successfully used on some of the earliest railroad bridges near Boston. The patent dates from 1830. Long's Peak in the Rockies is named for the same engineer. Long's trusses, finally, were followed by the Pratt and Howe trusses. These were widely used on the new railroad bridges, and played an important role in the transition from wooden to iron bridges.

William Howe was one of the three members of a Spen-

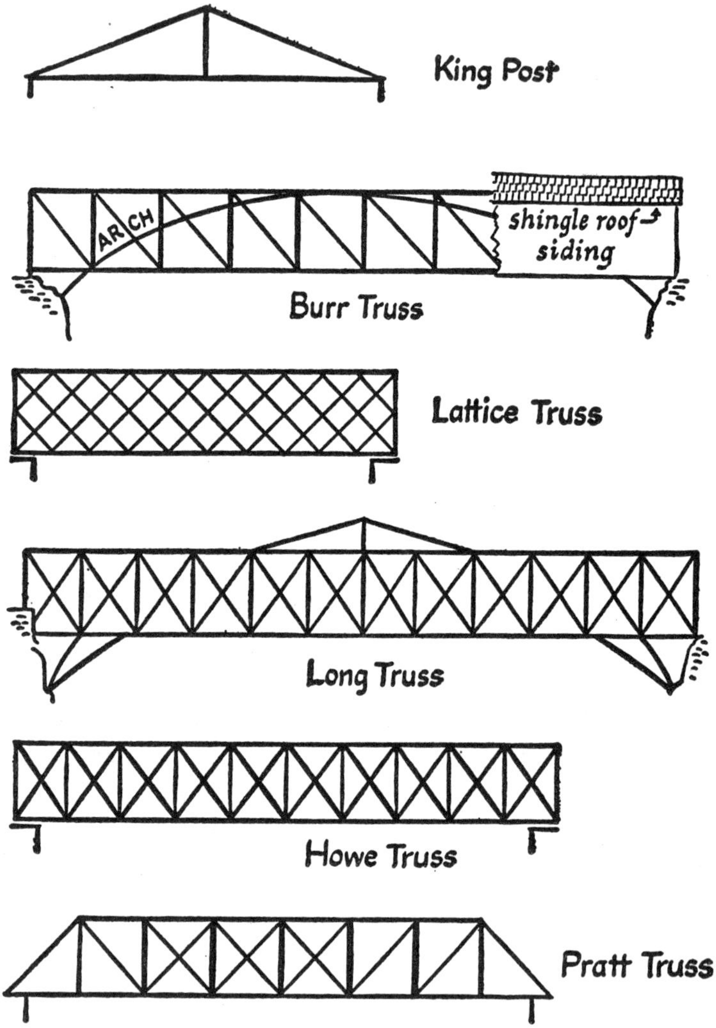

TYPES OF TRUSSES

Based on a drawing by Emily Watson in *Bridges and Their Builders,* by David B. Steinman and Sara Ruth Watson, G. P. Putnam's Sons, New York, 1941. Courtesy of G. P. Putnam's Sons.

cer, Massachusetts, family, who became famous far beyond the boundaries of their native Worcester County. A monument near Spencer Town Hall has bronze medallions of Tyler Howe, who invented the spring bed, of his brother William and of their nephew Elias, the inventor of the sewing machine. William Howe, born in 1803, took out patents in 1840 and 1850 on a pure lattice truss, in which the diagonal bars were linked by wrought-iron tension rods. The story is that the idea came to Howe while he was studying the roof truss of an old church near by—which may also have been the origin of the Town truss. He sketched his trusses on the plaster walls of a tavern in Spencer, where they were for many years on exhibition. The first Howe truss bridge was built in 1838 near Warren, not far from the Howe homestead. It was during this period that the Western Railroad was built from Worcester across the Connecticut to Albany, passing through Spencer and Warren.

The Howe truss appealed to the builders of the railroad, since it was quite adaptable to the needs of railroad traffic, which demanded larger and heavier spans than ordinary road transportation. In 1841 a firm specializing in the construction of railroad bridges bought the Howe patent. In the same year this firm built a seven-span railroad bridge—with a double web of the Howe design—across the Connecticut at Springfield. This structure was a success and the Howe truss was used for many other railroad bridges, often combined with other designs and modified for iron. It still figures prominently in modern books on steel and timber structures.

The Pratt truss is often mentioned with the Howe structure in the works on early railroad building. It was patented in 1844 by two Boston men, Thomas Willes Pratt and his father Caleb, both of whom were architects and engineers. The Pratt truss was the inverse of the Howe truss, since the vertical bars were made of wood and the diagonals were iron tension rods. The Pratt trusses, like the Howe trusses, were mainly used for iron bridges.

6

With the Howe and Pratt trusses we have arrived at a period in American engineering when wooden bridges were reserved for places with simple horse traffic and heavy traffic was led over cast-iron structures. These, in turn, were replaced by bridges made of wrought iron, and later by steel. Some iron bridges, however, date back to an earlier period. These first iron bridges in America were suspension bridges.

In 1809 the southern span of Timothy Palmer's arch bridge across the Merrimack near Newburyport was sold at auction as unsafe. It was decided to replace this by a suspension bridge, of which some already had been built elsewhere under a patent held by a Pennsylvania judge, James Finley. A Washington engineer, John Templeman, was called to Newburyport to build the bridge. It was completed in 1810 with the help of local craftsmen. The suspension bridge had a main span of 244 feet center to center of the truss, 10 chains, each 516 feet long, comprising the cable on one side, with a flooring about 40 feet above water. The chains were arranged in 3 tiers, 3 in the top and bottom ranges, and 4 in the middle; the material was Norway iron. The bridge received considerable fame in its day, even outside of America. It only withstood the elements until 1827, was rebuilt, and in 1909 replaced by a new wire-cable suspension bridge built to carry heavy electric cars. The stone gates of the old bridge are still standing; the busses now pass under them.

However, for some decades iron bridges remained an exception in the United States, while in Europe great advances were made in their construction. One of the most spectacular successes was Telford's suspension bridge spanning the Menai Straits in Wales, a structure 1710 feet in length, completed in 1825. George Stephenson, the inventor of the locomotive engine, was one of the first to build iron railroad bridges. Only after many years did America begin to catch up with Europe.

With increasing heavy railroad traffic the "try and fail" methods of local craftsmen became more and more outdated. Theoretical investigation of the strength of trusses and arches could no longer be neglected. Squire Whipple, who hailed from the Worcester County which produced the Whitneys and the Howes, was the first to publish a book on the computation of trusses. His *Bridge-building* appeared in 1847, when the author was manufacturing levels and transits at Utica, New York. Now other books and papers on the mechanics of bridge design appeared at regular intervals, until they were superseded by W. J. M. Rankine's *Applied Mechanics* of 1858. Bridge building in the second half of the nineteenth century became more and more the work of theoretically trained engineers.

7

One group of enterprising men had turned to road or bridge building; another group was doing some serious thinking about canals. Much had already been done—and still more dreamt—about this method of easy transportation, as old as Rameses of Egypt and Xerxes of Persia. However, in the colonies it had never really been tried out on any large scale. Both lack of interested capital and ignorance of the principles of civil engineering were responsible for the delay, and the colonial status of the country acted as a general deterrent. When the colonial shackles were removed, the country began to give serious consideration to canals as a means of opening up the country.

There were more practical inspirations for the enterprising Americans of this period than the example set by Xerxes or by the French and Italian engineers of the Renaissance. The old mother country itself had at last become canal conscious after the great opening venture of 1761, the Duke of Bridgewater's canal. It was this British example that the new American republic intended to follow. There was little local tradition, although small canals had been constructed occasionally even in the earliest days of colonization. Near Boston there were at least two canals dug before 1640, one at the head of tidewater on the Charles River in Watertown, the other connecting the Charles and Neponset Rivers in Dedham, at present known as the Mother Brook. Such canals were used to provide power for water mills, especially for gristmills. Canals as a means of transportation were discussed, but not constructed. As early as 1690 William Penn made a plea in Philadelphia to unite the Schuylkill and the Susquehanna Rivers. In 1762 actual surveying took place for such a canal, a project in which the astronomer Rittenhouse took a lively interest. The Revolution intervened, and only much later was Penn's idea realized in the Union Canal.

The most ambitious dream of colonial New England canal enthusiasts was to pierce the eight-mile shoulder of what Thoreau has called "the bare, bended arm of Massachusetts," Cape Cod. The first proposals to dig the canal date back to 1676 and 1697, but nothing was done until the War of Independence impressed upon the colonies the military value of a Cape Cod canal. Both General Washington and the General Court of Massachusetts were interested. A committee was appointed to decide on the practicability of the project (James Bowdoin was one of its members) and an engineer was assigned to it. This engineer, Thomas Machin—a Staffordshire man, who had worked on the Duke of Bridgewater's canal and in America had joined the continental forces (he was at Bunker Hill), was recalled almost immediately by Washington for other work, thus effectively interrupting a project which was not resumed until a much later date. During the nineteenth century many resolutions and plans were made, but the canal was not constructed until the years 1909-1914 and, with its two high bridges, is now the glory of the landscape.

After the Revolution a change set in with the construction of small canals to improve the transportation along the Connecticut and Merrimack Rivers. There were rapids and waterfalls in these rivers, where they gathered in season to fish. It is not unusual to find an Algonquin name like "Pawtucket" attached to these places. The new canals were ditches dug around these rapids to facilitate communication between the inland districts and the coastal markets. The same business principles which guided the road and bridge constructors also guided the canal planners. They formed private corporations working under a charter from the legislature and tried to make their venture pay by raising tolls. A Massachusetts company chartered in 1792 constructed the South Hadley and the Montague canals to promote traffic on the Connecticut. They were only two and three miles long respectively. The money for the South Hadley Canal came largely, it seems, from bankers in Amsterdam.

The adventurous men entering this field had only a very slight notion of engineering. The builder of the South Hadley Canal, Benjamin Prescott of Northampton, had to plan and work without any experience. The result of his ingenious improvisation, for which he had to blast an opening of 300 feet in length and 40 feet depth in solid rock, was a canal with levels connected by an inclined plane of timber, 230 feet long. The boats were moved up and down the incline in a large tank car filled with water and were propelled by cables set in motion by two water wheels. This type of water transport was continued until 1805, when a simple lock was introduced, which made the curious old contraption superfluous.

Some of these canals were built on the site where factory towns later appeared. Prescott's canal became part of the canal system of Holyoke. The canal system of Manchester, New Hampshire, grew out of a canal constructed under the guidance of a Samuel Blodget of Woburn, a judge of common pleas in New Hampshire. Blodget projected his canal around what were known as the "hideous waterfalls" of Amoskeag on the Merrimack. He accomplished this feat after much personal trouble between the years 1794 and 1807; his difficulties included imprisonment for debt. Another canal, built in 1796 by Newburyport merchants around the Pawtucket falls in the Merrimack near Chelmsford, after 1820 became part of the Lowell water system.

8

The real triumph of early New England canal construction was the 27-mile artery known as the Middlesex Canal. This was no longer a small venture of local contractors, judges or merchants, but rather an enterprise backed by members of the Boston ruling class. These men did not think in terms of two- or three-mile ditches to circumvent waterfalls in rivers, but in terms of the European canal builders, who connected whole regions by a system of waterways. Such ideas were shared by many leaders of the republic, including George Washington, who took an active part in projecting the Chesapeake and Ohio Canal from the Potomac to the Allegheny Mountains. We have seen how deeply James Bowdoin was interested in the Cape Cod Canal. Other merchants in Boston believed it possible to connect the Merrimack River above the Pawtucket rapids with the tidewater of the Charles River at Boston, and so open southern New Hampshire to Boston and the surrounding markets. This led to the chartering in 1793 of the "Proprietors of the Middlesex Canal," which numbered several important Bostonians among their sponsors. The financial projector was James Sullivan, brother of the hero of Trenton and Brandywine, and later a governor of Massachusetts; with him were Christopher Gore and other men of wealth, as well as James Winthrop, son of the Harvard astronomer. The leading technical spirit of the plan was Loammi Baldwin, whom we know as the lad from Woburn who in pre-Revolutionary days accompanied to Harvard College the boy who later became Count Rumford. After his return from military service under Washington, Baldwin, now a colonel, became a prominent man in the community of Boston. He resided in a large house in North Woburn, which he remodeled with pilasters and an elaborate doorway in the style of his time. The house is still standing, with a statue of the gallant Colonel in front. Here he indulged in political activities and in his old hobbies of surveying and experimentation. He became known as the cultivator of the Baldwin apple, which he is supposed to have grown from a sapling found in the fields. He played an active part in the technical work of the Middlesex Canal, which passed by his doorstep. In this way he developed into one of the earliest American promoters of civil engineering. His son and namesake became the leading civil engineer of New England.

The sponsors of the Middlesex Canal soon found that chartering a canal company and surveying the 27 miles of its projected site were two very different things. Local surveying talent was still singularly incompetent. The promoters were, in the words of a historian of the canal,

> met by an almost unsurmountable difficulty; the science of civil engineering was almost unknown to anyone in this part of the country. They were, however, determined to persevere, and appointed Mr. Samuel Thompson, of Woburn, who began his work . . . etc.

The directors felt a justifiable distrust in the results of the local surveyor. Loammi Baldwin was instructed to go to Philadelphia to see if he could obtain the services of somebody trained in Europe, preferably of William Weston, an English canal and bridge builder of considerable experience. Weston, who had arrived in Philadephia in 1793 and was engaged in working on the very Schuylkill and Susquehanna Canal forecast by William Penn, received for his services a salary of $1500 per annum. In response to Baldwin's urgent request and with the promise of a generous fee, William Weston arrived in Boston.

Weston's survey took place in the summer of 1794. He found that the canal would not have to ascend 16½ feet between the Concord River at Billerica and the Merrimack at Chelmsford as the Woburn surveyor had found, but that actually it had to descend 25 feet. He also corrected an error of 35 feet in 104 feet between the Concord River and the Mystic at Medford, but this was not so serious, as everybody knew that Boston was lower than Billerica. The path followed was the preglacial valley of the Merrimack, which used to run south through what are now the Mystic Lakes.

Weston asked $2200 for his job, a sum the proprietors paid without complaint. He maintained contact with the company after his return to Philadelphia. A number of his letters to Baldwin are extant; they deal with canal and lock cross sections, with watertight masonry walls and with the gate-opening mechanism of locks. As we have seen, English literature on civil engineering was still in its infancy; in this period the Robison articles appeared. Thus, lacking other resources, the Colonel secured his engineering training by correspondence course.

Work on the canal began in 1795. It started at Billerica Mills on the Concord River and continued to North Chelmsford on the Merrimack, near where Lowell was later built. It also ran from the Concord River through Wilmington, Woburn and Medford to the Charles River at Charlestown. The opening of the completed canal was celebrated in 1803 with a handsome entertainment at the Baldwin home. It was an occasion well worth celebrating. The canal was laid through swamps and across brooks, with the primitive hand tools of the day, "by main strength and awk'ardness"; it had 20 locks, 8 aqueducts, 48 bridges, safety gates, culverts, sluiceways, water weirs, and was all in all an impressive piece of engineering, the greatest construction work completed in the country up to that time. The experience gained in constructing and operating the canal was later put to good use in the building of other canals, especially the Erie Canal. It has been claimed that the first leveling instrument used in the United States was employed in the location of the Middlesex Canal.

The golden days of canal transportation occurred before and just after the opening of the Boston and Lowell Railroad. As a paying proposition the canal was as unprofitable an investment as most turnpikes, since its construction, and until 1817 its maintenance, constantly absorbed the money pledged by the proprietors without bringing them any return. In 1817 the total cost amounted to $1,164,200, a large sum for the period.

The first dividend was not declared until 1819, and the dividend from 1819 to 1843 amounted on the average to 1.39 per cent per annum. However, the economic importance of the canal was considerable, since it connected Boston with New Hampshire by means of the Merrimack, where additional canals were laid around rapids. The Middlesex Canal later also served the budding textile industry of Manchester and Lowell. For many years timber went down the canal to Boston, and gave cheap access of raw material not only to carpenters and shipwrights, but also to shoe factories, at that time working with wooden pegs. The canal also served as a means of traveler transportation, and an atmosphere grew up around it similar to that which we find in books on the good old Erie Canal. Here is a sample, from Edward Everett Hale's childhood memories:

> In the summer of 1826 it was announced one day that we were going by the canal. I have no recollection of the method by which we struck the Middlesex Canal; I suppose that we had to drive to East Cambridge and take the "General Sullivan" there. The "General Sullivan" was what was known, I think, as a packet-boat, which carried passengers daily from Boston to the Merrimac River, where

the name "Lowell" had just been given to a part of the township of Chelmsford. By way of escape from the heat, father had arranged that the whole family should go down to the tavern at Chelmsford and spend a few days.

The present generation does not know it, but travelling on a canal is one of the most charming ways of travelling. To sit on the deck of a boat and see the country slide by you, without the slightest jar, without a cinder or a speck of dust, is one of the exquisite luxuries. The difficulty about speed is much reduced if you will remember, with Red Jacket, that "you have all the time there is." Fullum, the dog, would spring from the deck of the "General Sullivan" upon the tow-path, and walk along collecting wild flowers, or perhaps even more active game. I have never forgotten my terror lest Fullum should be left by the boat and should never return. When he did return from one of these forays he brought with him for us little children a very little toad, the first I had ever seen.

Henry Thoreau, spending with his brother a week of September 1839 in a boat on the Concord and Merrimack Rivers, took the canal from Billercia to Chelmsford in order to pass from one river into the other. At that time the canal was already becoming a natural part of the landscape; and Thoreau philosophized:

> Nature will recover and indemnify itself, and gradually plant fit shrubs and flowers along its borders. Already the kingfisher sat upon a pine over the water, and the bream and pickerel swam below. Thus all works pass directly out of the hands of the architect into the hands of Nature, to be perfected.

Soon afterwards the canal passed even more directly into the hands of nature. It could not compete successfully with the Boston and Lowell Railroad and became more and more a liability to the stockholders. In the forties of the century attempts were made to transform it into an aqueduct for the Boston water supply. Nothing came of this and in 1852 the last boat passed through the locks at Billerica. The canal was sold for $130,000 and the money divided between the stockholders. The abandoned right of way was gradually absorbed by the growing towns at the Boston end of the canal. It can still be traced through the woods of Middlesex County, a long ditch, at some places dry, at others still filled with water, the overgrown towpath a good trail for a rough hike. The aqueduct across the Shawsheen River is a picturesque ruin, the tollhouse at Wilmington a well-preserved relic of the past. The canal can most easily be inspected beside the Baldwin mansion at North Woburn.

9

The technical difficulties which the Middlesex Canal proprietors had encountered in constructing the canal and its poor financial success had a discouraging effect for many years on New England canal enthusiasts. The Middlesex Canal Company tried to improve traffic on the Merrimack by constructing more canals around rapids and by taking a financial interest in the Amoskeag Canal, but this was all the inspiration emanating from this side. A change set in after the glorious completion of the Erie Canal in 1825. The fame of this achievement swept the country, and the new canal became a subject of national pride. At the time it was the longest canal in the world—363 miles from Albany to Buffalo—and had a determining influence on the opening of the Middle West and the spectacular growth of New York City. Moreover, it had been constructed under the leadership of native men, many of them frontier judges—Benjamin Wright, James Geddes, David Bates, Nathan Roberts—who without formal training had accomplished what England or France would have been proud to claim. And now there were even trained canal constructors in the country, who had worked under Benjamin Wright and his associates, and had learned civil engineering in the process of canal building itself. The country, moreover, was now economically more developed than twenty years before, markets had expanded, prosperity had increased. The result was a revival of the old enthusiasm for canals in New England, a kind of Indian summer, which was fated to cool off a decade later, when railroads appeared.

Several ambitious schemes were projected, three of which

were carried out. The Blackstone Canal, constructed between 1826 and 1828, connected Providence and Worcester; it was 45 miles long and had 48 locks of cut granite. Even more ambitious was the Farmington Canal in Connecticut, continued in the Hampshire and Hampden Canal in Massachusetts, which connected New Haven with Northampton over a total distance of 76 miles. It was built between 1825 and 1831 and cost $2,000,000. And in Maine—proud of its newly gained independence from Massachusetts—Portland was connected with Sebago Lake by a canal of 20½ miles. All these canals went the way of the Middlesex Canal and are now abandoned ditches, with here and there a romantic ruin to keep the memory of the New England canal period alive. The pleasure steamers on Sebago and Long Lake in Maine still use the Songo River passage, with a new lock replacing the old one.

The most ambitious of all canal schemes was never carried out, since the railroads appeared before its Boston promoters could make up their mind. It was a plan for a canal from Boston to the Connecticut River and thence through the Berkshire mountains to the Hudson. Here the connection with the Erie Canal could be accomplished, and the possibility opened for successful competition with New York, which now had sole access to the market of the miraculously expanding Middle West.

The canal was projected by Loammi Baldwin, Jr., New England's most outstanding civil engineer, the son of the promoter of the Middlesex Canal. Loammi, Jr., was an experienced canal builder who had carefully studied the British canals. Following Brindley's example in England and his own example on the Union Canal in Pennsylvania, Baldwin, after a survey in 1825 proposed a tunnel through Hoosac Mountain in the Berkshires. It was to be a tunnel of unusual length, four miles through solid rock, requiring eighty minutes in passage, if furnished with a towing path. This project was audacious enough, but eventually Massachusetts, in Professor Morison's words, "wisely accepted the veto of her topography." "With such means as are at command for ventilating," wrote the *New American Cyclopaedia* as late as Civil War days, "it would seem to be almost a hopeless undertaking to endeavor to penetrate this mountain." The tunnel was actually built as a railroad tunnel, but only in the years after the Civil War, and after a long and financially involved struggle of the interested parties.

10

All this technical ingenuity was applied by following essentially traditional lines of development, long known from England. Was it possible to break more completely with the past and to use the modern technical medium of steam power? The trouble was that England was not too keen to share its knowledge with the new republic. Already in colonial days, in 1765, an act prohibiting the emigration of trained operatives had been passed. Other statutes of a similar nature followed, and many of these enactments lasted well into the nineteenth century. But knowledge of the new force spread just the same, and gave many a good Yankee visions of power as fantastic to men of his age as those of Jules Verne or H. G. Wells were to men of later generations. The very fact that few of them had a chance to study the actual operation of the steam engine only stimulated their imagination. Since during the first decades of the republic so much interest in technological improvements was concentrated on waterways, interest in steam engines centered on their use in navigation. The fact that after years of experimenting with steam engines, even in England, no successful attempt at steam navigation had been made did not deter the American inventors. The back country had to be opened—an empire was waiting. Why not conquer it the modern way—by steam?

Since New England was geographically self-contained, and not connected by great rivers or main trails with the newly opened western territories, it never played an important role in the early attempts at steam navigation. New York and Philadelphia were the logical experimenting grounds for the steamboat visionaries. One of the best known of them, John Fitch, was a farmer's son of Windsor, Connecticut, and trained as a clockmaker, but his experiments were carried out in Philadelphia and on the Delaware. His patent for the application of steam to navigation dates from 1788, and Fitch actually organized a boat service on the Delaware between

1790 and 1792. The boat had a beam type of engine, with an 18-inch cylinder, which drove paddle boards at the stern; it ran a total distance of between 2000 and 3000 miles. Fitch's eventual failure was typical of American technology around 1790. Though he actually reinvented some of the English innovations, such as the condenser, it was all done in a cut-and-try way, without standard equipment, without tested material, without any standard precision work. Every part of a steam engine could afford—and actually did afford—the greatest surprises when put to work. No satisfactory steam engine could be made in a country where even shovels were made one at a time in the local smithy.

Other experimenters were James Rumsey, who in 1787 demonstrated a steamboat on the Potomac, and John Stevens of New York, who after years of work lived to see one of his steamboats enter the ocean (1809), and even to construct a steam locomotive (1825). There were more of such steamboat experimenters, but two of them belong entirely to New England—Nathan Read in Salem and Samuel Morey in Fairlee, Vermont. We have already met Nathan Read as a man of many ideas, who in 1789 experimented with paddle wheels to test his steamboat ideas. As to Captain Samuel Morey, he experimented in 1793 for a while with a little craft, loaded with a steam engine, on the Connecticut River and on a pond near his home at Fairlee, but never obtained much recognition. On misty moonlight nights, we are told, a tiny ghost steamboat still slips through the fog of the lake. But in 1808 the gallant captain tried again, now with more success, and had the pleasure of launching his own steamboat on Lake Champlain. In 1826 he was one of the first to invent an internal combustion engine.

The first commercially successful steamboat was Robert Fulton's boat, later known as the *Clermont,* after the estate of his sponsor, Robert Livingston. The *Clermont* sailed the Hudson from New York to Albany in 1807, twenty-five years after the experiments of Fitch and Morey. During this quarter of a century, at last a few Americans had been able to catch up with some of the developments in European engineering. Fulton's success was due to the fact that he was able to combine full theoretical knowledge with typical American daring in practical work. Most of his education was obtained in England and France; while in England in 1796 he published *A Treatise on the Improvement of Canal Navigation*. For his steamboat experiments he systematically studied not only the works of previous inventors, but also theoretical works on ship resistance, notably the works of Bossut and Beaufoy in France. In this way, Fulton became the earliest example of the theoretically trained American engineer-inventor. But he was only successful because of powerful backing. His principal sponsor was the wealthy and broad-minded New York jurist, Robert R. Livingston, into whose family he subsequently married. Livingston, a remarkable man of Jeffersonian cast—he negotiated the Louisiana purchase in 1803—was a patron of science and engineering, an experimenter with new farming methods as well as with steam engines. Thanks to his acceptance in Livingston's exalted circle, Fulton was able to remove many obstacles which had beset other inventors. The success of the steamboat venture was due not only to Fulton's ability to obtain a Watt and Boulton engine, but also to the circumstance that Livingston was able to press through the New York legislature a bill giving him and Fulton complete control of steam navigation in New York rivers. This regulation repealed the rights of the late unhappy John Fitch.

Fulton has left us many insights into his approach to science and invention. He absorbed the advanced teachings of the eighteenth-century reformers, defended the French Revolution and saw with misgivings the rise of Napoleon. He was a friend, not only of Livingston, but also of Joel Barlow, a Connecticut-born Jeffersonian, radical poet of *The Columbiad* and with Washington a supporter of the idea of a national university. He constructed the plans of a submarine around 1797 and submitted them to the French Directoire, in order to make it possible to destroy the British Navy, make England a republic and thus promote the cause of liberty. He saw in science a means to destroy human oppression, and believed in the possibility that science could promote peace and free trade. Such ideas were not uncommon among the inventors and scientists of the first generation after the

Revolution, although they were more common outside New England than within. New England's conservative merchant class was always very reserved in its adherence to the principles of the Enlightenment. Through his marriage into the Livingston family Fulton himself in his later life became increasingly identified with the wealthy patroon class on the Hudson. His and Livingston's monopoly of the Hudson River steamboat traffic became the target of many public attacks.

The successful performance of Fulton's first steamboat encouraged the sponsors to organize a regular service between New York and Albany. Already by 1807 the *Car of Neptune* was in service, and by 1812 three boats were operating on the Hudson, ranging in tonnage from 160 to 331. In 1814 New York possessed seven steamboats for commercial and passenger service. The new form of transportation rapidly became the popular method of travel on the large inland waterways, especially in undeveloped territories. The Hudson, the Ohio and the Mississippi Rivers became the testing places for steamboats.

Since New England did not lie along the main traffic arteries to the undeveloped West, it did little pioneering in the new form of communication. Here oversea trade remained the main interest, and for this sailing ships were for many years the only possible form of transportation. Opposition to the introduction of steamboats was successful for some time; it was supported by stagecoach and turnpike proprietors, who, for obvious reasons, welcomed the general distrust of that new and radical thing, the steam engine. Not until 1816 did a steamboat penetrate northeast of New York. On a breezy September day of that year the *Connecticut* arrived in New London from New York after a journey of twenty-one hours, running against wind and tide. Then, in October of the same year, a regular transportation line was opened in Long Island Sound between New York and New Haven. It was still a decade of experimentation. In this year 1816 the *Alpha,* of fifteen tons, ascended the Kennebec as far as Augusta, with steam power applied to a steam propeller. The owner, however, was dissatisfied with the performance and a short time afterwards sold his boat.

In June 1817 steam navigation was introduced into Boston:

> We understand [wrote the *Boston Daily Advertiser* of July 4, 1817] that the elegant steamboat "Massachusetts" will be here this day at ten o'clock, and will take a few gentlemen and ladies for a few hours to sail about the islands in the harbor.

This 230-ton boat, with an engine of 30 H.P., made a few trips between Boston and Salem, but its financial success was indifferent and it was transferred to the South the same year. In 1818 the *Eagle,* an excursion boat, ran from Nantucket to New Bedford for six months. Regular service between Providence and New York did not begin until 1822. From 1834 on, regular service was established on the Merrimack between Nashua and the new town of Lowell.

Both engines and boats improved considerably during these years of trial and error. The service between Providence and New York, once it was established, boasted of model boats. David Stevenson, in his book of 1837, described the *Narragansett* and the *Massachusetts.* The *Narragansett* was the finest seaboat he saw in the United States, with a length of 210 feet, a maximum breadth of beam of 26 feet and paddle wheels of 25 foot diameter which made 24 revolutions a minute. It plied the sea with speed and regularity despite the fact that between New London and Newport it was exposed to the roll of the ocean. The *Massachusetts,* of 200 foot length and 30 foot breadth of beam, received special praise for its comforts. It had 112 fixed berths ranged around its spacious general cabin, and 60 more in the ladies' cabin, while 175 persons could dine in the cabin without overcrowding. "The scene," wrote Stevenson, "resembles much more the coffee-room of some great hotel than the cabin of a floating vessel." In 1836 steamboat navigation already had a modern aspect.

11

The leading New England engineer of the canal period was Loammi Baldwin, son of the colonel in Washington's army who supervised the building of the Middlesex Canal. The history of these two Loammi Baldwins well illustrates the great strides which civil engineering made in the United States within one generation. The younger Baldwin was born in 1780 and graduated from Harvard College in 1800, after which he settled as a lawyer in Groton and later in Cambridge. He inherited his father's interest in tools and instruments, and was happier at the mechanic's bench than at the bar. The older Baldwin tried to interest his friend Rumford in his son, and Rumford actually persuaded a London instrument maker to offer young Baldwin an apprenticeship. Nothing came of this plan, perhaps because of the terms, which required a period of servitude lasting from two to four years and a premium of not less than sixty to a hundred pounds a year. At Groton, in 1802, the young lawyer provided the town with a fire engine, which, as "Torrent No. 1," functioned for many decades and is now the property of the local Historical Society. By 1807 young Baldwin could no longer endure the legal profession. He closed his Cambridge office and went to England to study the engineering works of that country. Upon his return he settled in Charlestown, opened an office and commenced at last the career he had always wanted to follow. This was no longer a case of dilettantism, but of increasing experience based on a growing theoretical knowledge.

During the war with Britain, in 1814, he built a fortification on Noddle's Island in Boston Harbor, named Fort Strong and situated in what is now East Boston. It was built with volunteer labor, the various trades and crafts taking special days for their part of the labor. After this Baldwin definitely established a name as an engineer, when he undertook the supervision of various public works in Virginia and in Boston itself. In 1814 the Boston and Roxbury Mill Corporation was chartered with the specific object of building a dam through the Back Bay, south of the Boston peninsula, in order to use the water power for tide mills. When Uriah Cotting, the director of the undertaking, died, Baldwin took over and completed the projected Mill Dam in 1821. It was one and a half miles long and built from stone quarried in Brookline; for those days and for America it was a remarkable undertaking. "This Vast Design," said the plan, "in the Wide Range of its Utility, will undoubtedly exceed any undertaking hitherto accomplished in this Quarter of the World." Not all citizens were equally enthusiastic, and a contemporary communication voiced their protest in some well-chosen words:

> Have you ever inhaled the Western breeze, fragrant with perfume, refreshing every sense and invigorating every nerve? What think you of converting the beautiful sheet of water which skirts the Common into an empty mud-basin, reeking with filth, abhorrent to the smell, and disgusting to the eye? By every god of sea, lake, or fountain, it is incredible.

The Mill Dam never succeeded in providing sufficient water power, and the indignant citizens of 1814 saw their prophecy fulfilled. We have already mentioned in Chapter 3 how later in the century the Back Bay was filled in. It is a now pleasant residential and shopping section of Boston.

In 1821 Loammi Baldwin planned the construction of the Union Canal in Pennsylvania, a structure of 79 miles. Here he proposed to practice all the famous devices of the English canal builders, including a tunnel of 739 foot length. After his untimely withdrawal from this project he spent a year in Europe, mainly in France, for practical as well as theoretical study. His later life was filled with important engineering work, mainly canal and dam building, both in New England and in other parts of the country. The two great works of his life were the naval dry docks at Charlestown, Massachusetts, and at Norfolk, Virginia, built between 1827 and 1834, which attracted considerable interest. They were among the works in America highly praised by David Stevenson. The Boston dock was 306 feet long, 86 feet in breadth and had a depth of water of 30 feet. Its masonry, of Quincy granite, was, according to Stevenson, "the finest specimen of masonry which I met in America."

Though a Harvard graduate, Baldwin was entirely self-made as an engineer. During his European trip of 1824 he laid the foundations of his engineering library, which in its time was the largest and the best in the United States. It was long preserved in the old Baldwin house at North Woburn and is now at the Massachusetts Institute of Technology with other Baldwin relics. He has been called the father of civil engineering in this country. During his days, which were essentially the days before the railroads, and in which engineering was hardly recognized as a profession, he was one of the leading men in his field. His career can be compared to that of his ten-year-younger contemporary, Major David Bates Douglass, an army engineer who taught at West Point and later went into consulting work. During the first part of the last century few public works in the United States were undertaken without some advice from Baldwin or Douglass. As to Baldwin, the Boston gentry respected him. When in 1825 the Bunker Hill Monument was projected, he was on the different technical committees with Daniel Webster, Gilbert Stuart, Washington Allston, George Ticknor and Jacob Bigelow. We can imagine the gentlemen walking on the Mill Dam, from where Baldwin pointed to the outline of Bunker Hill, fastening against the railing of the sidewalk small models of the obelisk he had prepared, to see how the monument would look from different distances.

The many canal plans from Baldwin's hand included the investigation, made in 1825, of the practicability of a canal from Boston to the Connecticut River. Baldwin also reported upon the introduction of pure water into Boston from ponds west of the city (1834), and it was his work, combined with that of Daniel Treadwell and others, which eventually led to the great system of Boston waterworks. In a similar way, Douglass advised on the Morris canal in New Jersey and on the Croton river waterworks of New York.

Loammi Baldwin died in 1838. He never became familiar with the steam engine, belonging in both training and temperament to that first generation of engineers, of which Brindley and Smeaton were the British prototypes. He accepted assistants in his office who passed through an apprenticeship, for which they usually paid $200 a year for two years while receiving some compensation for work in the field. In a period when technical schools were nonexistent, many engineers received their education in this way. Alexander Parris and Uriah Boyden were Baldwin pupils.

Baldwin's commanding figure—he was over six feet tall

—his unusual energy and ability of leadership, his social connections with Boston's aristocracy, gave him a prominent standing in the community. This helped to bring the engineering profession prominently before the public and gave it an intellectual and social prestige which it had previously been denied. Two of Loammi's brothers, James Fowle and George Rumford Baldwin, also made names for themselves as engineers. They survived their brother by many years and continued to impress a younger Boston generation with the increasing importance of modern engineering.

Chapter 5

The Beginnings of Mass Production

Les Yankees, ces premiers mécaniciens du monde, sont ingénieurs, comme les Italiens sont musiciens et les Allemands métaphysiciens,—de naissance.

—JULES VERNE, 1866

1

THE TRANSITION from tool to machine is not simply a change in mechanical complexity. The essential element in this transition is the replacement of the tool in the worker's hand by the tool as part of a mechanism. The machine proper is a mechanism that with its essential part performs the same operations that were formerly done by the workman with his tools. Whether the motive power is derived from man, from animals, wind, water or steam, makes no real difference. From the moment that the tool proper is taken from the worker's hand and fitted into a mechanism, a machine takes the place of the hand. The spindle and distaff are incorporated in the spinning machine, the fingers of the hand weaver in the power loom, those of the cotton worker in the gin, the body force of man and animal in the steam engine. The industrial revolution begins the moment the tool changes into the machine.

The fully developed machine consists not only of the tool or operating mechanism, but also of two other essential parts —the driving mechanism or motor, and the transmitting mechanism. Whenever the operating part of a machine is set to work, it influences the two other parts of the machine. Improvements are constantly made. The motor mechanism changes from man, animal, water or wind into steam and later into electricity, and the motor power of water or wind is itself stepped up. The transmitting mechanism with its gears, shafts, cams, pulleys, straps, and pinions, becomes more efficient; and again, the development of one type of machine is conducive to that of another; a process culminating in the building of machines to build machines. Often these different types of development can be quite well distinguished in a study of the industrial revolution within a country as a whole. This is especially true in the beginning of such a period where men are tackling tasks in fields that are still unrelated. In New England these fields were the textile and the arms industry.

To what extent were the elements necessary to the development of the machine in existence in the early years of the republic? On the social side there existed a merchant aristocracy, which was not originally too enthusiastic about manufactures, since shipping promised ever-increasing returns. Alexander Hamilton had to crusade arduously to convince his followers of the possibilities implicit in the industrial revolution. There were few proletarians, and an always open frontier existed to lure away the impoverished mechanic or farmer. Until the period of the Irish immigration the industrial revolution could only make progress through the wholesale use of women and children.

As to the ability of New England craftsmen to construct the many parts of machine and its transmitting mechanism, their greatest handicap lay in their inability to achieve uniformity of product. Machine design requires standard equipment and standard production methods. The blacksmith, the gunsmith, the carpenter, turning out products one at a time, gave to each artifact the stamp of individuality. This may be highly desirable from an artistic point of view, but it is often a highly undesirable quality in mass production. One of the main reasons for the failures of the early steamboat inventors was the fact that every new boiler had its own unexpected properties, which sometimes worked favorably, but just as often contributed to poor performance. Whitney's invention of interchangeable parts and Blanchard's invention of the copying lathe were steps essential to the change in craftsmanship required by the machine age.

Cheap motive power was not difficult to obtain. Yankees were not afraid of hard work, and there were oxen and horses to perform the tasks too difficult for men. Wind and water power were available in abundance, and New Englanders were accustomed to their use, if only on a primitive scale.

Colonial New England had many windmills, especially near the coast, where some are still standing. There were water wheels beside many brooks—heavy structures with a horizontal axis, deriving their motion from the momentum of streaming water falling on palettes placed along the rim. Until steam became available, water wheels were by far the cheapest and most reliable source of power. New England's industry developed along the falls of its rivers.

2

The political revolution which created the United States was not the result of an industrial revolution in the British colonies, but it created an atmosphere favorable to the introduction of the new economic system. During the colonial period hardly any machines were in use. Perhaps the most formidable one was the clumsy Newcomen steam engine at the Schuyler copper works in New Jersey, which functioned from 1755 to 1768, and again in 1793–1794. But inventive ingenuity set to work almost immediately after the Peace of Paris, to improve manufactures, and thus stimulate experimentation with machines. Some industry, notably that of firearms, developed during the Revolutionary War. After the war many men tried to improve the production of nails and tacks, others worked on cards and looms, others on clocks and muskets. Glass factories were started in several places, but many were shortlived despite occasional state protection; a $3000 lottery was organized to start a plant in Boston. The type of industry developed, its chances of success and expansion into new and revolutionary ways, depended primarily upon objective political and economic conditions.

Whatever their individual philosophy might have been, the Founding Fathers agreed that the Constitution itself should provide for the financial protection of inventive genius. Congress, in Article I, Section 8, was empowered "to promote the progress of science and useful arts, by securing for limited times to authors and inventors the exclusive right to their respective writings and discoveries." This provision of the Constitution, passed in 1787, was followed by the first Federal Patent Act, passed by Congress as early as 1790. This act provided that an examining board might grant a patent for fourteen years, to be signed by the President of the United States, the Secretary of State, and the Attorney General, at the cost of $3.70 plus the copying of the specifications at 10 cents a sheet. The first man to avail himself of the new privilege was a Vermonter by the name of Samuel Hopkins, who received a patent in 1790 for an improved method of "making pot and pearl ashes."

Other early patentees were working on the improvement of transportation. They were only following the trend of the times, since the merchant class had begun to spend a good deal of inventive and business energy on the improvement of inland transportation. The early days of the republic saw men engaged in many other economic activities, though they were sorely handicapped by lack of enterprising capital, lack of business experience, and insufficient knowledge of natural science and engineering. We have seen that even simple tasks, such as the mass manufacture of spades and picks, required a good deal of study and patient experimenting.

After the war, therefore, technological invention was attempted where men—and, by the way, not women—felt most handicapped in transportation and the more manly of the homespun crafts, such as candlemaking, glass blowing and nail cutting.

The leaders among the pioneers of invention and enterprise were outstanding political figures, signers of the Declaration of Independence, enactors of the Constitution and members of the Congress. They were motivated by the revolutionary belief in progress and by their conviction that inventions can be used for the improvement of human welfare. This philosophy could take many forms, and led either to the Hamiltonian belief in the progressive value of manufactures, under the aegis of a moneyed aristocracy, or to the belief of Jefferson in the basic importance of the small property owner, the farmer, tradesman, artisan and mechanic. We do not always know what the specific motivation of each particular inventor was, a fact interesting to study from whatever papers he left. However, it is well to keep in mind the fact that we are dealing with a generation which took its revolutionary mission very seriously, and which held the misery and the wars in Europe, and those brought about by Europe, to be the direct result of monarchical and feudal mismanagement.

These men envisioned the United States of America as the bulwark of progress and the hope of humanity. Both the merchant and the mechanic, each in his own way, looked upon his new republic in this fashion. Robert Fulton, in a letter to William Pitt written circa 1800, expressed this in the words:

> It is such men as Arkwright, Wedgwood, Boulton and Watt, Bridgewater, your iron-masters and manufacturers, whose inventions multiply the produce of labor, that England owes her prosperity, and not to an idle monarch, or a set of dissipated nobles and corrupt legislators.

Even those men who in our minds are most prominently connected with the introduction into America of the factory system with all the misery it involved were often men of progressive and humanitarian outlook. For example, Moses Brown, the Providence merchant-adventurer, who brought Samuel Slater to Pawtucket to build the new textile industry, was a Quaker and an abolitionist, with a strong bent toward public service. Many other early pioneers of industry were men of similar mind.

3

It is an interesting and instructive fact that in early republican New England most scientists came from the coastal towns, while many inventors and manufacturers came from the farming towns of Connecticut and Worcester County, Massachusetts. The main reason for this was the peculiar atmosphere of the Yankee farm, which created the "whittling boy," growing up into the general handy man. The poor condition of the soil drove many of them to emigrate to virgin territory, while those who remained added to their income by shrewd bargaining. Since the unlimited right to private property was accepted as an axiom, the typical conditions for the creation of a new bourgeoisie were present, a bourgeoisie which took its place beside the older merchant bourgeoisie of the coast. With the opening of the country through turnpikes and canals and the disappearance of colonial trade restrictions, this budding bourgeoisie had the opportunity to expand.

A characteristic aspect of this development was the traveling salesmen from New England—the "Yankee peddlers"—who gradually became welcome visitors all over the country. These tradesmen carried the products of the new manufactures from Maine to Georgia. Many of the more inventive and enterprising traders built factories of their own along their native brooks and rivers and relied upon their own efforts and those of their fellow townsmen to distribute their wares throughout the villages and farms of New England, the South and the newly opened West.

There was another cause for the growth of manufactures, especially of tools and household goods, in the interior parts of Connecticut and Massachusetts. These regions were old mining grounds. During the eighteenth century Connecticut was the leading iron producing colony, and its deposits of copper and other metals were widely exploited. The Salisbury mines provided iron, the Simsbury mines copper for numerous manufactures, producing household goods such as pots, kettles, spades or nails, as well as anchors, guns and cannon. Both the iron and the brass industry of Connecticut date back to colonial household industry. With the expansion of the markets this traditional ingenuity was put to excellent use. The brass industry of Waterbury, one of the largest of the United States, arose in the valley of the Naugatuck River. Again, the success of the older industries encouraged newer ones; Waterbury became a center for the manufacture of timepieces. Small textile mills appeared on many streams shortly after the Revolution. Worcester County began to specialize in iron industries. All these industries concentrated on small tools, household articles and precision instruments, and peddlers undertook the sale of all these "Yankee notions."

> With a small horse and a cart laden to the horse's capacity, he [the pedlar] would start out from the pedlar's center in his own or a neighboring village. These centers were in the larger communities and in them would gather the pedlars from the smaller villages and surrounding farms. Brisk trading, conducted with the Yankee cunning which every year grew more acute, went on before the departure.

Then, one by one, the little jingling carts pulled out, each headed on a long journey. A pedlar might be gone six months; he might come home only for the harvest, but as business prospered it drew more and more men forever from the farms. All through the South where what we now call "gadgets" were unfamiliar and exciting, the visit of the pedlar, bringing each year a host of new inventions, was anticipated with delight.[1]

This combination of peddler, whittling boy, inventor, manufacturer was typical of the leaders of early New England's mass industry. Self-made, original, hard working and resourceful, these men formed a class which set its stamp upon the character of the nation's industry. The revolutionary fervor of the early leaders had been characterized by a semipractical, semitheoretical approach to science. These men of a later generation had a more matter-of-fact approach, but in their own way they remained close to the people from whom they had sprung. Many of them had little use for college education with its emphasis on the classics, or even for a natural science which emphasized astronomy and theoretical studies of which they did not see the use. Invention and engineering

[1] The literary backwoodsman from Tennessee, David Crockett, in his description of Job Snelling, a Massachusetts peddler who had set up shop in the South, shows that this "delight" was also mixed with other feelings:

"The whole family were geniuses. His father was the inventor of wooden nutmegs, by which Job said he might have made a fortune, if he had taken out a patent and kept the business in his own hand; his mother Patience manufactured the first white oak pumpkin seeds of the mammoth kind, and turned a pretty penny the first season; and his aunt Prudence was the first to discover that corn husks steeped in tobacco water would make as handsome Spanish wrappers as ever came from Havana, and that oak leaves would answer all the purposes of filling, for no one could discover the difference except the man who smoked them, and then it would be too late to make a stir about it. Job himself bragged of having made some useful discoveries, the most profitable of which was the art of converting mahogany sawdust into cayenne pepper, which he said was a profitable and safe business; for the people have been so long accustomed to having dust thrown in their eyes, that there wasn't much danger of being found out." (1836)

were something to be learned through practice, with a minimum consultation of books. This close connection with the people led several of these enterprising men to an interest in welfare work for their employees. For a long time Samuel Colt's Hartford armory was known for the care it took of its workers. The same reputation was enjoyed by the South Boston foundry of a manufacturer-inventor named Cyrus Alger, established in 1809, which drew its profits primarily from the production of cannon and cannon balls. Eli Whitney, at an early date, tried to take care of his workers. Proximity to the people, from which they sprang, and among whom they often continued to live, created at the same time a typical patriarchal attitude. The employer took, or liked to think he took, personal care of his employees. This led to charity and welfare work, and also to a hostile attitude towards attempts of the workers toward independent action, especially towards the budding labor movement and the democratic reforms for which it stood.

4

One of the first and most sensational inventions of a machine occurred in the South. The inventor was a Massachusetts Yankee, who was destined to revolutionize Southern agriculture as well as Northern production methods. The name of this man, one of the most influential men in American history, was Eli Whitney.

Whitney's story has often been told. It is a tale typical of the inventor in a competitive society. The son of a farmer of some substance living in Westboro in Worcester County, Eli Whitney began his career as one of those characteristic products of the New England farm—a jack-of-all-trades. As a boy he made and repaired violins, worked in iron, manufactured nails and hatpins. Handicapped by his family's orthodox prejudice against sinful college life, he made up his mind rather late in life to study, and graduated from Yale in 1792 at the age of twenty-six. After graduation he followed the example of many a Connecticut lad of those days, and traveled south in the hope of becoming a tutor in some gentleman's house. He thus became acquainted with the widow of the Rhode Island general, Nathanael Greene, a gentleman of parts who had been an amateur scientist and agronomist and had settled after the war on a plantation near the Savannah River in Georgia. It was here that Whitney heard some planters discuss the state of the cotton culture. The industrial revolution in England had created an enormous demand for cotton, and the prices had risen sharply. Eli Whitney, stimulated by the talk of the planters, realized that a new type of ginning mechanism was necessary to improve the marketing possibilities of Georgian cotton.

Whitney set to work and in a short time designed just such a mechanism. By April 1793 he had built a gin with which one worker—of course a Negro—could produce fifty pounds of cleaned cotton a day. The invention discarded the ancient rollers and substituted for them a cylinder with bent spikes sharpened to form hooks. They were set in a ring and revolved through slots in a bar. When the cotton was pressed against the opposite side of the bar, the teeth pulled away the lint, which was then cleaned from the teeth by brushes. These brushes were made to revolve in a sense opposite to that of the teeth and several times as rapidly. A hand crank operated the whole machine. Almost overnight, Georgia cotton marketing became profitable. Rarely, if ever, was such a social revolution promoted by so simple a machine.

In May 1793 Whitney drew up a partnership agreement with Phineas Miller, another Yale man, who managed the Savannah plantation, a "man of respectability and property," who subsequently married Mrs. Greene. The partners were to engage in the patenting and manufacturing of cotton gins and to conduct a ginning business. Eli, in a letter to his father of September 1793, after his return to New Haven, described his work and his hopes as follows:

> In about ten Days [after being encouraged by Miller] I made a little model, for which I was offered, if I would give up all right and title to it, a Hundred Guineas.—I concluded to relinquish my school and turn my attention to perfecting the Machine. I made one before I came away which required the labor of one man to turn it and with which one man will clean ten times as much cotton as he can in any way before known and also clean it much

better than in the usual mode. This machine may be turned by water or with a horse, with the greatest ease, and one man and a horse will do more than fifty men with the old machine. It makes the labor fifty times less, without throwing any class of People out of business.

Whitney, we see, was also concerned with the human side of his invention. He then explained his chances of obtaining a patent in the United States and in England, and felt sure that he had attained success: "ten thousand dollars, if I saw the money counted out to me, would not tempt me to give up my right and relinquish the object." But the plans must remain secret: "I wish you, sir, not to show this letter nor communicate anything of its contents to anybody except My Brothers and Sister, *enjoining* it on them to keep the whole a *profund secret.*"

The invention could not be kept secret, and Whitney's troubles were only beginning. He received his patent in 1794. Miller and Whitney decided to manufacture the gins in New Haven, Whitney was to supervise the production, Miller to furnish the capital and to attend to business in the South. They tried to obtain a monopoly in their field, but they overreached themselves; it levied too great a tribute on the Southern planters, and the partners failed in their attempted monopoly. The patent was being infringed upon on all sides. Other troubles arose, the factory burned down in 1795, there was lack of capital, ugly rumors came from England that the ginned cotton contained knots, a rumor which was dispelled only after two years. Lawsuits followed each other with indifferent success. In all, Whitney received from various sources about $90,000, but the more than 60 suits to establish his rights swallowed up most of it. Whitney decided to concentrate on the manufacture of firearms.

The cotton gin remained, even though its inventor quit. The cultivation of cotton increased enormously. England began to order cotton from America—the first to order it was Robert Owen (1791). The crop of the United States rose from 1,500,000 pounds in 1790 to 35,000,000 pounds in 1800 and 85,000,000 pounds in 1810; in 1860 it totaled more than two billion pounds. The social implications were even more formidable. It re-established Negro slavery on a solid foundation and thoroughly discouraged manufacturing enterprise in the South. The scientific prominence of the South gradually disappeared. A civil war was necessary to redeem some of the evil effects which were the unpremeditated by-products of the genius of a progressive and benevolent inventor. The story of the cotton gin is a classic example of the disastrous effects of uncontrolled economic change.

Modern gins have increased considerably in size, and use steam, gas or electric power. They use circular saws instead of the rows of hooks which Whitney introduced—a patent of one of Whitney's competitors. But despite their differences gins are still constructed on essentially the same principles which underlay Whitney's invention.

5

Whitney's gin factory was at New Haven, where as early as 1798 he decided to undertake the manufacture of firearms for the United States Government. For this purpose he introduced a principle new to American manufacture, one destined to revolutionize production methods as fundamentally as his gin had revolutionized the social structure of the South, namely the use of interchangeable parts for tools, for machinery and for their products.

At the time Whitney signed his contract with the United States Government, there were no factories in the country for the manufacture of military firearms. Individual gunsmiths were making squirrel and other sporting guns. Some of them were famous during the war, such as that terror of the Hessians, the Kentucky long-barrel rifles, but arms were nowhere produced in quantity. Whitney's contract called for 10,000 muskets at $13.40 each. He purchased land and built an armory near New Haven, at the site now called Whitneyville. To produce this large amount of firearms he introduced his new principle of standardization of production by means of interchangeable parts.

Whitney concluded that he had to "substitute correct and effective operations of machinery for that skill of an artist which is acquired only by long practice and experience." He would have to make the same parts of different guns "as much like each other as the successive impressions of a copper-plate engraving." This required division of labor and the transfer of human skill to special tools and machinery. It took two years to produce and install the machines, after which the production of muskets could proceed on a mass scale.

Whitney met with considerable opposition. When Washington officials became uneasy at the prospect of advancing money without a single gun yet produced, Whitney went to the Capital and showed the Secretary of War and other interested army officers ten separate piles each containing identical parts of a musket. To the amazement of the audience he was able to select parts indiscriminately from each of the piles, and to put ten muskets together. The fact that this story has anecdotic value shows how dissimilar the ancient handmade guns had been and how much for granted we now take mass production.

Jefferson was impressed. In a letter of 1801 to Monroe, he wrote:

> He [Whitney] has invented molds and machines for making all the pieces of his locks so exactly equal, that take 100 locks to pieces and mingle their parts and the 100 locks may be put together by taking the pieces which come to hand.

The principle of mass production through a series of machines turning out the finshed product was carried out in ample detail in Whitney's armory. Although several operations, including assembly of the parts, had to be done by hand, many others were performed by machines. There was a machine to shape and bore the barrels of the muskets, another to shape the stocks, others to make the parts of the locks. One of Whitney's milling machines, claimed to be the first of its kind ever made, can still be seen in the New York Museum of Science and Industry in Rockefeller Center. One of the devices to introduce uniformity was the jig, a metal pattern equipped to guide a tool into the proper operation, so that every time the jig is applied the tool cuts an exact replica of the piece it cut out before. This insistence on uniformity was one of the most important contributions of Whitney to the introduction of modern industry into the United States. Whiney's armament manufacture and the growing textile industry delivered a decisive blow to old-fashioned handicraft —a blow which, although it destroyed some delicate esthetic and social values, was necessary for the industrial revolution to succeed in the United States.

In 1812 Whitney obtained a new contract from the United States Government for 15,000 muskets and one for a similar number from the State of New York. Whereas it had taken him eight years to fill his first contract, he could now deliver his order in two years. The armory in Whitneyville continued to be successful and allowed Whitney to recoup the fortune he had lost in his cotton gin litigations. The United States Government turned to him for advice in establishing its own armories in Springfield and in Harpers Ferry. Eli Terry was influenced by Whitney to apply his system of interchangeable parts to the manufacture of wooden clocks. Samuel Colt, in the beginning of his career as an inventor of revolvers, had them manufactured for several years in the Whitney armory.

Eli Whitney died in 1825, a well-to-do man of standing in his community. His portrait by the young artist Samuel F. B. Morse reveals an intelligent, strong-willed and handsome face, expressive of a man whose work influenced the whole social and economic history of the United States.

The business was continued after his death, first under Whitney's nephews, later under his son and grandson, both also called Eli Whitney. The second Eli Whitney, a Princeton graduate, was himself an inventor of no mean ability, with many patents improving gun manufacture to his name. Among his inventions were a method to make gun barrels of steel, and another to manufacture percussion-cap guns to replace the ancient flintlocks.

The armory continued under the management of the Whitney family until it was leased to the Winchester Repeating Arms Company in 1888. Several of the buildings and some of the company houses are still standing.

Another remarkably ingenious man was Whitney's nephew, Eli Whitney Blake. He came from the same town in Worcester County as his uncle, and like him grew up as a farm boy and general handy man. Times had changed since the early Whitney's days. The uncle had met with firm opposition from his family when he wanted a college education; young Blake was allowed to prepare for Yale, where he graduated with the class of 1816. He had his uncle's sixth sense for the invention of new tools, and though his patents protected reforms somewhat less revolutionary than those of his uncle, they nevertheless opened up new possibilities in several fields. He was one of the pioneers in the manufacture of domestic hardware, a field on which the American mind has expended so much of its genius. Blake established a hardware factory in Westville near New Haven and remained as its head for thirty-six years. He devoted his spare time to theoretical studies and shortly before his death, in 1882, he collected several of his papers on aerodynamics in a volume which shows him to be one of the first Americans to be interested in this field. His name is connected above all with the Blake stone crusher, patented in 1858, whose principles still dominate quarrying machinery.

6

Simeon North must be mentioned with Eli Whitney as one of the men who introduced standardization in the production of arms. He came from Berlin, Connecticut, and had the same farm background as Whitney and Blake. He opened a factory in Middletown, Connecticut. He undertook a contract for 500 horse pistols for the government in 1799, only a year after Whitney's contract for muskets, and produced them in one year in his shop in Middletown. Other contracts followed, and by 1813 he had manufactured at least 10,000 pistols and employed a crew of 40 to 50 men. In that year he built a three-story armory, where he continued production along the same principles as those of Whitney. A short time after the younger Whitney, in 1848, had shown the possibility of producing gun barrels from steel, he began to produce them himself. All the present great arms companies of Connecticut date back to the establishments of North and Whitney.

The history of the Connecticut arms industry is extremely interesting, but too long and complicated to follow here. It later again became famous through another Yankee jack-of-all-trades, Samuel Colt, the inventor of the revolver. Colt was born at Hartford in 1814 and had an adventurous career. He went as a sailor to Calcutta; later he embarked on a "lecture tour" throughout the country, demonstrating the effects of laughing gas to an amused public; he fought in the Seminole War, and finally settled in Hartford, where, after 1848, he established an armory. His revolver first became "popular" with the Texas rangers, and helped to open the West to the white man. In 1862, at the time of Colt's death, his factory produced 1000 firearms per day, and was manufacturing machinery for other armories as far away as Enfield, England, and Tula, Russia. Colt was also one of the inventors of the submarine telegraph cable, laying one in New York Harbor in 1843, long before Cyrus Field's day. One of the most dashing and adventurous of all the machine shop pioneers of America, he carried, it has been said, "Whitney's conceptions to transcendent heights."

Arms were made not only in private factories, but also in armories owned by the government. These armories held an important place in the mass production of arms. The Springfield Armory was established as a national institution in 1794 by act of Congress, continuing an establishment laid out during the Revolution. Production of muskets was started in 1795 and the armory soon took steps to improve the manufacturing methods of arms. Robert Orr became master armorer in 1804; he was the son of Hugh Orr of Bridgewater, one of the early pioneers of textile manufacture in New England, and was himself one of the first to try his hand at the mass manufacture of iron shovels. Under his direction attempts were made to introduce greater standardization of the finished product. About 1810 experimentation was in progress to turn out musket barrels with a uniform external finish. The superintendent's attention was drawn to a young man from Sutton, a town in that same Worcester County from whose farms had come Whitney and Blake. Thomas Blanchard, a timid boy who stammered badly and whose early history also reads very much like that of Whitney and Blake, had already made several inventions. The superintendent of the Springfield Armory offered Blanchard a contract for one of his machines, which Blanchard accepted. So began Blanchard's connection with the Springfield Armory, which led to several discoveries that contributed greatly to the standardization of manufacturing processes. The best known is Blanchard's "copying lathe."

The story is told that a worker at the armory remarked that although the work of grinding the barrels was done away with, his own work, that of making wooden gunstocks, could not be replaced by a machine. Blanchard, challenged, answered by solving the whole problem of turning irregular forms from a pattern. The idea of such a lathe struck him, we read, as he was driving home through Brimfield, and in his emotion he, like Archimedes, shouted: "I have got it, I have got it!" This occurred about 1818.

The principle of Blanchard's "copying lathe" is that of an imporved jig. First a pattern was made in the shape of the object which had to be reproduced, then every part of it successively brought into contact with a small friction wheel. This wheel regulated the motion of chisels upon a cutting wheel acting upon a rough block. From this block it pared off the superabundant wood while the pattern rotated. The machine produced not only gunstocks, but also shoe lasts, wheel spokes, hat blocks, handles and many other objects. The principle has been continuously applied until the present day, and has made machine production possible in many new industries. The original machine, used over fifty years, is still preserved in the Springfield Armory. It was discarded when work on metal could be performed with greater accuracy than work on wood.

Blanchard worked for the government for five years, during which time he saw his invention extensively pirated. When the patent expired in 1833, he petitioned Congress for a renewal, which was granted on the grounds that his machine was original and one of the first American inventions. Litigation followed, but Blanchard's rights were upheld in the courts.

Blanchard took out many other patents and was also one of the early promoters of steamboats and railroads. In his later life he became a patent expert. He lived in rather comfortable circumstances on the royalties of some of his patents, and died, seventy-eight years of age, in 1864.

7

The New England merchant class was slow to convert part of its interests into manufacturing. In the early days of the republic shipping and the overseas trade, the traditional occupation of the New England ruling group, expanded greatly and absorbed almost all energies. Merchants of the type of Moses Brown in Providence and George Cabot in Beverly, who shortly after the Revolution invested some money in textile industry, were rare. There was considerable distrust in the prospects of manufacturing. This attitude changed with the Embargo of 1807–1809, which prohibited an American vessel from sailing from any American port destined for any foreign country; it changed even more with the War of 1812.

Once started, the textile industry, even more than the arms industry, became the prototype of all mass industry. By 1790 Alexander Hamilton, in his *Report on Manufactures,* had already observed concerning cotton that "there is something in the texture of this material which adapts it in a peculiar degree to the application of machines." As to linen and wool manufacture, New England had a long experience. From the time of the earliest settlements a considerable quantity of linen and wool was spun and woven. The bulk of this work was done on the homesteads, while the fuller, who finished woolen cloth, had his mill on the town brook. Most products were for home use, and a considerable amount of woolens and cottons had to be imported from England, which jealously guarded its technological secrets.

The first sign of the interest of the merchant class in the use of textile machinery was the experimenting done by George Cabot, the highly successful Salem merchant, who during the Revolution owned at least forty privateers in partnership with his two brothers. He was engaged in banking, bridge building and Federalist politics, and later developed into the sage of the Essex Junto. In 1787 this enterprising adventurer set up a mill in Beverly near Salem with horse-driven carding machines and spinning jennies. Despite subsidies (even a lottery) from the legislature, the mill could not be made to pay. The machines of Cabot—and of the Bridgewater blacksmith Hugh Orr—were not good enough and they knew it. Too few details of the English machines were available.

The situation changed when the firm of Almy and Brown (Moses Brown's firm) in Pawtucket, which had also experimented with jennies without much success, took Samuel Slater into their service. Slater, a young Englishman from Derbyshire, was as a boy apprenticed to the cotton spinning business of Jedediah Strutt, the partner of Arkwright. An intelligent lad possessing a highly mechanical turn, he gained his master's confidence and friendship to such a degree that before he came of age he was entrusted with the supervision of a new mill and the construction of its machinery. Instead of continuing his promising career in England, Slater decided at the age of twenty-one to seek his future in America; in order to avoid a conflict with the laws of England he took no drawings of machinery with him. The fact that he looked more like a young farmer than a mechanic seems to have helped him at the customs. He arrived in New York in November 1789. Hearing that Almy and Brown were seeking a manager for their cotton mill, he wrote to Moses Brown that through his work for Arkwright and Strutt he knew the method of running the spindles the English way. "We hardly know what to say to thee," wrote back the Quaker merchant, "but if thou thought thou couldst perfect and conduct them to profit, if thou wilt come and do it, thou shalt have all the profits made of them over and above the interest of the money they cost, and the wear and tear of them." With Slater's arrival in Pawtucket, the textile industry in the United States began to be successful.

Slater knew enough of textile manufacture to attempt to build the machines. Pawtucket was an old ironmaking center, and the experienced local blacksmiths, Orziel Wilkinson and sons, were at Slater's disposal. The result was not only the making of the whole series of machines on the Arkwright plan, but also a little romance in the old-fashioned way:

> Moses Brown introduced Mr. Slater to Orziel Wilkinson of Pawtucket, R. I., as a suitable place for him to board; as the stranger came into the house, the two daughters, as is not uncommon, ran out of sight; but Hannah lingered

with curiosity, and looked through an opening in the door; Samuel saw her eyes, and was interested in her favour. He loved at first sight, but it was sincere, and it was permanent, nothing but death could have severed the ties which endeared him to Hannah Wilkinson.

Thus finally established in Rhode Island, by both occupation and marriage, Slater built up his cotton spinning business with great perseverance. By December 1790 he had carding, drawing and roving machines and seventy-two spindles in two frames. Power was obtained from the water wheel of an old fulling mill. Business prospered; "The Old Factory" was built in 1793, a second mill near it in 1799. In 1807, after Slater's brother John had come from England, another cotton mill was built; later mills were constructed in what is now the town of Webster, Massachusetts, where also woolens were manufactured. Here Slater settled in his later days, and here he died in 1835.

Mechanics of Slater's soon set up little mills of their own in many towns of Rhode Island and elsewhere, wherever there were falls in the rivers. Merchants were willing to back these mechanics financially, and in 1809 there were 62 spinning mills in operation in the country, with 31,000 spindles, while 25 more mills were being built or projected. Cotton came from the South, from Whitney's gins, and the yarn was sold to housewives or to professional hand weavers to make cloth for sale.

Slater solved the difficult problem of labor supply by moving whole families of impoverished farmers into his mill villages. All members of these families worked in the factories, even the small children. It was a method with which Slater had become familiar in England. This "family" system of employment became universal throughout Rhode Island and Connecticut and very general in southern Massachusetts. The families occupied single homes or tenements, owned by the companies. Despite Slater's personal benevolence, which he tried to show in the creation of Sunday school instruction, an innovation in America, conditions soon became miserable. Like Whitney's gin, Slater's jennies were transforming society at a great cost in human suffering. As so often in class society, this suffering had not entered into the speculations of the originators of the scheme. No control over social forces existed in those early days. When the misery created by the factory system became apparent, social philosophies were created to moralize about them. The *Memoir of Samuel Slater,* published in 1836, has a long chapter on the "moral influence of manufacturing establishments," with such sentiments as the following:

> In the present happy condition of the manufacturing districts, there are no advantages enjoyed by the rich, that are not reciprocated by the poor. Labor was never better paid, and the laborer more respected, at any period, or in any part of the world, than it is at present among us.

Business fell not infrequently into the hands of persons interested solely in profits. This again sharpened the incipient class struggle between the operators and the workers, and led to the beginnings of the modern labor movement. Thus a new element entered into American social history, and one which had considerable consequences upon the development of America's political and cultural life.

By 1810 the value of "goods manufactured by the loom" in the United States was about $40,000,000—and the bulk of this came from New England mills. It was almost one third of the total value of manufactures produced in the United States. Massachusetts had 54 mills, Rhode Island 26, and Connecticut 14, but there were few woolen factories. The dearth of industrial products caused by the War of 1812 compelled the Americans to extend their manufactures, and many millions of capital were invested in the establishment of woolen and cotton factories. More and more Americans became willing to risk their luck in the new mass production methods, not only to make a handsome profit, but also to show the world what the new republic across the ocean was able to perform.

8

The time had arrived when leading Massachusetts merchants began to think of cotton manufacture on a large scale. Among them were Nathan Appleton, a Boston importer, and

Patrick Tracy Jackson of the Newburyport mercantile aristocracy. With this newly awakened interest of the powerful ruling class, the infant industry entered into a new financial and technical phase.

Francis Cabot Lowell was one of the Newburyport merchants who had settled in Boston to enlarge their business interests. During a visit to England he met Nathan Appleton and both decided to enter into the manufacture of cotton. Upon returning to Boston they joined with their colleague Jackson, who was Lowell's brother-in-law, and made preparations to set up a new and technically advanced type of cotton factory. Lowell was particularly interested in the power loom, Cartwright's invention, already widely used in England, but he had not been able to obtain drawings or models in England. In Boston he reconstructed such a loom with the aid of Paul Moody, an expert mechanic of Amesbury. The team of Lowell and Moody was as successful as that of Slater and Wilkinson had been twenty years before. Looms were built and set up by the Charles River falls at Waltham. When completed, the mill had carding, drawing and roving machines as well as looms, and was the first complete cotton factory in the United States. The Slater mills had never done any weaving. Here in Waltham, however, cotton fiber was made into cloth in a process performed under one roof. Capital was furnished not by one man or one firm, but by a combination of men. Their company, the Boston Manufacturing Company, was established in 1813, during the war with England.

This mode of cotton goods production was an important development in the budding corporation type of business of America. The wealthy merchants who initiated the business invested large amounts of capital, petitioned the legislature for an act of incorporation and applied systematic organizing and selling methods. From the beginning their business was organized for mass production; it integrated all processes from the raw material through the finished product under one management and, if possible, in one plant. After Lowell's death the Waltham mills grew prosperous under the guidance of Jackson's management and Moody's inventive genius.

New large-scale enterprises with considerable capital investment were soon in the making. In 1821 a small group of Boston capitalists, led by P. T. Jackson and Appleton, guided by Moody, purchased a tract of about 400 acres near the Pawtucket falls on the Merrimack. They founded new manufacturing establishments, which expanded into the new town of Lowell. Mills and boardinghouses, which were to become typical of the northern New England factory towns of the nineteenth century, were built. The Merrimack Manufacturing Company, incorporated in 1822, started with a capital of $1,200,000. Other large corporations followed. The period of modern industry had set in.

Other factory cities, like Manchester and Dover in New Hampshire, and Saco in Maine, date from the same period. The workers were originally mostly young farm women recruited from all over New England. They were subjected to a rigorous system of factory supervision in boardinghouses, which was widely advertised by the manufacturers as a remarkable humanitarian scheme, in contrast to the "family" system in Rhode Island, where the workers—often woman and children—were left entirely on their own. It must be said that this paternalistic system did not lead to the excesses of poverty which were typical of the English system, but it did deprive workers just the same of health and civil liberties. The labor movement encountered as many difficulties in the towns owned by Boston financiers as anywhere else in Europe or America.

The "Waltham system," as this scheme of corporation controlled industry combined with welfare work was called, extended its control also to engineering, science and invention. It introduced from its very start a development which assumed gigantic proportions as the century grew older. Invention in spinning, weaving and dyeing proceeded regularly. Mechanics' contracts always specified that their inventions were to become the property of the company, and usually forbade them to disclose the mechanical secrets of their employers or to put these to their own use. In 1820 the Boston Manufacturing Company's agreement with Allan Pollock, their mechanic, contained the clause:

Should I be fortunate enough to make or suggest any

improvement for which it might be thought proper to obtain a patent, such patent or patents are to be the property of the Company.

Such methods also existed in smaller shops, but the very nature of the system gave the corporations the greatest chance to promote technological advances. Thus by their systematic training of specialized mechanics did the textile factories furnish a labor market for mechanics in other fields.

9

The phenomenal growth of the cotton industry influenced the entire development of American industry. An unending series of inventions revolutionized the technique of cotton manufacture and that of the related fields of woolens, rugs, carpets and silk. The production of textile machinery became in itself a new type of industry. One of the earliest producers of textile machinery was Amos Whittemore of West Cambridge, now Arlington, Massachusetts. Whittemore, a farmer's son, had (as he always claimed) a dream, in which he saw the possibility of a machine for puncturing the leather and setting the teeth of the cards, an operation always previously done by hand. He actually invented the machine and had it patented as early as 1797, then sold the patent for $150,000. His brother Samuel later repurchased it; a factory was founded and for some generations Arlington became a center of the manufacture of cotton and woolen cards. The business was subject to many ups and downs; in 1862 the plant burned down and was never rebuilt. The beautiful Amos Whittemore house is still standing near the library.

The technical methods of cotton manufacture influenced inventions in many other fields. Elias Howe worked for a while in a Lowell factory and may there have been inspired with some of the ideas which later led to his sewing machine. Early cotton manufacture also stimulated the imagination of Ichabod Washburn, an expert mechanic of Worcester, who in 1831 was the first in the United States to produce wire of good quality. This venture was partly responsible for the establishment of the large Worcester metal industry. It is significant that this new technological development in turn influenced Jonas Chickering, a piano manufacturer of Boston, a craftsman as well as a businessman. It led him to experiment with steel wires for pianos. In 1837 he introduced the first iron frame in the grand piano. Chickering collaborated with Washburn through years of trial and error, but he ultimately achieved such a remarkable perfection that the name of Chickering still remains famous in piano construction.

The influence of the new improvements in the manufacture of arms, and especially the principle of interchangeable parts, was equally broad. One of the first industries to receive the full impact of these innovations was that concerned with the manufacture of timepieces. New England, especially Connecticut, was the home of expert clockmakers in colonial days. In the early years of the republic these craftsmen, such as the Willards of Grafton, Massachusetts, often master carpenters, continued to produce beautiful examples of their skill. From their ranks came the men who were able to bring the change to mass industry. A Connecticut peddler by the name of Eli Terry blazed the trail. He came from East Windsor to settle near the colonial industrial settlement of Waterbury, where at about the turn of the century he began to use machinery in the fabrication of his wooden clocks, influenced by Whitney's methods. He set up his instruments in an old mill. The installation of machinery enabled him to produce interchangeable parts, with which he could produce clocks in lots of one or two hundred, which he sold himself, traveling about the countryside on horseback. At first his neighbors shook their heads when they heard of this mass production, for where was he to find a market? In 1807–1808 Terry undertook to make 4000 movements and completed the contract in three years. In 1814 he invented a 30-hour shelf clock, which he patented in 1816; it revolutionized the clock business and was produced in large quantities for many years until it was superseded by brass clocks. This clock sold originally for $15, and finally for about $5. Before Terry died, in 1852, he was making and selling from 10,000 to 12,000 clocks a year, thus earning a handsome fortune His hired man, Seth Thomas, soon set up for himself, buying, in partnership with Silas Hoadley, the original Terry factory and continuing mass production. The name "Seth Thomas" has appeared on so many dials that it is one of the best-known names in early American clockmaking. We can still inspect these curious wooden clocks with their simple and often semi-Gothic designs in the Smithsonian Institution in Washington, in Old Sturbridge Village and in other museums, as well as in private homes.

The third early Connecticut clockmaker of importance to engage in mass production methods was Chauncey Jerome, born in 1793. He also started as a carpenter and began his career as a worker for Terry and for Thomas. Jerome set himself up in business after the War of 1812 and was the first to break away from the custom of making the clocks of wood. After the establishment of brass foundries in Waterbury and Bristol, Jerome designed a brass one-day timepiece in a wooden case, small enough for easy transportation and cheaper than any clock made up to that time. The early Jerome clocks cost from $5.00 to $6.00, a low price for the time, the later ones even less. In 1840 Jerome established himself in Bristol, where he turned out brass clocks by the thousands, and sold them as far away as England. He continued to produce his clocks despite amazing ups and downs. He managed to reduce the cost of labor to about 20 cents a clock, which he sold for 75 cents; three men could make and cut all the wheels for 500 movements in one day.

In 1850 he joined with two other men to form the Jerome Manufacturing Company in New Haven, which was highly successful for some years, but later was manipulated into bankruptcy. It was taken over by the New Haven Clock Company. Jerome, in the crash, lost everything and died in relative obscurity in 1868. In 1860 he described his methods and his adventures in a curious book, which showed how he was done in by rascals all through his life—a story which could be told by many an inventor of the nineteenth century.

Even more directly related to Whitney's ideas was the development of American watchmaking. This is the story of Aaron Lufkin Dennison and the Waltham watches. There again it was in Waltham that the early pioneering found its start.

Chapter 6

Lecture Hall and Textbook

> Colleges and books only copy the language which the field and work-yard made.
>
> —RALPH WALDO EMERSON, 1837

1

WHILE SAILORS plied the oceans to the Indias and to China, while self-made engineers built canals, bridges and turnpikes, and manufacturers, inventors and merchants laid the foundations of mass production, the colleges remained very quiet. Lecture halls and professorial chambers produced no great ideas, and conceived no important projects. There was some progress in the medical schools, but the development of science had to wait till a new generation was well under way.

When New England became a part of an independent republic it had few institutions of higher learning only and these few were strongly dominated by the clergy. It had no laboratories, no observatories, and only small scientific collections and libraries. Advanced textbooks were scarce, and all were written abroad. There was not even much choice in grammar school texts. Generally speaking, science was the privilege of a few wealthy men and their associates, and was extremely one-sided.

John Winthrop's textbooks at Harvard, during the early part of his career and probably also later, consisted of Gravesande's *Mathematical Elements of Natural Philosophy,* at that time a modern introduction into Newton's theories, Euclid's *Geometry* and the *Young Mathematician's Guide* by John Ward, an English scholar of indifferent merit, whose book was also used at Yale, at Brown and at Dartmouth. Physics, chemistry and geology were hardly known, with the exception of Newton's optics and Franklin's electrical experiments, which were even the topic of popular performances. The number of professors of science was small, and the teaching was mostly in the hands of tutors, usually college graduates in the ministry in possession of some haphazard information on the laws of nature and logic. Common schools confined themselves to the teaching of writing and spelling, and did not even include arithmetic. This state of affairs continued for several decades. Bronson Alcott, born in 1799, describes the Massachusetts day schools of his boyhood as places where "until within a few years no studies have been permitted but spelling, reading and writing." "Arithmetic," he continued, "was taught by a few instructors one or two evenings in a week." In secondary schools there were some "ciphering."

During the Revolutionary War many institutions had to reduce their activities considerably or even closed their doors. Many teachers went away to engage in the war. Harvard had to abandon Cambridge for some time; its buildings were used as barracks. Troops occupied the premises of Rhode Island College for five long years. Recovery was slow, and scientific work continued at a low level.

It is not difficult to find the reasons for this slow recovery. New England's science remained for a long time mainly the science of a mercantile class, interested primarily in navigation. This meant Newtonian science, and with the sympathies of the ruling merchant group definitely pro-English—despite a slight pro-French break during the Revolution—this science was bound to be unproductive. Bowditch and Farrar, who finally broke the pro-English tradition in the exact sciences, did so only in the second and third decade of the nineteenth century. Interest in natural science, geology, zoology, botany and mineralogy was small in these circles of merchants and skippers; their indifference was the more pronounced since there was a French, or even a democratic, flavor to this knowledge, which a good Federalist resented. The clergy, still powerful in the colleges, often felt the same way, and had reason to combine this democratic spirit with freethinking or with atheism. For a long time the paramount interest in shipping and the aristocratic and traditional character of science in New England also made any connection between college science and the new thoughts in engineering and technology impossible. Benjamin Thompson, the only New Englander who thoroughly understood the necessity of refreshing college curricula with the study of the new technology, stayed far away in Europe, and his activity did not begin to bear fruit until 1819, when Bigelow was appointed to the Rumford chair in Harvard College. The earliest break with the old tradition came through the medical schools. It was followed by the appointment of Silliman at Yale in 1802 as professor of geology, mainly because of the worries of orthodox President Dwight concerning the atheistic character of natural science, and the necessity to do something about it, not by neglecting it but by appointing an acceptable man.

The prime need was the writing and publication of elementary text books, and during the early years of the republic there were authors willing to fulfill this need. Three New Englanders, two from Yale, one from Harvard, were outstanding and won acclaim throughout the nation in this field. One of them, the lexicographer, remains famous to this day. They were Noah Webster, Jedediah Morse and Nicholas Pike. Noah Webster's spelling books, of which the first appeared in 1783, had a lasting effect on American education, combining elementary instruction with the building of a feeling of nationalist unity. Of almost equal influence, if only for one generation, were the geography readers of the Reverend Jedediah Morse, like Webster a Yale graduate from a small Connecticut town, who ministered his flock in Charlestown, Massachusetts, from 1794 to 1820. His *American Geography* appeared in 1791 and made Morse's name a household word. Morse interpreted geography in a very wide sense, and included in his works some astronomy, botany, zoology, as well as lists of snakes and insects. Morse's texts, like Webster's, were builders of that strong American patriotism which had grown out of the exaltation of the Revolution.

Nicholas Pike, whom we met as a teacher and a magistrate in Newburyport, in 1788 published what was believed to be the first North American English arithmetic, though there had been an arithmetic written by Isaac Greenwood of Harvard College and published in 1729. Pike, like Webster, took care that prominent authorities endorsed his adventurous enterprise. The *New and Complete System of Arithmetic* was prefaced by presidents and professors of leading colleges and even by George Washington himself in a cautiously worded approbation. A Harvard recommendation was based on the patriotic sentiment that Pike's book would "save much money in the country, which would otherwise be sent to Europe." The text was quite advanced for an American school book and contained logarithms, trigonometry, algebra and conic sections, a sign of the progressive character of Newburyport's merchants and sailors. It passed through many editions; after the fourth edition in 1822 the advanced part was left out.

2

These early native school texts on grammar, geography and arithmetic originated outside of the colleges. The college staffs themselves had, for the time being, less to offer. Nehemiah Strong, of Yale, after having left the college for the bar, produced an *Astronomy improved* in 1784, which seems to have had little influence. Harvard had little to offer beyond a few uninteresting compilations from English authors by Samuel Webber, who succeeded Samuel Williams in 1789 as the Hollis professor of mathematics. They are of interest only as early efforts to provide the colleges with native texts.

At this time Priestley, Mitchill, Rush, Cooper and other scientists in New York and Pennsylvania were working on new problems in geology, botany and chemistry, inspired by the new French school, by several English scientists and by Jefferson, whom many of them also followed politically. The general atmosphere in Boston, despite the presence of well-intentioned academicians and the beginnings of literary and scientific interest in the Anthology Club (about which later), remained singularly indifferent to this newer type of research. This traditionalism struck young Silliman when he visited Boston in 1807, as a young Yale professor fresh from Europe. There was not much of a spirit of science, he declared, only some taste for literature. He found that Harvard's collection of minerals, a gift partly from the French Republic and partly from Dr. Lettsom in London and donated by Waterhouse, was hardly used.

This was indeed the heyday of Bostonian Federalism, the political creed of the established merchant circles, politically narrow-minded, snobbish and violently antidemocratic.

Though it expressed the ideals of a class which had played and still was playing a revolutionary role, it undermined its own progressive position by opposing the political aspirations of farmers and mechanics during Shays's Rebellion and long afterwards. In this way the Federalists blocked for many years those forces of progress on which Jefferson tried to build and which, in the thirties and forties of the nineteenth century, formed the background and the moving power for the flowering of New England's letters and science.

Still, in its own way, Federalism with all its limitations was a progressive force, promoting manufacture and navigation and to a degree even agriculture. Many outstanding Federalists, notably the members of the so-called Essex Junto and real die-hard Federalists like George Cabot, were patrons of science, learning and invention, as they conceived these arts. This was sometimes hard to reconcile with their bitter sentiments against Jeffersonianism. Thomas Jefferson was prominently identified with the advancement of learning in the republic, and in his circle moved such outstanding democrats and scientists as Benjamin Rush, Joseph Priestley and Thomas Cooper.

Gouverneur Morris, a New York Federalist, had remarked in the Federal Convention that he wanted none of "those philosophical gentlemen, those citizens of the world as they called themselves, in our public council." When the French Revolution sent scientists to the United States—Gallatin, Priestley, Cooper, Adet, Du Pont de Nemours—their presence was agreeable to Jefferson and odious to the Federalists. This led the conservative Boston gentlemen occasionally to a position which bordered on obscurantism. Timothy Pickering, Secretary of State under John Adams, was restrained only by the President himself from prosecuting Priestley under the Alien and Sedition Acts, under which Thomas Cooper, Priestley's friend, was sentenced to six months' imprisonment and a fine of $900. Such fanaticism was typical of many leading New England aristocrats, especially of the ponderous bigwigs of the Essex Junto. John Adams himself had a deeper understanding of the dignity of science as transcending party boundaries; in 1796–1797 he went openly to hear a lecture by Priestley, his old London acquaintance. But he drew the line at the scientists from France: "We have," he declared, "too many French philosophers already, and I really begin to think, or rather to suspect, that learned academies . . . have disorganized the world, and are incompatible with social order." Men of lesser stature went much farther in their fanaticism, and Jefferson's interest in the mastodon bones of Ohio, or his "mountain of salt" in Louisiana, was ridiculed in the Federalist papers. During the days of the Embargo a lad of thirteen in the Berkshires became so incensed against his President that he rhymed:

Go, wretch, resign thy presidential chair,
Disclose thy secret measures, foul or fair,
Go, search with curious eyes for horned frogs,
'Mid the Wild wastes of Louisiana bogs;
Or where the Ohio rolls his turbid stream
Dig for huge bones, thy glory and thy theme.[1]

This was William Cullen Bryant, who later voiced far nobler sentiments. Feelings ran so high that they penetrated into every sphere of life. It is reported that there were even Republican and Federalist packet sloops runing between Hingham and Boston. It took courage to be a man of learning and a Jeffersonian in New England. The Federalists controlled the positions in the colleges and in other outstanding places and could make it difficult for the occasional Jeffersonians in their midst, even when their political heresy was somewhat offset by their scientific merits.

Vernon Parrington, writing of this period, believes that this narrow attitude delayed the New England literary renaissance a full generation, and colored it with certain provincialisms

[1] Another young poet of this day, in Wallingford, Connecticut, rhymed similarly:

. . . to guard
To keep our People from the Statutes hard
Of Cursed Jefferson Son of the Devil
Whose thoughts are wicked and whose mind is evil.

This was James G. Percival, who later glorified the Greek rebellion.

when it did appear. This is probably true as well of the scientific development of the period. The Federalist scientist, despite all his merits, locked himself up within a rather narrow circle. The political and social pressure considerably reduced the number of men of learning in New England who, like Dr. Mitchill in New York, "supported the republican party because Mr. Jefferson was its leader, and supported Mr. Jefferson because he was a philosopher." The Reverend William Bentley of Salem, the Democratic philologist, kept his ministerial position, but the circle of his friends was restricted. Professor Josiah Meigs, freethinking Democrat, meteorologist and mathematician at Yale, could not stand the strain and accepted the presidency of the University of Georgia. When William Plumer, the Democratic governor of New Hampshire, tried to reorganize Dartmouth College as a state university after the model of Jefferson's University of Virginia, in "a wise plan for changing a narrow sectarian seminary into a broad university," his attempt was defeated by strong Federalist opposition, backed in final instance by Justice Marshall of the United States Supreme Court (1819). We have it on Sanborn's authority that this decision "seems to have delayed for half a century that cordial interest of the State in the affairs of its sole college." And Benjamin Waterhouse, certainly the most distinguished New England scientist of the period, lost his position at the Harvard Medical School because, as he himself always claimed, of his militant Jeffersonianism.

3

This brings us to one of the most outstanding scientific achievements of early Republican New England, the introduction of cowpox vaccination into the United States. Boston had pioneered with inoculation in the days of Mather and Boylston; it pioneered again with vaccination in the days of Benjamin Waterhouse.

Waterhouse, born in Newport in 1754, was influenced by Scotch physicians who practiced there and by the reading of medical books in the Redwood library. Young Benjamin decided to become a physician. After an apprenticeship at home he sailed in 1775 to London to study medicine, where he met the well known Quaker, John Coakley Lettsom, physician, collector, horticulturist and friend of embattled America. He also visited the universities of Edinburgh and Leyden. In Holland he lived in the home of John Adams, then ambassador of the rebellious colonies. Waterhouse, an ardent patriot, at his matriculation called himself "Liberae Republicae Americanae Foederatae Civis"—citizen of the free American Federated Republic. This frightened the cautious Dutch authorities, who insisted on a less provocative title, since His Majesty's ambassador might be prejudiced. Waterhouse's dissertation *De Sympathia Partium Corporis Homini* appeared subsequently in 1780 with the simpler title "Americanus" added to his name.

After his return to America in 1782 Waterhouse accepted the professorship of the theory and practice of physic at the newly established Harvard Medical School, where he taught from 1783 to 1812. He was not only an able physician, but a pioneer in the teaching of botany and mineralogy. He lectured on these subjects first at Rhode Island College from 1784 to 1786, in a series of public lectures designed to help that impoverished school, and repeated these lectures from 1788 on, in Cambridge. They were later published in the *Monthly Anthology* of Boston, as a pamphlet in 1810, and in part as *The Botanist* in 1811. This work brought Waterhouse again into contact with Lettsom, who sent him a cabinet of minerals which Waterhouse later donated to Harvard College. Through Lettsom Waterhouse also became acquainted with Edward Jenner's work on vaccination in England.

In the beginning of 1799 Lettsom sent him a copy of Jenner's *Inquiry into the Causes and Effects of the Variolae Vaccinae*, published during the previous year. By that time inoculation against smallpox had become fairly popular, even fashionable, in this country, but it was not without danger. Jenner's cowpox vaccinations held an advantage in that only a mild disease resulted, though the immunization against the disease was at least as great as in the older form of inoculation. Waterhouse made Jenner's results known and in 1800 received some vaccine from England, with which he experimented on his five-year-old son and on a servant boy.

The experiment was successful. "One fact in such cases," he wrote, "is worth a thousand arguments."

The news of his work soon spread, and aroused the inevitable controversey. One of Waterhouse's first supporters was Lyman Spalding, a Harvard M.D. and a physician at Cornish, New Hampshire, a countryman and friend of Professor Nathan Smith of Dartmouth, later of Yale. Like Smith, Spalding was a man of modern mind, and had even translated some of Lavoisier's work. The opposition was strong and encouraged by the fact that nonprofessional men soon began to dabble in vaccination. This activity by "stagedrivers, peddlers and in one instance the sexton of a church" resulted in a serious epidemic, in the course of which a number of people died. In 1802 Waterhouse was successful in having a committee of seven outstanding physicians appointed, which concluded that "the cow-pox" was a "complete security against the small-pox." Jefferson was interested and had about two hundred persons vaccinated with vaccine sent by Waterhouse. Through his insistence on maintaining the purity of vaccine virus Waterhouse finally managed to place vaccination upon a solid scientific basis in the United States.

Waterhouse's troubles with his colleagues and the authorities were of long duration and ended in dismissals. In 1809 he was dropped from the staff of the Marine Hospital at Charlestown, in 1812 from the staff of Harvard Medical School, and this after twenty-nine years of service. There is reason to believe that personal motives also played an important role, but Waterhouse himself attributed his dismissals exclusively to politics. In a memorandum of 1811 he wrote that his "appointment to the hospital called forth all the forces of the opposition of the Democratic party in Boston. . . . When Jefferson went out of office, I went out." A more detailed report can be found in a letter from Waterhouse to Tilton in 1815, where he writes:

"In 1806 this ancient University was taken possession by the notorious Essex Junto, who consider it their castle, or stronghold, and have accordingly manned it with picked men of their own cast." The fact that Waterhouse "had collected a very considerable cabinet of natural history, especially minerals," was of no avail. "I was," he claimed, "gradually stripped of the fruits of all my labour," and in 1812 "set adrift with the loss of everything but my honour."

> When Mr. Jefferson came into office, the late Judge Lowell, a leading man of the Junto . . . gave us, of the college, to understand, that the church and all our other sacred Institutions were in danger, particularly the University, that therefore it behooved us Professors to rally with the clergy, and together form the front-rank in Massachusetts army of federalism, in opposition to infidelity, Jacobinism and Jeffersonianism.

Waterhouse's colleagues accepted it, but Waterhouse refused to alter his position. Lowell wanted Harvard to produce Federalist politicians steeped in "true principles—hatred to France, adoration of England and contempt of their own country." Waterhouse believed that even John Quincy Adams, at that time professor of oratory and rhetorics, would have been dismissed if he had not been appointed Minister to Russia.

Waterhouse at last received a commission from President Madison as medical superintendent of the military ports in New England, an office which he held until 1825. He lived till 1846 on his "small but handsome seat with ten acres of land, on the Cambridge Common," as he described it, and which still exists, though without the acres of land. In his later days he dabbled in general literature and with increasing age became more and more a picturesque old-time Cambridge figure. Oliver Wendell Holmes used to know him as a boy, and noted "his powdered hair and queue, his gold-headed cane, his magisterial air and diction." A man of action, he not only fought epidemics and Federalists, but also opposed intemperance and the use of tobacco, as we can learn from the following lines sometimes ascribed to him:

> Tobacco is a filthy weed,
> That from the devil does proceed,
> It drains your purse, it burns your clothes,
> And makes a chimney of your nose.

It is conceivable that Waterhouse's political enthusiasm in Jefferson's day made him somewhat of a trial to his Federalist colleagues, so that men of quality like Warren and Jackson found it impossible to establish harmony in organizing the medical work at the school. There were bitter conflicts, especially when Waterhouse, who had little clinical interest, opposed the transfer of the Medical School to Boston so that it would be near the contemplated hospital. Waterhouse, his enemies claimed, published "false, scandalous and malicious labels upon the other professors," and sponsored a Massachusetts College of Physicians, rival to the Massachusetts Medical Society. Dismissals for political reasons are often defended as dismissals for other more "social," or "ethical" reasons. Waterhouse's case was certainly complicated by professional conflicts.

4

The great new science of the day was chemistry, fascinating not only because of the wealth of its results, but also because of the controversial issues involved. The older chemistry was built around the so-called "phlogiston" theory, the result mainly of the efforts of J. J. Becher and G. E. Stahl, German chemists of the seventeenth and early eighteenth century. It implied that combustible substances, including metals capable of "calcination" (oxidation), contained a substance called phlogiston or "the principle of inflammability." When metal was recovered from an oxide by burning with charcoal, the metal was held to have absorbed phlogiston from the charcoal. The more complete the apparent diminution of a substance in combustion, the more phlogiston it contained. Phlogiston was supposed to escape from substances upon the application of high temperature; oxides of metals were due not to the absorption of a substance as we believe today, but to escape of phlogiston from the metals. The theory was accepted for about a hundred years, although it paid little attention to the quantitative side of reactions, and had no scruples in assigning a negative weight or "levity" to phlogiston, since its loss meant increase in weight. Yet it was the way in which chemistry emerged from alchemy, and explained so many facts known at the period that many men of great learning supported it, notably Priestley and Cavendish in England, Kirwan in Ireland, and Joseph Black in Edinburgh.

The phlogiston theory itself helped to build up the science which discarded it. The very persons who backed the old theory of Becher and Stahl were not unfrequently those whose researches contributed most to its destruction, such as Priestley and Cavendish. In 1775 Priestley discovered oxygen, calling it "dephlogisticated air." The name "oxygen" was coined by Lavoisier, who made it one of the principal elements of his new chemistry, which rejected the phlogiston hypothesis. Lavoisier's *Traité élémentaire de chimie*, published in 1789, established quantitative chemistry, and explained combustion not as an escape of phlogiston but as an absorption of oxygen. Metallic oxides and sulphuric acid were compounds, not elements, while sulphur, phosphorus and the metals, assumed to be compounds, were probably elements. With the gradual disappearance of the phlogiston theory chemistry emerged as a new and modern science.

Lavoisier still had a hard fight on his hands to assure victory to his new ideas. Priestley remained faithful to the old ideas, and when he came to America in 1794 he waged a friendly fight with his admirers Rush in Philadelphia and Mitchill in New York, who become convinced of the correctness of Lavoisier's theories. With Priestley's death in 1804 the phlogiston theory lost its strongest defender, and American scientists were ready to accept without controversy the new discoveries of Dalton, Davy, Gay-Lussac and the other European chemists.

In geology, another new field of science to North Americans, controversial theories also vied with each other for acceptance. Nearly all geologists of the early decades of the nineteenth century were divided into Wernerians and Huttonians. The followers of Abraham Werner of Freiburg in Germany advanced the idea of the aqueous origin of rocks, which were held to be precipitates of a primeval ocean. For this reason they were also called Neptunists, and for them volcanoes were rather abnormal phenomena, perhaps caused by burning subterraneous beds of coal. Opposed to them were the followers of James Hutton of Edinburgh, who were disposed to consider the structure of the earth's surface to be the result of subterranean heat, for which they were called Vulcanists.

Neptunists and Vulcanists were both catastrophists; they believed that elementary forces worked on the earth's crust in past periods which did not operate in the present period. This was in accordance with Biblical opinion as it was usually interpreted, and the result was that popular prejudice played a more considerable role in the arguments of the geologists of both schools than in those of the chemists. Both theories were not definitely abandoned until in the thirties Lyell proposed his theory of gradual development; after this modern geology emerged as modern chemistry emerged with Lavoisier.

Early American geology, as well as chemistry, did not participate productively in the solution of the controversial problems. The natural scientists of the United States confined their work mainly to observation and classification. Modern chemistry was accepted on the authority of the great English and French chemists, and as to geology, even the inhabited parts of the American continent were geologically unknown. Only with the accumulation of recorded observations did it become gradually possible to carry conclusions beyond the point of mere observation. At this stage, American scientists usually took a conservative position. Catastrophism found little active interest, but was quietly accepted, and Lamarck's early refutation in his *Philosophie zoologique* of 1809 was little studied. This also accounts for the fact that the ideas of evolution, as suggested by Lamarck, were entirely discarded. The full impact of the continental controversies on geology and biology did not come until the thirties and forties, when Lyell and Agassiz paid their visits to the United States.

5

Waterhouse's introduction of the study of natural history into New England's college halls was only a very small beginning. The first New England college to devote serious thought to the inclusion of the newer forms of natural science into the curriculum was Yale. There President Timothy Dwight, an orthodox minister and a stanch Federalist, showed himself a man of wider vision and broader interests than many of his contemporaries. He knew of the tremendous strides which chemistry and geology had taken in Europe, he knew that America was still geologically unexplored and may also have been aware of the importance of chemistry for the industrial revolution. He was at the same time concerned with the way in which science affected religion and how often scientists were sympathetic to democracy. The political difficulties with the Jeffersonian Professor Meigs affected him painfully. Rather than turn his back upon natural science, he decided to have it taught by a man of reliable religious and political background. His choice was easier, because he had some scientific understanding himself, being a founder of the Connecticut Academy of Arts and Sciences and a pioneer statistician. Looking for a promising teacher, he found a young tutor, Benjamin Silliman, whose father he befriended, and with remarkable foresight saw in the young Silliman the person he was looking for. In 1802 he had Silliman appointed to a newly created professorship of chemistry and natural history, although Silliman had not really decided to make science his lifework until Dwight's acumen mapped out his future career. The president could not have made a better choice.

At the time of his appointment at Yale Silliman was twenty-three years old. He came from a family which had settled in what is now Trumbull, Connecticut, and had entered Yale under President Stiles. Both Stiles and Professor Meigs stimulated his interest in science, Stiles in astronomy and Meigs in chemistry through an occasional lecture on the works of Chaptal, Lavoisier and other French investigators. After graduation in 1796 Silliman studied for the law, and he passed his law examinations just before his appointment to the new chair. To become a worthy incumbent Silliman began to take the study of natural science seriously, with means supplied by the university. Philadelphia at that time was the leading North American center of natural science. Here Silliman attended the lectures of Rush, Barton and Wistar, and met Robert Hare, destined to a great career as a chemist. At Wistar's house he met Priestley, the exile from England. At Princeton he learned chemistry from John Maclean.

Back at Yale he gave his first lecture in 1804, and received funds to establish his new laboratory. For books and apparatus he received an allowance of $9000. Since he wanted to put this appropriation to the best possible advantage, he went to Europe, where he learned from the great men in London and Edinburgh, from Dalton, Accum, Davy and others. At Edinburgh, under Thomas Hope, he received his first taste of geology, a relatively unexplored field for an American. After his return in 1806 he decided to remain a geologist, and began to explore the surroundings of New Haven with hammer and blowpipe. A visit to Boston and Cambridge convinced him even further that Yale had a pioneering task to perform in geology and mineralogy. Looking for materials for a cabinet of minerals, he obtained $1000 from Yale to purchase a collection of minerals—the New York collection of Benjamin Perkins of "metallic tractors" fame. About this time, on a visit to Newport, he met "Colonel" Gibbs and saw his main chance.

We have already met George Gibbs as a member of a wealthy Newport family. He had traveled extensively in Europe as a collector and amateur mineralogist, and obtained an excellent cabinet of minerals, mainly by purchase. When, in 1807, it was brought to Newport, it was supposed to be the best American collection of its kind; Gibbs loved to show the 10,000-odd specimens to his friends. Silliman, who understood its pedagogical possibilities, persuaded the colonel to place the collection on public exhibition at Yale College. George Gibbs consented and allowed the collection to be transferred to a room of the college, where it remained from 1811 to 1825 as a loan exhibit. By 1825 a new spirit had gradually come over America, and Silliman was able to raise the large sum of $20,000, mainly by public subscription to purchase the collection for the college.

Thus Yale had the finest collection of minerals in this country, as well as a brilliant teacher to explain its uses. Silliman was a great teacher, a great organizer, and a great popularizer of science, the first example of a modern scientist in New England. He knew how to attract pupils, and how to attract the attention of the public. As early as 1808 he arranged a course on "affinities of matter," which "ladies attended." Later, to stir up sentiment in favor of the purchase of the Gibbs collection, he organized a public propaganda meeting. This contributed materially to the success of his venture. Visitors from all over the country came to see the exhibit, and anecdotes grew up around it. A reverend gentleman worried and asked Silliman: "Why, dominie, is there not danger that with these physical attractions you will overtop the Latin and the Greek?" and Edward Everett complained that he had been unable to persuade Harvard's friends to secure the collection. When he hoped "the affair would give a useful lesson to our people against delay," Silliman declared that he was welcome to any moral benefit to be derived from the matter, "we, meanwhile, will get what good we can from the Cabinet."

A New Yorker, Dr. Archibald Bruce, had started an *American Journal of Mineralogy* in 1810. The venture did not flourish and seemed about to come to an end with Bruce's death in 1818, when Silliman stepped in. Encouraged by Gibbs, and by Bruce himself shortly before his death, Silliman took the journal under his own wings and, in 1818, started the *American Journal of Science,* which became a lasting monument to his talents and his enthusiasm. Edited for many years by Silliman, later by his son and namesake, and then by his son-in-law, James Dwight Dana, it has continued to exist and is now the oldest scientific journal in the country not published by a society. The history of natural science in the United States in the earlier part of the nineteenth century is largely identified with the history of the *American Journal of Science,* or *Silliman's Journal,* as it used to be called.

In Silliman's school—the earliest "school" in New England in the sense of a body of followers of a teacher—many young scientists found their early guidance and training. Two of them came from his own family. Silliman had married the daughter of a Connecticut governor, a Miss Trumbull. His son, Benjamin, Jr., followed in his father's steps and became his colleague both in the chair at Yale and on the editorial board of the *American Journal of Science.* Another of his pupils was James Dwight Dana, who married a daughter of Silliman's and later became one of the nation's leading geologists. Other pupils of Silliman at Yale were Edward Hitchcock, Charles U. Shepard, Oliver P. Hubbard, Thomas S. Hunt and Josiah P. Cooke, all of whom distinguished themselves in fields opened to them by their teacher.

Silliman also found time to organize at Yale a geological society, the first in the country, and to write a series of widely used textbooks. Beginning, as was customary in the early days of the republic, with an American edition of an English text, William Henry's *Epitome of Chemistry* (1808–1814), he himself later wrote a two-volume *Elements of Chemistry* in 1830–1831. He wrote about his trip to Europe in 1805–1806, and when, after his retirement at Yale, he paid a second visit to Europe, he described his experiences in a two-volume *Visit to Europe in 1851* (1853). He died in 1864, at the age of eighty-five.

Silliman's biography, compiled after his death by G. P. Fisher, contains large sections written by Silliman himself. They reveal his intense interest and his boundless devotion to his chosen fields of science, and the unfolding of an activity directed in the main toward organization, classification and teaching, which prepared the way for research of a more penetrating type. At the same time we see a certain narrowness developing, not so much the narrowness of a Bowditch, which essentially represented the point of view of a proud member of the ruling caste, but rather the anxious care which exists in the mind of certain scholars eager to avoid antagonizing the powers that be. Like so many American and British scholars of his generation, he was constantly trying to pacify the orthodox clergy by asserting that nothing in the teachings of modern science was antagonistic to the Bible. It should, however, be remembered that at critical periods in American history Silliman did not forget that he was only one generation removed from the men who made the Revolution, and that he himself helped to defend New Haven against the British in the War of 1812. At the time of the Civil War in Kansas he courageously braved public opinion to collect funds to help the Northern settlers with firearms—in 1856—and during the last years of his life he was an ardent crusader for the cause of the Union.

With such a capable organizer and teacher active at Yale, Harvard College was placed rather in a secondary position. The Harvard Medical School, however, set the pace for the college. One of its early appointees was the physician Aaron Dexter, who began to teach chemistry and materia medica

6

in 1782, at the very begining of the school's history, and gradually extended his work. He was not a man of research, and not, it is said, a great teacher, yet among his pupils were Lyman Spalding, Nathan Smith and Parker Cleaveland. In 1791 he induced his friend and patient, Major William Ewing, to endow his own chair for chemistry and mineralogy with a thousand pounds, thus establishing the Ewing chair which exists till the present. Benjamin Waterhouse introduced some botany and mineralogy into the college, and obtained money to be used for a botanical garden, which was established in 1805. But the real backbone of the Harvard Medical School was supplied by the three stalwart physicians, John Warren, his son, John Collins Warren, and his lifetime associate, James Jackson.

John Warren was presumably in the Boston Tea Party and was a brother of the revolutionary physician whose handsome statue adorns Warren Square in Roxbury. He held the chair of anatomy and surgery and at his death in 1815 was succeeded in this chair by his son John Collins, himself the father and grandfather of other medical Warrens. James Jackson, one year John Collins's senior at college, was the son of a Newburyport merchant, a pupil of old Dr. Holyoke in Salem, and for sixty years, from 1800 to 1860, a practicing physician in Boston. We possess a loving picture of this wise practitioner in his later days from the pen of Oliver Wendell Holmes. Through the hard labor of the Warrens and of Jackson the Harvard Medical School began to flourish and to achieve the leading position which it has since established throughout the world. Dartmouth followed Harvard, in 1798, with its own medical school, led first by the able Nathan Smith and later by Lyman Spalding, the physician and chemist, both Harvard men; Yale followed when it obtained Nathan Smith for its professor of medicine. The first great piece of scientific work associated with the Harvard Medical School was Waterhouse's introduction of the cowpox vaccination in 1799. In 1808 J. C. Warren and Jackson pioneered with a *Pharmacopoeia,* written for the Massachusetts Medical Society, the first in this country. This led Spalding at Dartmouth to prepare a U. S. Pharmacopoeia, eventually printed in Boston in 1820, reprinted and enlarged many times since. Then in the same year, 1810, in which Jackson obtained the chair of clinical medicine, he and John Collins Warren placed before the city a plan for a hospital.

Even at that late date, 1810, there was no permanent hospital in New England, though Philadelphia already had a hospital for almost half a century. Warren and Jackson's plea resulted first in the erection of an insane asylum at Charlestown, an innovation of great importance for the future, and at last, in 1820, in the foundation of the Massachusetts General Hospital at Boston. Bulfinch designed the building, which still stands in mellow dignity next to the massive modern structures. Jackson became the first physician, Warren the first surgeon of the new hospital. They were laying the cornerstone of an institution where startling experimentation was to occur. In 1846 John Collins Warren himself, in the halls he helped to erect, gave Dr. Morton the opportunity to administer to a patient the first anesthetic for surgical operation—one of the greatest events in an eventful century.

The college itself at Cambridge, in the meantime, was registering some progress. In 1794 the trustees of the recently formed Massachusetts Agricultural Society had, under the influence of Waterhouse's botanical lectures, appointed a committee "to consider the expediency of procuring a piece of ground for the purpose of agricultural experiments." In 1801 the committee voted an appropriation of $500 towards the foundation at Harvard of a professorship of natural history. These measures were steps in a drive which culminated in an appropriation of some thirty or forty thousand dollars for the establishment of a botanical professorship and a botanical garden. In 1804 William Dandridge Peck was appointed to the chair, and in 1805 the Botanical Garden was established. Friends of the institution contributed "exotick plants" from their own greenhouses, other plants were sent from the tropics and from Europe. The land was given by Dr. Andrew Craigie, who lived in what is now the Longfellow house, and had been apothecary-general to the Continental armies. Peck laid out the grounds and built a house on it, together with a greenhouse.

Peck was the son of John Peck, the shipbuilder of Kittery, Maine, and had lived a retired life on his father's farm, where he had built up a reputation as an entomologist. His work, performed in the Jefferson-Adams utilitarian tradition, concerned mainly agricultural pests, such as the cankerworm which lives on apple trees, or the slug worm, or beetles living on pear and pine, on oak and cherry. He received awards for some of these papers from the Massachusetts Agricultural Society. He was a man of excellent craftsmanship, who made his own microscope and his own beautiful drawings. At Harvard he is best remembered as the founder of the Botanical Garden, in behalf of which he visited Europe. It was in his care until his death in 1822, at seventy-nine years of age. In the White Mountains, where so many names keep the memory of New England's naturalists alive, a rosaceous herb preserves his name: *Geum Peckii.*

When Peck died, his professorship was vacated. As curator of the Botanical Garden he was succeeded by Thomas Nuttall, an able English naturalist, a botanist as well as an ornithologist, who had traveled widely in the United States. We shall discuss his work in a later chapter. He left in 1833, and after an intermission his place as head of the Botanical Garden was taken by Asa Gray.

Many men who pioneered in natural science during the early decades of the nineteenth century had been pupils of Warren and Jackson at the Harvard Medical School, two scholars who were not only men of action, but excellent teachers as well. One of the pupils was John Gorham, a Boston physician who graduated in 1801 and who developed ideas similar to those of his contemporary, Silliman. He met Silliman in London, where he had gone to study chemistry with Accum. Back in Boston he became Warren's pupil, married his daughter, and obtained an M.D. in 1811. When Dexter resigned from the Ewing chair of chemistry and mineralogy, Gorham was appointed his successor. This gave Gorham a chance to introduce modern chemistry to Harvard, which he did with great teaching and organizing ability. But for his death in 1829, at fifty-six, he might have accomplished for Harvard what Silliman was accomplishing at Yale. There were differences of education between the two men, since Silliman was a chemist and mineralogist, and Gorham a chemist and a physician, but in those days such differences were not as fundamental as they are today. With Warren and

Jackson, Gorham projected the *New England Journal of Medicine and Surgery,* which he edited for fifteen years. His two-volume *Elements of Chemical Science* (1819–1820) was the first systematic treatise written by an American chemist and was long a standard text. With this book we have finally entered the stage where New England began to produce its own academic textbooks.

We have also arrived at the stage where more and more chemistry was being used in industry, and in which even the colleges began to pay some attention to industrial applications of chemistry. At Harvard Gorham was succeeded, in the Ewing chair, by John White Webster, also an accomplished teacher of chemistry and geology, though as little an original mind as Gorham. His *Manual of Chemistry,* published in 1826, proclaimed itself openly a compilation of the works of European chemists, such as Brande, Henry, Berzelius and others. It paid, however, full attention to practical applications; it was up to date, and even quoted Faraday on the liquefaction of gases. Equally electric was the *Boston Journal of Philosophy and the Arts,* which he edited with John Ware and Daniel Treadwell from 1823 to 1826, but which showed a wide interest in English and French scientific work. Webster was a famous lecturer in his day—Edward Everett Hale speaks of "his brilliant power of experiment"—but is now remembered chiefly as the defendant in one of the weirdest murder trials in American history. He lived above his means, got into debt and in a moment of rage killed his creditor, Dr. George Parkman, at the Chemical Laboratory.[2] In his panic he chopped the body up and burned it, but eventually paid for his crime at the gallows (1850).

[2] Since last evening, our whole population has been in a state of the greatest possible excitement in consequence of the astounding rumor that the body of Dr. Parkman has been discovered, and that Dr. John W. Webster, Professor of Chemistry in the Medical School of Harvard College, and a gentleman connected by marriage with some of the most distinguished families, has been arrested and imprisoned, on suspicion of being the murderer. Incredulity, the amazement, and the blank, unspeakable horror have been the emotions, which have agitated the public mind as the rumor has gone on, gathering countenance and confirmation. Never in the annals of crime in Massachusetts has such a sensation been produced. (*Boston Evening Transcript,* December 1, 1849.)

7

The Medical School at Harvard not only devoted its attention to the study of chemistry, but showed interest in other fields of natural history. In the middle of the second decade of the century the power of old Federalist Boston began to wane, and with it came the decline of the staid Newtonian tradition. Nathaniel Bowditch was its exponent and high priest, established firmly in the respect of his citizens as a local Gauss or Laplace, but younger men began to turn their minds to other fields of science. Who, after all, knew anything about New England's nature, except perhaps old Manasseh Cutler in Ipswich and Peck in Cambridge? Who knew its animals, its plants, its minerals, its geological structure? Interest in such subjects was rife in New York, in Philadelphia, in the South, in the White House. Jefferson, as early as 1789, had written to President Willard of Harvard: "What a field have we at our doors to signalize ourselves in: The botany of America is far from being exhausted, its mineralogy is untouched, and its Natural History or Zoology totally mistaken or represented." The New England treasures had nevertheless remained untouched. To be sure, a few foreigners had penetrated into New England as if it were a barbaric country, to study its nature and to take some of its treasures to more civilized sections for exhibition. Early explorers, like Peter Kalm of Sweden, had carefully avoided it, and even the older Michaux did not do much better. Alexander Wilson, the Scotch-born ornithologist and poet, came to Salem in 1808 as a guest of Dr. John Prince, the minister and inventor, but when he set off for Newburyport, he saw only "a rocky, uncultivated and sterile country." He returned to New England in 1812, where his exploring habits brought him under suspicion of being an English spy.

William Maclure, the wealthy Scottish geologist, the "father of American geology," came also to New England in 1808. He visited Professor Parker Cleaveland at Bowdoin College and Professor Silliman at Yale, traveling "in a private carriage with a servant, and a pair of horses which, as they transported loads of stone from place to place, were lean and dull." He might have had considerable influence in stimulating interest in geology, but he brought to New England a touch of European freethinking, which the local respectability did not too well appreciate. Maclure was a fighting social reformer with communist leanings, he had as his companion Thomas Cooper, the veteran fighter for human rights, democracy and freethinking, who shocked Silliman and Dwight by his aggressive talk.

At about this same time the younger Michaux, Francois André, came from France to the United States to study its trees. Michaux started his exploration in 1806 in Maine and traveled along the Kennebec to Hallowell and Farmington. On a second trip, from Boston to Lake Champlain, he crossed New Hampshire and Vermont, and in all made five trips through New England. He studied trees from the point of view of the farmer, of the lumberjack and of the botanist. Studying pine logging on the Kennebec, he was struck by the size of the pine trees. One trunk of white pine was 154 feet long and 54 inches in diameter, another 142 feet long and 44 inches in diameter 3 feet from the ground. A third was even 180 feet long and 6 feet in diameter. The destruction of these trees filled him with alarm and, like Peter Kalm before him, he called for protection and for reservations. This was all published in the magnificent *North American Sylva, or the description of the forest trees of the United States, Canada and Nova Scotia,* in three volumes, published at Paris in 1819. Michaux was full of admiration for the lofty American sylva with its 140 species of trees 30 feet or higher, as against the 30 species existing in France. It is a curious thing that despite his keen detective instincts, the mountain laurel escaped his notice in New England, and he only observed it on Long Island. From the botanist's point of view New England was still almost virginal territory.

After his return from England, Silliman had gone out around New Haven with box and hammer and later described the geological features of the landscape for the Connecticut Academy. After 1810 young men around Boston also went out to explore the outlying ponds, the marshes, the meadows and the rocks. It was fascinating to take descriptions of the flora from other parts of the world and see if the harvest of a day's hunting could be recognized in the books. There was always the thrill of discovering a *novum,* something observed perhaps a thousand times but never classified. In 1814 some of these amateurs formed for a while the Linnaean Society, in which they discussed their discoveries. A young physician, Jacob Bigelow, a minister's son from Sudbury and a Harvard graduate of 1806, assumed the leadership. In the hours between practice he collected flowers, demonstrated them in a class on botany and wrote a book about them. His *Florula Bostoniensis* appeared in 1814, with the description of plants found within a circuit of ten miles around Boston. This was the first systematic flora of New England flowers, and with its enlarged editions of 1824 and 1840 the only flora of its kind until Asa Gray began his work in the forties. Scanning the book allows us to follow the young physician around Fresh Pond, through the wet meadows of Roxbury and Cambridgeport, along Chelsea beach and on "Sweet Auburn" in Cambridge, finding delicate plants in places now often occupied by residences and dismal tenements. *Viola acuta* was found, "particularly about the pine trees on Craigie's road, in moderately damp soil," and the scarlet pimpernel, "a humble but delicate flower," in South Boston, where it was common. The second edition listed the results of trips to more remote places, including an expedition to the White Mountains with his friend Francis Boott in 1816, which was described in the *New England Journal of Medicine and Surgery*. Bigelow Lawn and Boott Spur on Mount Washington preserve the memory of this trip.

Bigelow's classification was the old Linnaean one, which he did not abandon even in the third edition. There is a personal touch to the text which gives the booklet a piquant old-fashioned flavor, as in this description of the Indian summer:

> Among the crimson and yellow hue of the falling leaves there is no more remarkable object than the Witch Hazel, in the moment of parting with its foliage, putting up a profusion of gaudy, yellow blossoms, and giving to November the counterfeit appearance of spring.

The doctor might have developed into an early Thoreau, or might at least have completed a flora of the whole of New England, as he seems to have planned, had he not found other things to keep him busy. In 1816 he was appointed to the chair which Count Rumford had endowed at Harvard for the instruction of "the application of the sciences to the useful arts," a first attempt to create a meeting ground for self-made inventors and academic scientists. There being no good name for such a field, Bigelow coined for it the name "technology," which has passed into common language—"the science of systematic knowledge of the industrial arts," according to Webster's Dictionary. The results of Bigelow's teachings were given to the world in 1829 as *The Elements of Technology*. In the meantime he had published a three-volume *American Medical Botany* (1817–1820) with descriptions and colored engravings of medically useful plants—it was still necessary to defend the study of plants by referring to their direct usefulness. The author himself made most of the engravings, and experimented on the chemical properties of plants. He was also one of the men who, at the initiative of Lyman Spalding, joined in order to continue the work of Warren and Jackson and compose an *American Pharmacopoeia,* which appeared in 1820. We shall meet this remarkable and many-sided man again in these pages. Sometimes brilliant, often original and always active, he lived long enough to become a trustee of the Massachusetts Institute of Technology.

The Dana brothers gave us the first description of the geology of Boston and its surroundings. James Freeman Dana and Samuel Luther Dana were sons of a naval officer who had seen action in the Revolution and had settled in New Hampshire. During their Cambridge days, the two Dana brothers, both Harvard graduates of 1813, went out into the fields with the geologist's hammer, in order to do something against "the coldness and indifference, with which mineralogy is here treated." The result of their work appeared in a little booklet of 1818 called *Outlines of the mineralogy and geology of Boston,* published, as was Bigelow's *Florula,* by the University Press—a private publishing firm in Cambridge which did distinguished work for those days. They found "that more than forty simple minerals and several rocks occur within a few miles of Boston." The book was written, as the authors took care to point out, in moments of leisure by men who did "not profess to be mineralogists," and its very elementary character was a token of the infancy of the science in Boston. They found several types of quartz like "petrosilex" on the Nahant, Nantasket and Chelsea beaches; some copper with iron in Woburn and a little lead in quartz at Medford and at Brighton. They ventured, like Bigelow, into the unknown, and called a "Schaalstone" near Chelmsford "Chelmsfordite." A geographical map gave a survey of their results.

The youthful Danas had for guidance a native authority. This was Parker Cleaveland, a Harvard graduate of 1799, a pupil of Dexter, who had been elected to a professorship at the new college in Brunswick, Maine, endowed by James Bowdoin. Bowdoin College was incorporated in 1794, but did not open its doors to students until 1802. In 1805 it added Parker Cleaveland to its staff, and Cleaveland remained at Bowdoin for more than fifty years. He became the college's outstanding scientist, and in his later life its most remarkable landmark. Cleaveland was appointed to teach chemistry, but soon became interested in mineralogy. He studied the countryside from the seacoast to the White Mountains, adding to his cabinet at the college and working on his book. When it was published in 1816 by the University Press in Cambridge —a private organization, not connected with the college—the *Elementary treatise on mineralogy and geology* was easily the best book on the subject in the country and one of the best in the English language. The basis of classification was the chemical composition of minerals. He made a thorough job of it, basing it mainly upon the French system of Brongniart and Hauy, and attracted the sympathetic attention not only of his colleagues in America, but also of scientists abroad, primarily because of its information on American sites of minerals. Davy, Brewster, Cuvier, Brongniart, Hauy, began to correspond with Cleaveland, and he received honors from several learned societies abroad. New England had not had such a scientific celebrity since the death of John Winthrop. The book was indeed a good and substantial piece of work, with its 668 pages and a geological map, and it is perhaps a pity that a man of such attainment as Cleaveland consented to remain cooped up in a small college in Maine, which at that time was a long distance from Boston. But he loved his college, its town, its fire company, his classes, his collection, his Maine landscape. He liked teaching, and during his fifty-three years of it taught nearly every subject in the curriculum at one time or another, during this period missing only three recitations. We do not know whether the good Brunswick townspeople set the clock when Cleaveland passed by, as they did in Königsberg when they saw Kant, but it is reliably reported that in sixty-seven years he never missed a meal. He saw Bowdoin grow and win a reputation through its pupils, Longfellow, Franklin Pierce and Hawthorne; with Longfellow he established a life-long friendship. The *Mineralogy* remained his magnum opus. A second enlarged edition appeared in 1822, which stimulated research to an even higher degree than the first, and he died while preparing a third edition.

8

Silliman, Gorham, Cleaveland, marked a definite step forward in New England's scientific life. Their work took shape in the period after the war with England, when new energies were released aimed at a greater development of the country's natural resources. Though leaning strongly on England, leading men of science were no longer afraid to take lessons from France. The old Newtonian tradition had discouraged natural science, and also the study of continental mathematics. Bowditch, the lone navigator and insurance pioneer, had attempted to break the ground in this field, but his work had not affected the colleges. The man who introduced continental mathematics to New England's college students was John Farrar.

Farrar, a Harvard graduate of 1803, succeeded to the Hollis chair of mathematics and natural philosophy in 1807. Mathematics in the United States had hardly developed beyond the stage in which it was left at the time of John Winthrop's death. It even lagged behind England, which itself was trailing behind France and Germany, partly through the lethargy of its colleges, and partly because of prejudice against France and its revolution. Farrar's Harvard education smoothed his path to culture in general and to continental mathematics in particular, a path which a generation before had been so thorny for Bowditch. His interests were also sharpened by his contact with the men around William Tudor, a well-to-do businessman of Boston and a man of considerable culture.

Tudor was a Harvard graduate; he had traveled on the continent of Europe, and, after his return to Boston, had joined with some others in 1805 and founded the Anthology

Club, a genteel center for the cultivation of the arts and sciences. The Boston Athenaeum, now a library of 200,000 volumes, is a living descendant of this club.

The club also published, from its beginning, a periodical, the *Monthly Anthology*. This literary magazine gave also some regular information on the gentlemanly type of science prevalent in Boston, on eclipses, on useful plants, on the Memoirs of the Academy. In 1815 it adopted the name of the *North American Review and Miscellaneous Journal*, and for many years remained the oracle of Bostonian gentility. Like *Silliman's Journal* it has survived until this day. Farrar, Cleaveland and Bowditch were regular contributors, and kept the readers informed on astronomy and meteorology.

European events were discussed in the Anthology Club and in the pages of the *North American Review*, as well as events in the scientific world. In the second volume of the *Review* abstracts of the *Transactions* of the French Institute for 1813–1814 appeared, probably prepared by Bowditch or by Farrar, with remarks on the great works of the French mathematicians and physicists, on Lagrange's mechanics, Laplace's probabilities, on Poisson, on Biot and others. By that time Bowditch had begun his translation of Laplace's books on celestial mechanics, and continued to impress his friends with the value of continental mathematics and astronomy. It did not escape the friends of the *Review*, despite their Federalist bias, that important educational and scientific changes had taken place on the Continent as a result of the Revolution and the Napoleonic period. A great center of education and research in engineering and the exact sciences had grown up in the Ecole Polytechnique of Paris. Here some of the best French scientists were engaged in systematic teaching of modern science and engineering, the results of which they published in remarkable textbooks, which, like Legendre's *Elements of Geometry* or Monge's *Descriptive Geometry*, have not lost their educational value even today and which are the prototypes of many of our present textbooks in the exact sciences and in engineering. Farrar decided to bring as many of these texts as possible to Harvard.

He began, in 1818, with an adaptation of Lacroix's *Elements of Algebra*. Then followed, in rapid succession, translations from Lacroix, Legendre, Bézout, Biot, and also from

Euler. He was eclectic, selecting those parts which he considered best "for the use of the students at the University of Cambridge, in New England." Through his fine abilities as a teacher he brought the new type of mathematics to many students. Among his innovations was the introduction of continental calculus in the school of Leibniz, to replace the antiquated forms of Newtonian algorithm, which were not only taught in the colleges, but also for many years presented in articles appearing in *Silliman's Journal,* in perfect innocence of what had happened in Europe during the past hundred years.

Farrar continued his educational work at Harvard until 1836, when illness caused him to resign. He died in 1853. Although as little of an original mind as Cleaveland, Gorham or Dexter, he has nevertheless left indelible traces upon the path of American science, as did all these men. These scientists could hardly have been different from what they were: pioneers in a new field, doing the first backbreaking work that others might harvest. With Farrar, and Bowditch in the background, Harvard took the leading role in mathematics, while Yale with Silliman led the way in the natural sciences.[8]

This is the place to introduce Nathaniel Bowditch in his

[8] A historical sketch of mathematics in early New England must have a word about 1810, when New Englanders were surprised to see a young boy of six years of age, Zerah Colburn, exhibit an astonishing ability for complicated arithmetical computations. It was a boy from a Vermont farm, without much education, who was led around by his father, in the way of a circus exhibit, to show his exceptional gifts. In answer to a request he found instantly that $247483 = 941 \times 263$. "The child is the greatest phenomenon I ever beheld," wrote a spectator. The history of this prodigy is a sad comment upon paternal pride and greed. An offer by men at Dartmouth and at Boston to give him an education was neglected, and the father continued to show his son around, also abroad. After his father's death Colburn taught at the Fairfield Academy, became an itinerant Methodist minister and a teacher at Norwich Academy. In his *Memoir* he indicates some of the methods he use for rapid calculation. He also found that $2^{32} + 1$ is divisible by 641.

Another Vermont prodigy was Truman Henry Safford, born in 1836, who before he was ten had calculated an almanac for Bradford, Vermont. His education was not neglected and he became a graduate at Harvard. For many years he taught astronomy at Williams College.

9

later years, after he had moved to Boston, in 1823, to become a well-paid actuary of the Massachusetts Hospital Life Insurance Company. This venture of the Boston merchant class was a noble attempt to combine humanitarian support of the new Massachusetts General Hospital with the new and promising business of issuing life insurance. The fifty-year-old mathematician had refused to leave Salem for a professorship at Harvard, for one at the new University of Virginia—offered in 1820 by Jefferson himself—and for one at the Military Academy at West Point. But having once decided to make Boston his new field of activity, he approached it with his usual energy. The Life Insurance Company, led by the conservative policy of some of Boston's most outstanding merchant manufacturers, and guarded by a nicely buttressed monopoly, became under Bowditch's supervision a model of solidity, and even weathered the crisis of 1837 with relatively small losses. It was as solid as a rock, but the premiums were high, so that many New Englanders took their insurance out in Pennsylvania. Popular opposition to this policy merged in the thirties with a general attack on the tight control over Boston by the merchant clique. The New England Life Insurance Company of Boston was founded, partly as a protest, in 1835. In 1843, after a long legal struggle, the virtual monopoly of the Hospital Life Insurance Company was broken and the excessive rates came down.

Bowditch's spare time was mainly dedicated to the translation of Laplace's *Mécanique Céleste,* that enormous four-volume classic of mathematical astronomy. Bowditch had begun the translation after he settled in Salem. Rather than teach continental science in America by means of elementary texts, he chose to do it by bringing within the reach of the English-speaking world the highest product of the science. It was a world in which even some of its very elements were hardly known. To overcome this handicap, Bowditch added a commentary, which exceeded the original in extent and which was itself a kind of extensive introduction to continental mathematics. It also gave Bowditch a chance to practice on the highest authority in the field his old game of finding errors. "I am sure," Laplace is reported to have said, "that M. Bowditch comprehends my work, for he has not only detected my errors, but he has also shown me how I came to fall into them."

The first volume of the translation was published in 1829, the second and third some years later, and the fourth volume was nearly completed at the time of Bowditch's death in 1838. It appeared in 1839. The publication was paid for by the translator himself and copies were sent to all the famous men in the field. Many of them responded in some way or another. Madame Laplace, in Paris, sent him a large bust of her late husband, which can be seen in Osgood's picture of Bowditch in the Peabody Museum at Salem. Gauss, in Göttingen, gave his son a recommendation to Bowditch when the young man came to America to look for a job in railroad construction. Young Gauss wrote home:

> Schumacher wrote about Mr. Bowditch, that he belongs to a gloomy and unsociable sect, but I did not find this confirmed. He [Bowditch] is a vigorous and cheerful man with a very pleasant family. It is not true that he has ruined himself financially by the translation of Laplace's *Mécanique Céleste,* which he had printed for $20,000 at his own expense and which only a few people buy. He is very well to do and has beside a salary of $6000—for his position as director of the Mutual Life Insurance Company, which occupies him only a few hours every day.

What has been the influence of Bowditch's labor of patience and love? Simon Newcomb, the astronomer, has expressed himself rather pessimistically. The translation might have helped others to study mathematics:

> . . . yet we cannot concede that this was the most advantageous form in which the help could have been given. . . . It was like furnishing a classical author with an interlinear translation, that the unfavored student might read him without studying the grammar of his language. This Commentary, as well as some other of Bowditch's writings, betrays the want of that inspiration which comes from immediate contact with the masters of the subject.

This judgment is probably correct, but we must not go too far in minimizing the importance of Bowditch's work. We know that his book was studied not only by Benjamin Peirce, a young friend of Bowditch's in Salem, but also by Maria Mitchell, the Nantucket astronomer, both persons of influence. Peirce assisted Bowditch in the preparation of the later part of the Laplace translation for the publisher, and used the book for a while in his classes, when he obtained a teaching position at Harvard. Since modern mathematical teaching began at Harvard with Farrar and modern research with Peirce, we may justifiably see in Bowditch's opus a formative factor in establishing modern mathematics in America.

10

A few words must also be said about the scientific development in the countryside, especially the western part of Massachusetts and Vermont. In colonial days this had been a country of hardy pioneer farmers on a high enough intellectual level to produce a setting for Jonathan Edwards, one of the keenest thinkers of the eighteenth-century English world, whose study of the relation of freedom and necessity is a remarkable piece of dialectical thinking. The towns along the Connecticut River, in the Berkshires and in the Green Mountains, shared fully in the partisan struggles of the early republic, and maintained a remarkable interest in whatever came their way in sciences and belles-lettres. Hartford was a cener of Federalist literati, the Hartford wits. Eventually at least one of these wits, Joel Barlow, deserted to the people's camp; another, David Humphreys, became a pioneer reformer in agriculture and in manufacture. Many a farmer took an intellectual interest in his rocks; when in 1844 the historian Francis Parkman visited the Berkshires, he was impressed by the knowledge of geology which he found among the old-timers. Elkanah Watson, in 1807, initiated an agricultural reform movement from Pittsfield, which inaugurated the popular country fairs. Williams College was founded in 1793 at what became known as Williamstown with a staff of good teachers, who remained faithful to the college despite its poverty.

The scientific tradition at Williams began with Gamaliel Smith Olds, who left in 1808 for the University of Georgia. After him came Chester Dewey, whose work was supplemented for a while by Amos Eaton. Dewey, a Williams graduate of 1806, was a Yankee science teacher typical of the early days of the republic. He was well versed in the Bible, held a license to preach, and had a sound foundation in chemistry, mineralogy and botany. At Williams he taught from 1810 to 1827. As a lecturer he was tireless; by the time of his death, when he taught in Rochester, he had delivered over four thousand lectures, and preached nearly as many sermons. He was not merely a teacher, but a remarkably good one, and not only a dabbler in science, but a serious specialist, who for many years contributed monographs on the sedge (Carex) to *Silliman's Journal.* Albert Hopkins, who later taught science at Williams, and did it well, has left this picture of Dewey's original ways of teaching (c. 1825):

> The professor enjoyed adding to the thrill by pretending that his experiment would not work, and then setting it off with a terrific bang. He would call up a boy and pour alcohol into his hand, and then touch it off with an electric spark, causing the youth to leap in terror to the great joy of the class. Or he would pass an electric current through the class and give them a terrific shock. One day he covered their faces with phosphorus, so that, in the darkened room, they shone like ghosts.

Amos Eaton, the geologist, lecturing at Williams in 1817, found so much response that "an uncontrollable enthusiasm for natural history took possession of every mind." The work of Dewey and Eaton was continued by Ebenezer Emmons and Edward Hitchcock, who both participated in the geological surveys of the thirties and forties, and by Albert Hopkins, who built an astronomical observatory at Williams before Harvard and Yale acquired one. This early scientific teaching at Williams also influenced the early growth of Amherst College and of the Rensselaer Institute at Troy, across the Taconic Mountains.

In these remoter parts of New England there also flourished for several decades a number of medical schools, in Pittsfield, Massachusetts, in Woodstock, Vermont, and in Castleton near Rutland, Vermont. They were the creations of local physicians, a kind of expansion and modernization of the old system of apprenticeship. The first of these schools was founded in 1812 at Fairfield, near Utica, New York. It was attached to the local academy, and placed under the University of the State of New York. Its most famous pupil was Asa Gray. Between 1812 and 1860 there were about thirty such schools in the United States, almost all the lengthened shadow of one man. The Castleton Medical Academy was founded in 1818, and had Dr. Selab Gridley as its central figure, a good Vermonter who was a physician as well as a poet, a politician and a general store proprietor. Dr. Theodore Woodward, a nephew of Nathan Smith, and occasionally other lecturers such as Amos Eaton and Ebenezer Emmons, addressed the students. Chester Dewey used to teach at the Pittsfield and Woodstock schools. These schools were far from bad for their day, the Pittsfield school competed for a while in attendance with the Harvard Medical School. The usual course consisted of an annual fourteen-week series of lectures, after which the students had to seek clinical instruction elsewhere, usually in country practice. It was this lack of clinical facilities and their general primitive character which were in the long run responsible for the disappearance of these schools, despite their honorable career and services. By Civil War days they had largely ceased to exist or were partly absorbed by the medical institutions of the colleges.

The University of Vermont was established at Montpelier as early as 1791. Under President James Marsh, a Dartmouth graduate of 1817, it played an important role in the educational reforms of the thirties. Marsh was a philosopher of uncommon merits for his period; his work reflects that independence of thought in early Vermont which occasionally also stimulated scientific work of more than home-town flavor. The Green Mountains produced not only a sterling geologist in the person of Zadoc Thompson, but also the inventor of the electric motor, Thomas Davenport.

PART III

THE JACKSONIAN PERIOD

Chapter 7

The Surveys

Let us not underrate the value of a fact, it will one day flower into a truth.

—HENRY DAVID THOREAU

1

BY THE END of the second decade of the nineteenth century the conditions for a new expansion of American democracy had matured. Production had increased with the growth of manufactures, transportation facilities had improved, educational facilities had widened and new methods of technology were constantly applied. At the same time the control of all these new and marvelous means of promoting human welfare had remained for the most part in the hands of a small ruling class. Large sections of the population began to feel that they had no share in the increasing opportunity to raise the standards of living and thinking. A new class, the urban working class, was developing. This new class, in alliance with the older mainstays of an embattled democracy, the mechanic and the backwoods farmer, began seriously to assert itself. The impact of the popular struggles overseas, the growing labor movement in England, the war of liberation in Greece, the July revolution in France and the Polish rebellion, gave an added impetus to the development of democracy on this continent. And where the earlier Jeffersonian democratic movement somehow never got well started in New England because of the stubborn strength of the seacoast merchant class, the lid blew off the kettle with the advent of the Jacksonian period. The decades between 1830 and 1850 became the years of the common man.

The struggle for social emancipation was fought along many fronts. All classes participated, even the top-rank bourgeoisie, which, despite the antidemocratic bias of many of its members, was after all a progressive class. These years of the "militant thirties" and "militant forties" witnessed the intensification of the campaigns for popular education, for the separation of church and state, for the ten-hour day, for greater civil rights, for the emancipation of women, for prison reform, for better institutions of higher learning and research, for improved medical care, and even for a revolution in the whole economic system. It was the period of Horace Mann, of Seth Luther, of William Lloyd Garrison, and of the Brook Farm visionaries. Popular interest in science increased by leaps and bounds, as a means to social betterment, as an intellectual avocation and as a means of education. The lecture system flourished, and audiences of thousands of people came to hear scientists lecture on their special subject.

The economic expansion of the country, the gradual opening of the West, the growth of its industries, its navigation, its transportation, its mining and agriculture, greatly increased the need for a deeper knowledge of the nature and the resources of this expanding nation. How did this great country look on the map? What treasures were buried in its rocks, hidden in its forests, lying beneath its waters? How could nature be controlled so as to offer better opportunities for industries, transportation, agriculture and men? Never in American history had so many people been interested in so many things. Questions were asked, and answers were expected, of which only few could be given. Private and public institutions were under pressure to devote their attention to these topics of general concern, and the existing institutions, notably the colleges, were singularly inadequate to the demand. New organizations, lyceums and public schools, societies and research institutions, sprang up to explore these various fields. The scientists in the colleges found themselves in high demand as lecturers and advisers, but they were limited in number and not all responded to the call to break out of their ivy-clad halls. The changing society demanded a new approach to science.

Interest in these problems came from many sides. It might spring from economic motives, from persons with utilitarian purposes, who wanted to know where to build a factory, a mine or a railroad, or how to construct a dam, a locomotive or a new piece of textile machinery. Others were eager enthusiasts for education, for knowledge, for the cultivation of the arts and the sciences. All these groups pressed the national and state legislatures to grant funds, and these public bodies now began to support research in the practical problems of natural history, geography and astronomy. This interest of public bodies was very uneven and the very fact that legislatures were called upon time and again to authorize grants brought unfortunate decisions by ignorant or prejudiced men, as well as decisions of great progressive value. When public initiative failed, the ancient American habit of private initiative afforded support through personal devotion and endowments. Time and money were contributed for the promotion of learned societies, of popular lyceums, of expeditions, surveys, the latter perhaps the most important of all for the direct stimulation of scientific work.

2

The idea of such surveys goes, in North America, back to Thomas Jefferson, himself the author of an early survey of Virginia. While President of the United States he sponsored the expeditions of Lewis and Clark (1803–1806) and of Zebulon Pike (1805–1807) into the West, he encouraged the search for minerals and fossils, and was active in establishing the United States Coast Survey (1807). It was not Jefferson's fault that the Coast Survey did not flourish until the thirties of the century. The country did not yet appreciate such careful survey work of the coast, its currents and its inlets, despite the excellent work done by the same European governments. There was some individual pioneering work in charting, like that of Bowditch in Salem, but government-sponsored scientific efforts were long a cause of opposition and ridicule in this land of individualism and republicanism. In European countries these ventures had usually been undertaken by princes, but in the United States such personages were nonexistent and the work of such undemocratic agencies was distrusted. In this respect Jefferson had a deeper understanding of the tasks of a democracy than many of his followers or opponents, even those during the period of Jacksonianism. John Quincy Adams, however, in a political camp different from that of Jefferson, had an equally fine understanding of the obligation of a republic to science. In

1816–1819 he pleaded for the establishment of a national astronomical observatory similar to such institutions abroad. His appeal was received "with shouts of ridicule" and the wits of the country had a field day wisecracking on the "light-houses in the sky," as Adams had named his cherished brain children. His learned report on weights and measures, presented in 1821, supplementing an earlier report by Jefferson, also shows both his interest in matters pertaining to government and science, and the fundamental agreement between his way of thinking and Jefferson's.

The lack of public encouragement did not dampen the ardor of the pioneers in scientific organization and surveying. Silliman, Bigelow, the Danas and others went ahead on as large a scale as was then possible. Denison Olmsted was the first who induced a public body to authorize a survey.

Olmsted was a Connecticut man, a Yale graduate who, like Whitney, Meigs and Baldwin before him, had gone to the South for a living. He became a professor of chemistry, mineralogy and geology at the University of North Carolina and later returned North to accept a Yale professorship. His interest in popular education and particularly in scientific education was deep and many-sided. In the years between his graduation and his departure for the South, he worked to improve the wretched grammar schools of his period and urged the establishment of normal schools for teachers, thus blazing the trail for Horace Mann.

In 1821 Olmsted proposed to undertake a geological survey of the State of North Carolina, which was authorized by the State Board of Agriculture in 1823. He agreed to perform the whole task free of charge except for a hundred dollars to defray traveling expenses. Olmsted set to work and published the result in 1824 and 1825. This free vacation work of a single individual remained for several years the only survey authorized by a public body.

All these surveys were chronicled in *Silliman's Journal*, which had a regular column with notices of miscellaneous mineral sites, to which local amateurs contributed. Olmsted's success was duly appreciated by these men. In reviewing Olmsted's reports in the *American Journal of Science* in 1828, Edward Hitchcock held them up as an example: "What an accession would be made to our resources, and to the knowledge of our country, were a thorough examination to be instituted into our mineralogical, geological and, even botanical riches!" He pointed out that Olmsted's report revealed the existence of mineral riches previously unknown. The timeliness of Hitchcock's remarks was borne out by the continuation of important surveys carried out by private persons, of great enthusiasm but naturally of limited resources. Two young Bostonians, Francis Alger and Charles Thomas Jackson, surveyed Nova Scotia, and published their results. Interest in such work grew wider and wider.

Hitchcock had become the main driving force behind the campaign for a systematic Massachusetts survey, sponsored by the legislature, of all natural resources. He met with generous support from many circles from all those groups and individuals interested in the country's geology, fauna, flora and geography. The governor of the Commonwealth at that time, Levi Lincoln, was originally a Jeffersonian Democrat and a member of a family long associated with the promotion of arts and science. The General Court responded in 1830. The governor was then authorized to appoint both a trigonometrical and a geological surveyor. He appointed Hitchcock to the last position, while the trigonometrical survey eventually passed to the leadership of Simeon Borden. Both the Hitchcock and the Borden surveys exerted a tremendous influence in the United States.

3

Edward Hitchcock belongs, with Bowditch and Silliman, to the grand old figures of New England science. He was born and reared in the Berkshires, a country which he loved and which he knew as few others have known it before him and few will know so well again. His father came from an old New Haven family and was a hatter in the town of Deerfield, which even now has preserved its old New English atmosphere more intact than most other neighboring towns. The Berkshires were intellectually alive, but poor financially, and hatter Hitchcock suffered with his fellow men. Edward, a boy of delicate health, obtained whatever education he could get at the local academy, where there were some astronomical instruments. His observations of the comet of 1811 were published in the *Memoirs* of the American Academy through the efforts of his uncle; this attracted the attention of Bowditch, who had also observed the comet. In the meantime young Hitchcock, who resembled Bowditch in his search for errors in existing tables, made life miserable for the publisher Blunt in Newburyport, who had offered prizes for the discovery of errors in the *Nautical Almanac* he published. It is recorded that Blunt acknowledged the errors, but did not pay the reward.

The influence of Chester Dewey, Amos Eaton and Ebenezer Emmons, all naturalists at Williams College, led Hitchcock to the search for plants and minerals in his locality, which in turn brought him into correspondence with Silliman. For young Hitchcock this was an added inducement to go to Yale, though he came to study not for science, but for the ministry. This combination of science, geology and the ministry was typical not only of Hitchcock but, as we have noted, of many other scientists of this period in which natural science was hardly a profession in New England. After a pastorate in Conway, Massachusetts, he accepted in 1825 the chair of chemistry and natural history at near-by Amherst College. Amherst was at this time a young and struggling institution founded in 1821 mainly to provide free education for the ministry. From 1825 till his death in 1864, Hitchcock's name became as intimately associated with Amherst as that of Cleaveland with Bowdoin. The college gratefully preserved anecdotes about him, of his mania for nostrums, of his passion for renaming mountains, and of his attempts to reconcile science with orthodox religion. Geological commissions entrusted to him by Massachusetts, New York and Vermont drove him away from his college, but he returned to serve as its president from 1845 to 1854, aiding it during a difficult financial period. His collection of fossil footprints is still a pride of the college. Like Silliman he had a son and namesake who continued to do distinguished work in the fields selected by his father.

Hitchcock spent three years exploring Massachusetts, traveling 4550 miles and collecting 5000 specimens of rocks. He was greatly encouraged by the way his mission was received by the people. A "universal disposition," he wrote, was "manifested by all classes of the citizens of the Commonwealth, and in every part of it, to do all in their power to forward the objects of my commission. . . . The excursions," he confessed happily, "have greatly exalted my opinion of the kindness, intelligence, and happy condition of our population, and sensibly increased my attachment to my native state." This was a far cry from the indifference which Bigelow had encountered less than twenty years before. But such popular interest also had its drawbacks. There was pressure on him to publish his results quickly, and Hitchcock had to report his findings as early as 1832 and 1833. The final report, published at Amherst in 1833, was a book of 700 pages and, though the result of enormous labor, left Hitchcock still very much dissatisfied.

Such a geological report of a century ago was a classification of minerals for the professional, a guidebook for prospectors, a textbook for the student, and a Baedeker for the tourists. Hitchcock, in Part I, called "Economical Geography," indicated the places where metals could be found, and went out of his way to announce a deposit of gold in southern Vermont. In his chapter on "Topographical Geology" the "principal object will be to direct the attention of the man of taste to those places in the State, where he will find natural objects particularly calculated to gratify his love of novelty, beauty and sublimity." He gave descriptions of mountains, of the view from the Boston State House—with moral and political contemplations—and of autumnal scenery. Part III, "Scientific Geology," which comprised the bulk of the book, was a textbook on geology with field examples. It compiled the opinions of foreign auhors, applied De Beaumont's theory of mountains to Massachusetts, and struggled with the Mosaic records. The last part of the book was a catalogue of animals and plants of the state, compiled from contributions by different specialists.

There is an evaluation of the technical strength and weakness of this report in Merrill's excellent history of American geology and we can safely refer to this account. A nontechnical aspect of Hitchcock's work lies in his eagerness to see the biblical records confirmed by geological evidence.

Hitchcock, like Silliman, was a spokesman of what was called natural theology, which tried to find in nature the confirmation of theological tenets. A pious Christian of the old Congregational school, Hitchcock looked for evidences of the deluge and found them in the drift of erratic blocks distributed so freely over New England. Opinions like those of Charles Lyell, whose *Principles of Geology* appeared in 1830 and which emphasized the uniformity of the action of geological agencies in the earlier periods of the earth's history, could not curry favor in Hitchcock's eye. He was a catastrophist:

> The resemblances between the plants and animals in each of the divisions of the strata, that have been mentioned, even to the very limits of each division, and the suddenness of the change that then takes place in their characters, preclude the idea, so much of a favorite with certain philosophers, that all was the result of a gradual metamorphosis. Now if we thus ascertain that God has specially interfered with the operation of natural laws in the instances under consideration, the presumption is, that he may interfere again, whenever the good of the universe demands. Thus do we get rid of a host of atheistical objections, with which the student of natural theology finds his path encumbered.

One of the reasons why the earlier geological agencies were considered as being of greater intensity than the later ones was the fact that the older rocks are more distorted and metamorphosed than the younger ones.

4

The action of the Massachusetts Legislature in 1830 set the example for the whole country. It was the breaking of a hole in a dam. Suddenly a flood of new enterprises was set free. Hitchcock, in 1841, could state with some pride that it might "not be irrelevant to state that since Massachusetts began this geological exploration, no less than eighteen other states of the union have commenced similar surveys, while the Government of the United States, as well as some European Governments, especially that of Great Britain, have followed the same example." The list of the establishment of these surveys is impressive:

Massachusetts, 1830; Tennessee, 1831; Maryland, 1834; New Jersey, Connecticut, Virginia, 1835; Maine, New York, Ohio, Pennsylvania, 1836; Delaware, Indiana, Michigan, 1837; New Hampshire, Rhode Island, 1839. Alabama, South Carolina and Vermont (1844) followed, and there were also surveys in Nova Scotia and New Brunswick.

Indeed, the nation had become conscious of its treasures and of the necessity of exploring them. When in 1815 Bigelow and his friends tried to establish the "Linnaean Society" for the promotion of natural history, the attempt fell through after a few sessions and the lone publication of a story on the sea serpent. By 1830 the general situation had changed. The apathy towards agricultural improvement had disappeared, and interest in horticulture had increased considerably. The natural sciences had obtained a foothold in the colleges and had been widely accepted in schools and lecture halls. When the Boston Society of Natural History was founded in 1830, it proved a success from the beginning. Led by energetic and expert men, it soon acquired a number of influential members and considerable financial resources. The popular movement of the thirties stimulated the ardor of its membership, which contained well-to-do merchants as well as educators. Among the founders was Amos Binney, a wealthy merchant and an expert on mollusks, and George Barrell Emerson, headmaster of a private school for girls and an authority on the fauna and flora of Boston and surroundings. Both Binney and Emerson were men who got things done. Emerson, who, by the way, was not related to the Concord philosopher, and Josiah Holbrook were the moving spirits of an organization known as the American Institute of Instruction. Its many members, drawn largely from groups of teachers and other friends of education, were active in convincing the General Court of the necessity for founding a State Board of Education. When the board was established Horace Mann, president of the State Senate, resigned to become secretary of this board. He knew that he had a mission to perform, and that he could now count upon the support he needed to institute his reforms. Horace Mann was to set the stamp of his work on elementary education not only in the Commonwealth, but throughout the entire country and even beyond its boundaries.

In 1836 Amos Binney was elected to the legislature. In Whig Governor Edward Everett, a man of considerable culture, who had been Emerson's tutor at Harvard, the reformers found the support for their program which Hitchcock had found in Governor Lincoln. The General Court was willing to continue the surveys, and in 1837 authorized the governor to make appointments for a further and thorough geological, mineralogical, botanical and zoological survey "of the Commonwealth, with the special task to discover coal, marl, and ores and to analyse the various soils for agricultural benefit." Emerson, in that year President of the Boston Society of Natural History, was appointed chairman of the committee established to conduct the survey. The different fields were referred to experts. By this time a remarkable number of such experts was available.

Hitchcock now obtained a chance to do a better job than before on the geology of Massachusetts. His final report was printed in 1841; it consisted of two quarto volumes of 831 pages, and contained much of the older material. Again we can refer to Merrill's book for a critical discussion. Some twenty pages were devoted to fossil finds, especially to the famous fossil footprints of the Connecticut Valley. Mrs. Hitchcock had drawn the delicate sketches of fossils and of landscapes. As in the first edition a sketch of the distribution of animals and plants was again added to the report.

Such a book had become by this time the subject of popular interest. It naturally came to the attention of Henry Thoreau, who reviewed it in the *Dial,* the periodical of the transcendentalist group of Bostonian intellectuals. "Books on natural history make the most cheerful winter reading," he wrote, "I am singularly refreshed in winter when I hear of service-berries, poke-weed, junipers. Is not heaven made up of these cheap summer glories?" Thoreau checked up on the birds in the appendix, and missed the veery, though he had often heard the College Yard in Cambridge ring with its thrill.

By and by the other reports came in. George Emerson had taken as his subject the trees and shrubs of Massachusetts. His report appeared in 1846 and has long remained authoritative. He pleaded for the conservation of natural resources, as Michaux had done before him. Binney set to work on terrestrial mollusks, but died in 1847 at the age of forty-three, so that it was posthumously published by his friend Augustus A. Gould, between 1851 and 1855. Gould himself had contributed a study of invertebrate animals of Massachusetts (1841). Other reports were written by Chester Dewey (on herbaceous plants), Thaddeus W. Harris (on insects) and D. H. Storer (on fishes).

5

The Maecenas of this group was Amos Binney. His leisure hours were devoted to his early love for natural science, especially for mineralogy and conchology. He became one of the first conchologists of the country, a specialist in a field opened by Thomas Say in Owen's utopian colony of New Harmony. His work was continued by his son William, who became the chief American authority on mollusks. It is claimed by their biographer that the clarity and the high standard set by the work of the Binneys saved the literature of the land mollusks from the confusion apparent in American work on other molluscan groups done in this same period.

Among the members of the Boston Society of Natural History who participated in these surveys we meet the curious figure of Dr. Charles Thomas Jackson, one of the most talented and one of the most irascible scientists of his period. He was born in 1805 at Plymouth, Massachusetts, of old New England ancestry; became an orphan at twelve, and prepared himself for Harvard under the supervision of his guardian, in the meantime studying the new natural sciences, especially geology and chemistry. He continued these studies while taking courses at the Harvard Medical School under James Jackson and Walter Channing, devoting time to the new and startling discoveries in electromagnetism, which at that time began to be more widely known in America. He received his M.D. in 1829. With his quick perception of the new and original in scientific research and organization he went the same year to Nova Scotia for a geological survey, accompanied by his friend Francis Alger, and published the result in *Silliman's Journal.* His ambitions carried him to Europe, where he studied medicine at some of the best schools, in the meantime traveling on foot through large parts of Central Europe with the perceptive eye and hand of the chemist and the geologist. At Vienna he had a chance to assist in the dissection of the bodies of two hundred victims of a cholera epidemic, and in this way became better acquainted with some of the statistical methods which began to be introduced into European medicine, as well as into sociology. He returned to the United States in 1832 on the packet *Sully,* a man perhaps better acquainted with modern science than any other person in the United States.

Jackson had on board with him some electrical apparatus of the type Ampère and Faraday were using. He talked about it to a fellow traveler, the painter Samuel F. B. Morse, son of the Charlestown minister, who had received fame by his schoolbooks on geography, and who was returning from an artist's tour of Europe. He had picked up some physics from Silliman and Day at his father's alma mater, and was interested in Jackson's speculations concerning the future of electromagnetism and its bearing on communication at a distance. The result was that Morse, after landing, diverted his talents from painting to electricity, and began that series of investigations which led to the invention of the electric telegraph. He was aided by the knowledge of Joseph Henry's research at Albany. But Jackson—not Henry—promptly claimed the priority of the invention and started a bitter controversy with Morse. This was only the first of a series of similar priority fights which brought Jackson into conflict with a number of excellent men. They cost him valuable time and energy, as well as the respect of many of his peers, and eventually landed him in an asylum for the insane. The best known of these controversies was his bitter struggle with Dr. Morton over the use of ether for anesthetic purposes.

Jackson, a man of unlimited energy, settled in Boston and opened a private analytical chemical laboratory in 1836. This laboratory, and Booth's in Philadelphia, were the first laboratories to receive students for practical instruction in chemistry. Here he trained William Channing and many other chemists and gave the dentist Morton the information which changed the history of medicine. In this workshop he performed many practical and theoretical experiments which made his name known throughout the country and even abroad. The time had come when chemist and mineralogist began to find a wide market for their special knowledge. Jackson was one of the first men to engage in agricultural chemistry, in which field he studied the properties of sorghum and the economic possibilities of cotton seed. He discovered the presence of tellurium and selenium in America and the presence of chlorine in meteorites. He conducted extensive geological surveys of Maine, Rhode Island and New Hampshire and explored the mineral deposits of the Lake Superior region for the United States Government, on which he wrote a report in 1849. Without doubt, his was a wonderful scientific fertility and originality, greater perhaps than an American had ever possessed before. He was the first scientist in New England who lived up fully to European standards. Jackson, constantly overflowing with ideas, both concrete and vague, simply had no time in which to carry them out to see where they would lead; but he claimed them as his own when other men had worked them out to their ultimate results.

In 1835 Ralph Waldo Emerson married Lydia Jackson, Charles Thomas's sister. In the subsequent years Emerson often visited Jackson in his laboratory. The frequent allusions to chemistry in Emerson's works can probably be traced to this relationship. When, in his *Conduct of Life,* Emerson wrote that "Nature is a rag merchant, who works up every shred and ort and end into new creations; like a good chemist whom I found the other day in his laboratory, converting his old shirts into pure white sugar," we may be sure that he was describing the laboratory of his brother-in-law.

Jackson's report on New Hampshire, which appeared in 1844 and represents field work in the years after 1839, has been criticized as superficial, but it has remained quite readable as a kind of geological guidebook of the state. This was the period in which New Hampshire went in for mining on a modest scale, and Jackson's report may have stimulated some activity of this kind. He recommended the erection of glassworks, and of iron, copper and zinc smelting furnaces because of the cheap supply of wood. The opening of the West changed all plans based on these reports, but summer visitors to the White Moutains may remember Jackson, when they visit the Franconia iron mines which he held up as an example. Jackson also claimed to be the discoverer of the tin which was mined for a while near the village of Jackson—named in 1829 after the President—now an excellent rocky site for a rural picnic.

6

A large number of men connected with these geological surveys have left their stamp on American science. We have mentioned Hitchcock and Jackson. The Rogers brothers, William and Henry, worked in Virginia and Pennsylvania, in New York State worked Hall and Conrad, in Vermont Zadoc Thompson. One of the most curious, though not most influential, of these state geological surveyors was James Gates Percival of Connecticut.

Percival, in his day, enjoyed fame as a poet,[1] and for many years his works, last published in 1859, brought sentiment and inspiration to the more romantic souls of the country. He was born in 1795 in Berlin, Connecticut, the son of a physician, had what he called "a neglected orphanage" and was singled out at an early age as a poet of talent. He studied at Yale, where he not only wrote poetry, but also studied geology and botany. After his graduation in 1815 he turned to medicine and obtained in 1821 a Yale M.D. Later he held several jobs, as a teacher, a philologist, a linguist, but could not keep a position very long.

Some of his best known poems appeared in 1821–1822, among them "Prometheus," written in a style reminiscent of Byron, but with sufficient originality to make parts of it still readable. America was poor in literary production, and Percival easily became the leading poet. "God pity the man who does not love the poetry of Percival. He is a genius of Nature's making," wrote young Whittier in 1830.

[1] See Chapter 6, 2, and Chapter 8, opening motto.

In 1835 the legislature of Connecticut appropriated money for a geological survey and at Silliman's recommendation the governor appointed James Percival and Professor Charles U. Shepard as state surveyors. The legislature only expected a superficial examination, mainly a report on the available mineral deposits. Shepard in a short time made a 200-page mineralogical report which met the requirements. But Percival, in charge of the geological report, totally discarding the position the legislators would be likely to take, went around the country making a most elaborate survey. In 1842 his report, which Percival called "a hasty outline," appeared; it contained in 500 pages such a mass of unreadable details, with no discrimination between important and unimportant matters, that according to Shepard it never found a single reader among the persons for whose benefit it was written.

James Dwight Dana later analyzed Percival's findings, and showed that they contained much valuable material, above all a description of crystalline rocks and an investigation into the crescent shape of trap dikes. This was an achievement, since few geologists of that period had ventured into dynamical geology. Percival's studies amounted to a beginning of a theory of mountain formation—a field in which William and Henry Rogers later achieved considerable results through their study of the Appalachian system in Virginia and Pennsylvania.

Percival, a disappointed man, more erratic than ever, burdened with increasing debts, now began to live a secluded, melancholy, bachelor life at the State Hospital in New Haven. Alone with his ten thousand books and mineral collections, a recluse with an enormous stock of information, he continued to write in many fields. He wrote German poems, translated poems from several Slavic languages, and occasionally went surveying. After 1851 he traveled west, surveyed for the American Mining Company lead mining districts of Illinois and Wisconsin, became state geologist of Wisconsin, and died in 1856 in Hazel Green, Wisconsin, where he is buried.

From Shepard's descriptions of his adventures with Percival as state surveyor, we receive a vivid picture of the type of travel such work involved. They went together in a horse-drawn cart—later Percival went afoot—explaining to the inhabitants of the villages the meaning of their mission:

> Great was the wonder our strange outfit and occupation excited in some rustic neighborhoods; and very often were we called upon to enlighten the popular mind with regard to our object and its uses. This was never a pleasant task to Percival. He did not relish long confabulations with a sovereign people somewhat ignorant of geology; and, moreover, his style of describing our business was so peculiar, that it rarely failed to transfer the curiosity to himself, and led to tiresome delays.

State surveying, we see, gave the rural population a chance to get acquainted with geology, even if the instructors did not always relish the job.

7

We do not intend to give an exhaustive account of all the geological surveys of those days. Merrill's study gives all the further necessary information. But the people of New England wanted information not only on geology, but on other subjects as well. The agricultural survey of Massachusetts, undertaken by the Reverend Henry Colman in the Berkshires, was a token that the cause of better farming, in which so far only some merchant reformers had been interested, had now become the concern of the people as a whole. Another project of these days, and one which received wide attention, was the Borden triangulation of Massachusetts, for many years the only survey of this kind authorized by a state.

After the authorization in 1830 by the Massachusetts General Court of the appointment of "a surveyor well skilled in astronomy and in the art of surveying upon trigonometrical principles—to make a general survey of the Commonwealth and . . . to project an accurate skeleton plan of the state which shall exhibit the external lines thereof and the most prominent objects within those lines and their locations," the field work was begun in 1831 under Chief Engineer Robert Treat Paine, grandson of the signer of the Declaration of Independence, and the son of a poet and a wit. An able mechanic named Simeon Borden, who had already made a name as a surveyor and as the maker of an excellent surveying compass, was assigned as a topographical engineer to this survey. Borden devised and constructed a special apparatus for measuring the base line of the survey, fifty feet in length enclosed in a tube, and invariant with respect to all variations in temperature. Accompanied by four compound microscopes, it was at that time and for several decades to come the most accurate and convenient instrument for surveying purposes. Borden's zeal and enthusiasm made him the soul of the enterprise, which in 1834 was placed entirely in his charge, and with which he became identified to such an extent that the survey has become known as the Borden triangulation. The work was completed in 1841, not without great difficulties, since many local maps necessary for the survey proved to be extremely inaccurate. The map was published in 1844 and has been widely praised for its accuracy, despite some previous criticism by old Hassler, the Superintendent of the United States Coast and Geodetic Survey.

Borden continued his work as a surveyor in later years. He marked the boundary between Rhode Island and Massachusetts, which was the subject of a case argued in the Supreme Court in 1844. He assisted in the construction of several railroads, and in 1851 published a booklet with formulas for railroad construction. In that same year he supervised from the northern part of Manhattan the suspension across the Hudson of a telegraph wire, which was over a mile long and attached to masts 220 feet high. Borden died in 1856, only fifty-eight years of age.

8

The Federal Government only reluctantly followed the states in supporting scientific surveys. Congress, which in its early days had occasionally yielded to the requests of its revolutionary leaders for support of scientific enterprise, had now partly lost its belief in the dreams of enlightenment. John Quincy Adam's voice was for a time lost in the wilderness. The Jacksonian change did not improve matters, but rather was a factor in withholding governmental encouragement of science. The era of the scientifically inclined gentlemen of the North and South was disappearing, and the democracy of the frontier and of the workingman was not always able to understand the place of science in the new industrial economy. The uncertain political aspects of the forties and fifties, which were increasingly dominated by the slavery question, made the attitude of Congress erratic. The Presidents of the United States resulting from this uncertain balance of power were men of indifferent merit, too often inclined to yield to the pressure of a slave economy which was rapidly yielding the intellectual splendor of the great Virginian days to an anxious conservatism. Notwithstanding all this, increasing pressure from enlightened public opinion was manifested in favor of federal support of scientific enterprise. The Coast Survey, that neglected pet child of Jefferson, was reorganized in 1832 and the talented Swiss surveyor Ferdinand Hassler was returned to his position of superintendent. When Hassler died, in 1843, the survey of the coast had been extended from New York eastward to Point Judith in Rhode Island, and southward to Point Hinlopen in Delaware. The glorious days of the Coast Survey had begun. The new superintendent, Alexander Dallas Bache, who was steeped in the traditions of his great-grandfather, Benjamin Franklin, continued Hassler's work with remarkable vigor and efficiency, managing to attract the services of some of the best men in science. Both Louis Agassiz and Benjamin Peirce did work for the Coast Survey, the latter succeeding Bache as superintendent in 1867.

In 1834 the Federal Government began to interest itself in a geological survey. It also organized the realization of that old dream of the Stonington skippers, the exploration of the southern Atlantic and the Pacific Oceans. As in the case of the Coast Survey, Congress acted only after years of agitation, but the United States Exploring Expedition, under the command of Commander Charles Wilkes, finally sailed from Norfolk, Virginia, in 1838. It was launched primarily in the spirit of the men who had agitated for it, namely: "that commerce might be benefitted by surveying the coasts frequented by our hardy fishermen . . . and that new channels might be opened for commercial pursuits, especially in animal fur." However, in 1838 America had begun to realize the importance of "pure" science for utilitarian as well as educational purposes, and considerable care was bestowed on the selection of scientists who were to go on the expedition. By this time trained observers began to be available. The expedition has since become known as the Wilkes expedition; it consisted of five vessels accompanied by a store ship. It returned to New York in June 1842 after a journey which had led successively to Madeira, the Cape Verde, Rio de Janeiro, Cape Horn, Valparaiso, Tahiti, Samoa, Australia, New Zealand, Fiji, Hawaii and California, returning by the way of Borneo, Singapore and the Cape of Good Hope. A coral island in the Pacific was named "Bowditch Island" after the Salem astronomer, recently deceased. An enormous amount of material was collected, and it took American scientists many years to describe and classify it.

Among the scientists on this voyage was young James Dwight Dana, a Yale graduate and an assistant to Silliman, who, in 1837, at the age of twenty-four years, had published a *System of Mineralogy* which became the standard American textbook on this subject. For Dana this voyage with Wilkes was what the voyage of the *Beagle* had been for Charles Darwin only a few years before. If his observations on board the *Peacock* and the *Oregon* did not lead to the development of a theory as spectacular as that of evolution, they did lead him to his theory of cephalization, which pointed to the growth of the nervous system in animals during the eons of geological time and in which we now may see a contribution to the theory of evolution. He collected an enormous amount of data and specimens; during the thirteen years following his return he was active in publishing the various reports of the Wilkes expedition committed to his charge. It was largely this work which was responsible for Dana's emerging as the leading geologist of his period.

Asa Gray, then a young botanist from western New York and curator of the collections of the New York Lyceum of Natural History, had secured an appointment, but did not sail with the Wilkes expedition. Neither did Nathaniel Hawthorne sail, though he wanted to go as a "historiographer." But Charles Pickering, grandson of old Timothy, the Salem Federalist, did sail with Wilkes, in the function of a naturalist interested in comparative botany, zoology and anthropology.

He was at that time in his thirties, a Harvard graduate of the class of 1823; and a man of leisure who as a boy in Wenham had already instinctively "seemed to know the habits and resorts of flying and creeping things." Pickering's rambles eventually led him to the most remote corners of the earth, including India and Africa, and were recorded in books of wide scope, in which he described the geographical distribution of animals, plants and the races of man. He became a man with an incredible amount of information. The work of his last sixteen years was a *Chronological history of plants* published posthumously in 1879, an incredible hodgepodge of information, covering 1222 crowded pages. It was a veritable encyclopedia of the occurrence of plants in the literature of all ages, beginning with Egypt, 4713 B.C., and progressing chronologically via the Pentateuch up to 1872. It was somewhat of an apotheosis of classification in American literature on natural science, appearing when this phase gradually began to be abandoned for studies from a comparative and dynamical point of view.

9

The change which took place in twenty years in the attitude of the public, and the corresponding change in the scientific atmosphere of the country, were indeed astonishing. Charles T. Jackson, in a letter to Silliman in 1836 in which he suggested a plan for the making of "a universal collection of the objects of Natural History of the USA," was fully aware of it. The American people, he wrote, are liberal: "no other people in the world, I may safely affirm, have ever called on their governments, to furnish information of this kind; from which fact we may conclude that the American people are more enlightened respecting the application of science to the arts, than the people of any European state." The museum of the Boston Society of Natural History, he continued, is opened "freely to the public one day in the week; young persons throng to the cabinet for instruction and amusement, and . . . many a germ of science has begun to unfold itself in their minds, the fruits of which no man can calculate." We may see, in these words of Emerson's brother-in-law, a translation of the ancient Jefferson-Adams creed into the terms of a new generation. As a matter of fact, even at present the European tourist and scholar is agreeably surprised to see how easily, as a rule, public and semiprivate collections of America are made available to general use.

A member of the Boston Society of Natural History, writing some of his reminiscences at a later date, wrote that in 1830 there was

> not, I believe, in New England, an institution devoted to the study of natural history. There was not a college in New England, excepting Yale, where philosophical geology of the modern school was taught. There was not a work extant by a New England author which presumed to grasp the geological structure of any portion of our territory of greater extent than a county. There was not in existence a bare catalogue, to say nothing of a general history, of the animals of Massachusetts, of any class. There was not within our borders a single museum of natural history founded according to the requirements and based upon the system of modern science, nor a single journal advocating exclusively its interests. The laborers in natural history worked alone without aid or encouragement from others engaged in the same pursuits, and without the approbation of the public mind, which regarded them as busy triflers.

Even as late as 1838 Edward Tuckerman, the Boston botanist, complained to John Torrey in New York, at that time the leading botanist in the country, that "there seems little enough botanical feeling in this region: and Bigelow is still the manual of students. . . . Perhaps, indeed it is true, that New England is behind the age in science." When Tuckerman wrote to Torrey, Torrey's young friend Asa Gray was already writing the books which would replace Bigelow's. By the middle of the forties there were public and private funds available for the scientific investigation of nature, there was a growing body of scientists, and a public faithful in its devotion to the work of these men. De Tocqueville, the shrewd French observer, saw the change and remarked: "If the democratic principle does not on the one hand induce men to cultivate science for its own sake, on the other, it does enormously increase the number of those who do cultivate it." At the same time he philosophized on the form which this science took: "Permanent inequality of conditions leads men to confine themselves to the arrogant and sterile researches of abstract truths, whilst the social condition and institutions of democracy prepare them to seek the immediate and useful practical results of the sciences. The tendency is natural and inevitable."

It was natural that geology gained most from the new foundation laid for science in the United States. A large number of geologists were being trained, but many other scientists of the middle nineteenth century received to a large or small degree their education while working on the surveys between the years of 1830 and 1850. Their number includes not only geologists, but also chemists, botanists and zoologists. In 1840 the geologists organized themselves in the Association of American Geologists, the first professional scientific society in the country. Among the prime movers were Edward Hitchcock and Henry Darwin Rogers, a geologist from Pennsylvania, as well as his brother, William Barton Rogers of the University of Virginia, who later became the first president of the Massachusetts Institute of Technology. The first meeting of the Association was held in Philadelphia in 1840. It met in Boston two years later, and was received with great enthusiasm by a wide and influential public. Charles Lyell, England's geological celebrity, attended, and Nathan Appleton, Boston merchant and father-in-law of Longfellow, helped to defray the expenses. These geological meetings were such a success that the Association invited all naturalists to join them, which led the following year to the American Association of Geologists and Naturalists. In 1848 all natural sciences were organized by the establishment of a general professional society, the American Association for the Advancement of Science. It was tangible evidence that science in America had achieved a certain degree of maturity. The "Triple A–S" has continued to grow and has remained the over-all organization of American scientists. Its first president was William Redfield, a businessman and a distinguished meteorologist of Connecticut.

10

By changing the quantity of its work, American science was changing in quality. One of the new problems was the systematic training of professional scientists. Early scientific work was the work of isolated individuals, related to other men of their turn of mind at most by correspondence or an occasional scientific chat or conference. The new type of work began to demand teamwork, if only on a small scale, and required a moral responsibility not only to truth, or to the body of peers alone, but also to the public at large, which paid for it. It is a long way from the solitary vacation work of Olmsted in 1823 to the geological survey of the public lands in the Upper Mississippi Valley, accomplished in 1839–1840 and led by David Dale Owen, in which 139 assistants organized into 24 field parties participated. These assistants had to be instructed in "such elementary principles of geology as were necessary to their performance of the duties required of them," and the survey of 11,000 square miles was accomplished in two months and six days. The Rogers brothers, when surveying Virginia and Pennsylvania, also had to train their own staff, several members of which later made a reputation for themselves in American geology. Since very few opportunities existed to obtain a scientific education at any American college, a number of the leading scientists connected with the surveys were either self-made men or men trained for other purposes. Among those engaged in geological activity in New England, Edward Hitchcock started as a clergyman, James Gates Percival as a physician and poet, Charles T. Jackson as a physician, Zadoc Thompson as an almanac maker. They were supported by public funds, by academies, or by businessmen such as Teschemacher and Binney in Boston, who were themselves distinguished naturalists. Dana and the Rogers brothers were younger men, and had an academic background. Silliman's outstanding educational work was bearing fruit.

Thus the United States began gradually to emerge from the stage of exclusive collecting and classifying in which it had been forced to languish for so many years. For the first time since Benjamin Franklin, new and fundamental scientific contributions were made in North America. This is particularly true for geology, where the investigations of the Rogers brothers on the formation of the Appalachian Mountain system offered new contributions to structural geology. This contribution was, like Franklin's of several generations earlier, recognized as important work in Europe, and all scientific recognition still depended on European approval. America, however, was also able to recognize scientific merits: the work of Redfield on the direction of storms found its recognition by the election of its author as first chairman of the American Association for the Advancement of Science. European approval, as well as American ability to recognize scientific genius, made it possible for the Swiss geologist Agassiz not only to be invited to the United States, but also to find an attractive field of work in this country. With Agassiz a new chapter opened in the teaching and study of natural science in the United States.

Chapter 8

Humanists and Humanitarians

The whole machine of worlds before his eye
Unfolded as a map, he glances through
Systems in moments, sees the comet fly
In its clear orbit through the fields of blue,
And every instant gives him something new
Whereon his ever-quenchless thirst he feeds;
From star to insect, sun to falling dew,
From atom to the immortal mind, he speeds,
And in the glow of thought the boundless volume reads.

Truth stands before him in a full, clear blaze
An intellectual sunbeam . . .

—J. G. PERCIVAL, 1821

1

FROM THE TIME of the earliest town meeting, New Englanders had been accustomed to come together to deliberate on public issues, to acquire information and edification and, if circumstances permitted, to speak up. They had participated enthusiastically in debates on questions of theology and on moot points with a theological flavor in the arts and sciences, such as the pro and con of the lightning rod, the inoculation against the smallpox, or the ungodliness of the theater. Many persons of colonial Boston had flocked to hear Dr. Spencer and his imitators, to witness their demonstrations with electrical apparatus; it was at one of these sessions that Benjamin Franklin had received his first impulse to study the problems of the Leyden jar. Even during the relatively sterile days after the Revolution popular interest in science had never waned, with or without a touch of theology. When, in the early years of the nineteenth century, Silliman wished to strengthen his position as a pioneer in the new fields of science, he found it extremely useful to give popular lectures.

The American public of those years took kindly to lectures on chemistry or geology, despite an occasional disapproving frown from the pulpit. A well-known popular lecturer in the early decades of the last century was Amos Eaton, a Williams College graduate with some schooling at Yale under Silliman. He had adopted this new profession after a short but painful career as a lawyer in Catskill on the Hudson, which had ended in a scandal and a jail sentence. Bearing up courageously under a fate which many people recognized as unjust, he taught geology and chemistry for several years at Williams College in the Berkshires, at Castleton Academy and other places. He was a skillful and successful teacher, stimulating the interest of Edward Hitchcock, Albert Hopkins and other talented young men of the locality. Possessed by the true fervor of the evangelist, from 1817 to 1824 he wandered through the New England states and New York lecturing on several branches of his beloved natural sciences to all who were willing to listen. People from all ranks of life crowded around his lectern, the common man as well as the town aristocracy, ladies as well as gentlemen. He passed from Northampton to Belchertown and Worcester, ending up in Troy, staying several months in each place, occasionally lecturing on botany. He used the simplest kind of apparatus:

> A pewter sucking bottle [he wrote to John Torrey, the New York botanist] is my fluoric gas bottle, a stone pig and a tin tube my earthen retort, a teakettle with the cover luted on is my iron retort, etc. I have a complete pneumatic cistern with several of my own improvements. My glass retorts etc. are also regular. But much of my apparatus is *sui generis*. . . . I illustrate the most obtruse parts by a dishkettle, a warming pan, a bread-tray, a tea-pot, a soap bowl or a cheese press.

Eaton was an ideal teacher for that large group of people who were anxious to know about the newly discovered facts of nature, which were not to be found in college curricula or in libraries. His method perhaps was crude, but it answered a growing demand of the community. When Eaton ceased lecturing in New England, others took his place. The younger men had learned from him and from the few other Americans who recognized the need and understood the charm of modern science. Eaton taught them how to teach:

Never offer less than 24 lectures for a course. To save the public from imposition, make the fact known, as extensively as possible, that none but imposters will offer less than 15 lectures for a course; and that 30 ought to be given. These peddling swindlers, who offer to sell tickets for isolated lectures, ought to be despised. They are always contemptible quacks of no integrity; and they ought not to be allowed to sleep near traveler's baggage, at public inns.

[These] swindlers . . . are chiefly foreigners, and most of them are illiterate Scotch, Irish or English. But justice demands that we make many honorable exceptions.

It has been said that there is more real beauty in the chemistry of a distant star, or in the life history of a protozoon, than in cosmogony produced by the creative imagination of a prescientific age. Eaton and his followers could not lecture on astrophysics and microbiology, but the information they gave held as deep a fascination for their listeners as our modern science holds for us.

Eaton's personality subsequently drew the attention of the shrewd and wealthy old patroon of Albany, Stephen Van Rensselaer, who give him a chance to continue teaching and writing in a more organized form. As protégé of the patroon, Eaton in 1820–1821 helped survey some of the Hudson counties and in 1824 explored the region of the Erie Canal, then under construction. This work resulted in some of the earliest geological reports in this country. After the foundation of the Rensselaer Institute he became its leader, as well as its outstanding lecturer and instructor. Here he introduced the laboratory method of teaching, taking students out into the field to study animals, plants and rocks—a novel system even in Europe, and not widely accepted in America until Agassiz's day. However, the emphasis at Rensselaer was placed rather on the education of teachers and mechanics than on the training of scientists. Eaton's voluminous textbooks on geology and botany, for instance his *Manual,* indicated the trend in his teaching; they remained for many years a standard source of information for the younger generation.

2

From now on the lecture system became an integral part of the new emancipation movement. The people of the United States at last began to realize the enormous possibilities involved in building a new economy, expanding locally and toward the west. During the thirties and forties this resulted in a broad popular movement which affected all phases of life and thought, and all classes of the population. To the industrialists it meant expansion and profit; to the merchants new markets and more daring navigation; to the farmer and the mechanic greater civil rights and better educational facilities; to the factory worker organization and ten hours of work a day; to the women, participation in public activities; to the inventor and the scientist, new realms of nature to explore. It was accompanied by a search for new values of life and for a new outlook on nature, society and man, in an amazing variety of aspects.

All this inspiring, unfettered and often ruthless and utterly uncontrolled activity was truly a continuation of the American Revolution, bursting now into full flower after an incubation period lasting for more than a generation. The old foes of democracy and the friends of the *status quo* were there again; they wore new faces, just as they emerged and continued to re-emerge time and again in every period of American life under varying disguises, but at this period they lacked the earlier homogeneity of the Tories in the colonial period or the Federalists of the previous generation. They were centered in the old merchant class, now definitely turned industrialist, and spread throughout the Whigs as well as throughout the old Democratic party, which became more and more the stronghold of the southern slave economy. However, the new progressive trend was at the same time thoroughly in evidence in both Whig and Democratic parties. The Democrats were connected with the newly created industrial proletariat, and still remained under the influence of the older progressivism of the mechanics and dirt farmers. Those pillars of the Whig party, the merchant-industrialists, had their eyes opened to wider horizons than those of their ancestors; these men were more tolerant of change and less worshipful of England and the *status quo*. The same "broadcloth" gentlemen who dragged William Lloyd Garrison through the streets of Boston and tried to lynch him were also the builders of new industrial cities, the promoters of clipper ships and steamboats, who revolutionized the country by railroad construction and other seemingly fantastic engineering operations. By solid investment and by wild and often ruinous speculation they expanded the domestic market enormously. Their role in the general political field is well illustrated by their somewhat overrated statesman, Daniel Webster, who was perfectly willing to function at the bar and in Congress as a paid servant of the New England merchant-industrialists, but who rose to great statesmanship in his reply of 1830 to Hayne in which he extolled the unity of the American nation.

The turbulent social current of the thirties and forties carried the germs of deep social conflict, since expanding industrialism could not in the long run maintain its friendly attitude towards an increasingly arrogant slaveholding class in the South, and found itself for some time drawn into a reluctant companionship with the abhorred labor movement and the even more abhorred abolitionists. Some Whigs, though descendants of the tie-wig Federalists, moved hesitantly forward and graduated into the rank of Free-Soilers and eventually into Lincoln Republicans. The Democrats, heirs to the progressive traditions of Jefferson and Jackson, began increasingly to take orders from a Southern gentry, fast losing contact with its Jeffersonian tradition. New Hampshire born President Franklin Pierce, who started as a follower of Jefferson, and later of Jackson, prepared in his presidential career the road to Buchanan and Southern secession. The inevitable conflict of the sixties was in the making, to be followed in the next decades by the ruthless class struggles between workers and employers in the North, freedmen and Bourbons in the South. But even civil war and subsequent corporation control could not undo the great positive achievements of these earlier decades of optimism. Indeed, they were bound, in their own self-conflicting way, to offer their own particular contribution to the results of these early victories.

3

New England in those years maintained a leading role in the nation, and thus lost some of its hitherto typical provincialism. It contributed generously to the new literature and science, and in its own way helped give to the new culture a more purely native American form. Both science and literature still depended strongly on European models, and looked steadily across the Atlantic for inspiration and recognition. However, the approach to the arts was more universal than before and became increasingly independent. England now no longer set the example for Boston, Providence or New Haven; Germany and France were fully recognized as countries which could contribute greatly to American literature and science. Young men who went abroad to study went not only to London, Edinburgh, or perhaps Leyden, but also to Paris and Göttingen, to Rome and Vienna. German and French publications began to be widely circulated. They introduced new ideas, not only in romantic literature, but also in mathematics, in physics, in chemistry and in engineering. In this period the New World no longer adapted, it absorbed, transformed and digested, and began to turn out a home product of genuine merit. Thoreau was typical of New England, as were Redfield, Howe, Holmes and Hitchcock. Large masses of the population were willing to share with their intellectual leaders the results of new thought and new experiment; they came to listen by the hour to discussions on English authors, the periods of the earth's history, the care of the crippled, or the socialism of Fourier.

The backwoods of New England contributed their full share to the emancipation of an awakened democracy. Williams College, Castleton Academy, the University of Vermont, did remarkable educational work. The first Workingman's Association in New England was founded in Woodstock, Vermont, which was also the site of one of the early medical schools. New religious movements of a fundamentally democratic type were initiated by leaders from these remote regions, such as John Humphrey Noyes of the Oneida Community, Brigham Young of the Mormons, both of whom were Vermonters. And one of the most important advances in popular education was undertaken in 1826 by the farmers and mechanics of Millbury, a budding textile community in Worcester County, when they launched the first of the so-called "lycea."

4

Josiah Holbrook, the leader of this lyceum movement, was a man of considerable erudition, one of the many Yale graduates who had received inspiration from Silliman. Like Eaton he was a gifted popular teacher with a deep veneration for the natural sciences, and a believer in what we may broadly call the Jeffersonian concept of education. He taught school on his farm in Derby, Connecticut, and later tried to embody his ideals in an educational institution of broader scope, but this Derby Academy, also conducted on his homestead, lasted only from 1824 to 1825. It was an early experiment in agricultural education and as such had much in common with Vaughan's Academy in Gardiner. An attempt was made to combine manual labor with instruction in several branches of the natural sciences, including even the analysis of soils. Under Eaton Rensselaer favored somewhat the same approach. Like all these early experimental academies, Derby did not aspire to original scientific work, but concentrated on the training of teachers and able craftsmen.

After his Derby experiment had come to an end, Holbrook concentrated on general educational reform. The common schools were in poor shape, and had deteriorated seriously in the last decades, since the well-to-do had been sending their children to private institutions. At the same time the interest of the people in education and self-improvement had grown by leaps and bounds. There were popular lectures everywhere, discussions on popular science and on literature, accompanied by a demand for informative books. Holbrook, like Eaton, was a beloved traveling lecturer, and was keenly aware of the new trend expressed through this grass-root curiosity. In 1826 he summed up his conclusions in a paper printed in the new *American Journal of Education;* in 1828 he organized the American Lyceum or Society for the Improvement of Schools and Diffusion of Useful Knowledge. There was a definite connection between this movement and the new industrialism, expressing both the aspirations of the workers and those of more progressive employers. Daniel Webster, Edward Everett, Governor Lincoln and other influential men collaborated. In 1830 some of them met in Boston to found the American Institute of Instruction, which concentrated on the improvement of the public school system. In 1831 a National Lyceum was organized, which held annual meetings until 1839. In 1834 there were already nearly 3000 lyceums in the United States; in 1839 Horace Mann counted 137 in Massachusetts alone. Their character varied from place to place, but they were often large organizations. Most of them went in for discussion of such questions as corporal punishment, the teaching of ancient languages in schools, the necessity of teachers' training schools, the advantages of a manual labor system in schools both for boys and girls, and the need for visual means of education in the sciences, notably cabinets of natural history. The reform movement, which under the leadership of Horace Mann and Henry Barnard was able to reorganize the public school system, found its mass support in the lyceums. The reformers not only supported more modern methods of education, but also experimented with them in their own halls. Holbrook and Warren Colburn were always willing to try out new visual means of education, models, tools, apparatus, or an occasional magic lantern. There was, of course, influence from educational reformers abroad, but the whole lyceum movement had a thoroughly American grass-root flavor.

For the lyceum movement Holbrook constructed apparatus to promote the teaching of science, a relative novelty which even today might be more widely emulated. The program of the *American Lyceum*, published in 1829, advertises tools for the instruction in geometry, arithmetic, astronomy, geology, chemistry and "natural philosophy." The geometrical apparatus consisted of 2 sheets of diagrams, 15 geometrical cards, 4 transparent figures, 26 solids and a book with questions; the apparatus for astronomy contained an orrery, a tide dial and an instrument to show eclipses; that for natural philosophy had levers, pulleys, an inclined plane, wedge, screw, wheel and axle. The chemical apparatus, consisting of tubes, flasks, retorts, crucibles and compound blowpipe, is remarkable for the fact that it was introduced at a period when even

Yale and Harvard had barely enough instruments to accommodate the students of Silliman and Webster. Holbrook's work constituted a definite break with staid traditions, especially in the teaching of mathematics, which until then had looked for inspiration—or better, lack of inspiration—to the sixteenth-century humanists. Holbrook later concentrated his efforts upon the construction of his visual tools of education. At the time of his death, in 1854, he had a shop in New York for this purpose.

5

Textbook reform was now the order of the day. Farrar, a friend of the lyceum movement, had begun this reform on a college level in the field of mathematics already in the previous decade. Amos Eaton had tried to embody the experiences of his popular lectures in his texts. Now other men came along, who wanted books not only for adult or college audiences, but also for the younger generation. The publishers' catalogues began to advertise an increasing number of schoolbooks written by contemporary American authors. There were, for instance, the books by John Lauris Blake, for a while an Episcopal clergyman and the principal of a "young ladies' seminary." Already in 1814, two years after his graduation from Brown University, he published a *Textbook of geography and chronology,* and continued writing and compiling book after book, including a schoolbook on astronomy, which because of the simplicity of its language was often reprinted. The school textbooks of the thirties, indeed, are beginning to use a language which our generation can still appreciate.

This holds especially for the books of one of the best known of the school reformers of those days, a young Harvard graduate by the name of Warren Colburn, whose short career illustrates the connection between the educational reform movement and the new industrialism. Colburn was a farmer's son of Dedham, and obtained a good working knowledge of machinery while working as a boy in factories. From 1817 to 1820 he was at Harvard, where he studied mathematics under Farrar—even reading Laplace, which was quite an achievement in the America of that period. His main project became the improvement of the system of elementary instruction in mathematics. He opened in 1821 a select school in Boston, where he developed new methods in teaching arithmetic under the influence of Pestalozzi. When, however, Patrick T. Jackson offered him in 1823 the position of superintendent of the Waltham textile mill, he accepted, and exchanged this position the next year for that of superintendent of the Merrimack Manufacturing Company in the new town of Lowell. In this position Colburn had full opportunity of combining his technical knowledge with his educational ambitions. He became a popular lecturer on all kinds of subjects relating to nature, technique and astronomy, using models to illustrate hydraulical principles as well as the magic lantern. His fame, however, was mainly based on his excellent school textbooks, above all on his first, the *Lessons in Intellectual Arithmetic* published in 1821, and which became an immediate success. It teaches elementary computation by means of hundreds of short and pointed questions, emphasizing solution by mental methods; it was translated into most European languages; by Civil War days, more than two million copies had been sold in the United States alone, and missionaries had translated the book into several languages of India, as well as into Hawaiian. The edition of 1863 had a foreword by George B. Emerson, at that time one of the leading educators of Massachusetts, and a revised and enlarged edition appeared as late as 1891. The *First Lessons* were followed by a *Sequel* and in 1828 by an *Algebra.* Colburn might have become one of America's leading educators, but the hard life in the new industrial settlement was too much for his health. He died in 1833, only forty years of age.

The greatest triumph of the lyceum movement was the reform of the public school system, the foundation of state departments of education and the establishment of normal schools for teachers. This was largely the contribution of Horace Mann and his co-workers, George B. Emerson, Henry Barnard, and many others from all walks of life. It may be that this reform, which was accomplished in Massachusetts in 1837, had something to do with the decline of the National Lyceum Organization, which met for the last time in 1839. Locally, however, the lyceums continued their activity for many years, and some are still active today. They remained one of the most effective media for general adult education. Lecturers on all possible subjects of popular interests came to the lyceum meetings, some attracted by the noble purpose, others because they had a real message, others by generous fees, still others because of a combination of all these reasons. Daniel Webster, Edward Everett, Horace Mann and other public figures of national fame appeared on the lyceum platforms. Considerations ranged from $5 to $100. College professors were invited to conduct courses on science, on commercial subjects or on modern languages. A glance through the works of Wendell Phillips, Ralph Waldo Emerson, Theodore Parker, or Henry Thoreau shows that many of their essays were written as lectures for such lyceum or similar platforms. Theodore Parker, the radical Unitarian minister of West Roxbury, gave an average of forty to eighty lectures a season for ten years. Oliver Wendell Holmes also did a fair amount of speaking.

6

During the winter of 1837–1838 twenty-six courses were delivered in Boston, each consisting of more than eight lectures and attended by an estimated number of 13,000 people. There were many defects in this system, since the thorough preparation of good courses required more money and time than most organizations were willing to pay. Not all people were like Amos Eaton, willing and capable of lecturing on chemistry year after year with only fifty dollars' worth of equipment at their disposal. The lecture system could be substantially improved and prevented from degenerating by more generous endowments, which also could lessen some of its dilettantism. Such grants were occasionally forthcoming. The most generous of them was the Lowell bequest. John Lowell was the son of the ingenious Lowell who had reinvented the power loom and helped introduce the corporation system into the textile industry. He was of weak health and spent much of his short life traveling. He wrote his will among the poetic ruins of Thebes on the Nile, leaving approximately $250,000 for the maintenance of annual courses of free lectures in Boston. Shortly afterwards, in 1837, he died in Bombay, not yet thirty-seven years of age. The Lowell Institute opened its doors in 1839 and was successful from the beginning; it attracted the best talent available and is still functioning, though perhaps in a less spectacular way. Early scientific lecturers included Silliman, Nuttall, Lovering, Wyman, Asa Gray, Henry Rogers, Charles T. Jackson and Benjamin Peirce, all from the United States and many from the immediate neighborhood; Lyell and Tyndall came from England to lecture. The direct cause of Agassiz's arrival in the United States was an invitation to lecture for the Lowell Institute. If the Institute had accomplished no other good, this single fact alone would have justified its existence. Many lecture courses lasted several weeks; some had enormous attendance. Benjamin Silliman, then at the peak of his fame as a lecturer, opened the Institute in 1839 with a series on geology. When tickets were given out for his second course, on chemistry, the crowd filled the streets adjacent to the lecture hall—the old Odeon on Federal Street—and crushed into the windows of the Old Corner Bookstore. The number of applicants for a single course sometimes rose as high as 8000 or 10,000, so that tickets were distributed by lot.

The regular inclusion among the lectures of a theological or religious subject was symptomatic since it shows the care which was taken to avoid any appearance of atheism, then still widely considered inherent in natural science. But no minister was asked, as scientists were often requested, to repeat his lectures; interest lay overwhelmingly in science. When Lyell lectured in 1841, no less than 4500 tickets were given out. In his *Travels in North America* he gives his impression of the audience, which amazed him by its interest and decency: "persons of both sexes, of every station in society, from the most affluent and eminent in the various learned professions to the humblest mechanics, all well dressed and observing the utmost decorum"—an attitude reminiscent of that of some present-day visitors to the Soviet Union, who also have been led to expect barbarism and meet an unexpected degree of civilization. After all, it had not been so long ago since Englishmen, through their press and thanks to some of the visitors returning from the United States, had been made to believe that their cousins across the ocean were a most reprehensible species of semibarbarians. Lyell must have had his misgivings before he ventured into this disorderly wilderness.

Lyell also admired the way in which the Lowell Institute spent its money on lecturers rather than on buildings. The result was that lecturers in Boston were remunerated on a scale three times higher than that offered the best literary and scientific lecturers in London. Lyell's book also contained some excellent popular science, information on Boston's geology and fauna, on the rocks of Nahant, on Dr. Gould's collection of marine shells, and on the glacial theory, then in its rise. His observations on social conditions were less excellent; he was only too willing to accept at their face value the glowing stories of the Lowell manufacturers and the proslavery philosophy of the Southern plantation lords.

Lyell was amazed at the taste for good books prevalent in Massachusetts. Editions of Plutarch's *Lives* sold from 5000 to 20,000 copies, Froissart's *Chronicles* 16,000, Liebig's *Animal Chemistry* 12,000; all these were cheap editions, with no author's copyright. Similarly popular was Herschel's *Natural Philosophy*. Four thousand copies of Prescott's *Conquest of Mexico* were sold in one year at the rate of six dollars apiece. All this allowed him to "indulge very sanguine hopes of the future progress of this country towards a high standard of civilization." Holbrook and his friends, as well as the labor leaders, had not worked in vain.

In the outlying towns reports on the progress of science were received with the same ardent expectation. Audiences in college towns kept their professors busy, especially when they happened to be not only well informed, but also stimulating lecturers. Outside the college towns the audiences also listened eagerly when a good man came along. In 1845 Henry Rogers, the geologist from Pennsylvania, lectured before the lyceum in Portsmouth, New Hampshire, before 1000 people, and later addressed a private audience of 200.

7

With interest in the new sciences so widely spread, it was only natural that the new literature should reflect this vigorous trend of the times. Ralph Waldo Emerson always kept his eyes and ears open. His sympathy for chemistry and geology dated from his early student days at Harvard, where he had listened to John Gorham; he later taught science for a while in his brother's school. Occasionally, while philosophizing by the side of the road or on Sweet Auburn, he indulged himself as a naturalist. His interest was kept alive by his associations with his brother-in-law, C. T. Jackson, with Francis Cabot Lowell, his classmate at Harvard, with Thoreau, Holmes and Agassiz. During his European trip of 1832–1833 he visited John Davy, the brother of Humphry Davy, and the Modena microscopist Amici—"Emerson hardly knew his privileges, he may have been the first American to look through an immersion lens with the famous Modena professor," wrote Oliver Wendell Holmes in later days. At Paris Emerson attended the lectures of Gay-Lussac and a séance of the Institut, Classe des Sciences. He understood the leading role of France in science and later described it in his delightful reminiscent way:

> We remember, when arriving in Paris, we crossed the river on a brilliant morning, and at the book-shop of Papinot, in the Rue de Sorbonne, at the gates of the University, purchased for two sous a Programme, which announced that every Monday we might attend the lecture of Dumas on Chemistry at noon—For two magical sous more, we bought the Programme of the Collège Royal de France, on which we still read with admiring memory, that every Monday, Silvestre de Sacy lectures on the Persian Language; and other hours, Lacroix on the Integral Mathematics; Biot on Physics; Elie de Beaumont on Natural History; Thenard on Chemistry; and so on to the end of the week. On the same wonderful tickets, as if royal munificence had not yet sufficed, we learned that at the Museum of Natural History, at the Garden of Plants, three days in the week, Brogniart would teach Vegetable Physiology and Gay-Lussac Chemistry. With joy we read these splendid news in the Café Procope and straightway joined the troop of students of all nations, kindred, and tongues, when this great institution drew together to listen to the first Savants of the world without fear or reward.

Back home, Emerson read his paper on the American scholar and tried to grasp the function of that new phenomenon, the scientist in a growing democracy. This was in 1837, when Silliman and the Harvard Medical School had been modestly teaching science for a generation, when geological and other surveys were gaining impetus and when for the first time an increasing number of men were grooming themselves for a purely scientific career. Emerson saw very well that despite the overcrowding of popular lectures, American science was still in its infancy. Above all, it lacked originality of thinking. Emerson, however, felt confidence. "Our day of dependence, our long apprenticeship to the learning of other lands, draws to a close," he said. He did not draw an outline of research for his America, a task which neither he nor any of his more competent scientific contemporaries felt the need of—but he tried to delineate the ideal character of the emerging American man of science in his relations to nature, the past and society. Although in accordance with the existing trend in American colleges, he perhaps underrated the creative element of science—he spoke of scholars rather than scientists, and of the study of books rather than of the testing of new ideas—he did stress the necessity of intellectual courage, of standing on one's own legs, and of being a responsible member of the community. "Action is with the scholar subordinate, but it is essential. Without it he is not yet man. Without it thought can never ripen into truth. Inaction is cowardice, but there can be no scholar without the heroic mind." He warned against too much listening "to the courtly muses of Europe," which has made the scholar "docent, indolent, complaisant." This was a voice to which many a smug academician of 1837 could listen with profit, and the lesson has kept its timeliness to the present time.

Emerson was always more concerned with the social effects of science than with its content. He represented, not without a certain complacency himself, the balanced man of progress, who was willing to appreciate the creative function of science in an expanding democracy, but was keenly aware of the darker sides of the coming machine age. Science was brought to America as an essential part of an utterly uncontrolled capitalism in a wild orgy of expansion. It brought visions of power and of prosperity, but also misery, squalor, vice and degradation. Emerson admired the arts and manufactures, the inventions and the crafts, but he stressed above all the dignity of the individual human being. In former periods, he said, the doctrine of Reform had always respected something, but now all things heard the trumpet, and must rush to judgment—Christianity, the laws, commerce, schools, the farm, the laboratory; and there was not a kingdom, town, statute, rite, calling, man or woman, but was threatened by the new spirit. The love of foot rule and balance brought too often in its wake a bitter degradation of human dignity. Let there be worse cotton and better men, he pleaded. The fact that under favorable conditions better cotton might produce better men, as Jefferson and some of his own Unitarian and socialist contemporaries believed, was somewhat lost on Emerson.

William Ellery Channing represented more than Emerson the positive social approach to the new period of science and technology. Channing was a scion of a leading Newport family and a brother of a prominent scientist at the Harvard Medical School; he was for many years the outstanding Unitarian clergyman of Boston. In all these capacities he enjoyed full opportunity to watch science at work in the laboratory and study its effect upon the people for better or for worse. This great preacher felt optimistic: "Where is science now?" he exclaimed in 1841 from a Philadelphia platform:

> Locked up in a few colleges, or royal societies, or inaccessible volumes? Are its experiments mysteries for a few privileged eyes? No, science has now left her retreats, her shades, her selected company of votaries, and with familiar tone begun the work of instructing the race. Through the press, discoveries and theories, once the monopoly of philosophers, have become the property of the multitude. Its professors, heard not long ago in the university of some narrow school, now speak in the mechanic institute. There are parts of our country in which lyceums spring up in almost every village for the purpose of mutual aid in the study of natural science. The characteristic of our age, then, is not the improvement of science, rapid as this is, so much as its extension to all men.

Science, then, had become democratic. Had democracy, on

the other hand, a use for science? There were many reasons to believe that it had. There were the many geological, agricultural and topographical surveys, which were publicly sponsored and showed that science could help promote the economic interests of the people, not only of the rich, but also of many a small businessman, a professional, a mechanic, a factory worker or a husbandman. Science could also help bring succor to the sick and destitute. A new and better world was possible:

> It is conferring on us that dominion over earth, sea and air, which was prophesied in the first command given to man by his Maker; and this dominion is now employed, not to exalt a few, but to multiply the comforts and ornaments of life for the multitude of man.

This was a far cry from the utilitarianism of the Founding Fathers of the republic, however deeply this might have been tinged with the fervor of the Enlightenment, but it also lacked the healthy skepticism of Emerson. In our day, Channing's vision is shared by uncounted millions, who, often for the first time, see their path to freedom open.

8

The misery which the factory system brought to thousands of workers and the cheap materialism which often accompanied the new industrialism were certainly not lost on Channing. But he saw some of the essentials clearly, as we can now more easily perceive with the perspective of a passed century between us. Another, even more essential aspect of the truth was proclaimed by the boldest dreamers of their age, the socialists, the men and women who went out to build with their own hands an America devoid of the exploitation of man by man. The best known of these settlements was that at Brook Farm near West Roxbury, ten miles south of Boston, where Theodore Parker preached. Here, in rural surroundings which even today the growing city has not completely absorbed, a community of cultured people under George Ripley and Charles Dana tried to realize a Christian co-operative, and later a socialist colony according to Fourier. Fourier's visions of a world built on co-operation of all productive workers enjoying the benefits of all resources of science and industry had the mark of genius, even if occasionally he got excited over such epic notions as the possibility of disinfecting and perfuming all the waters of the sea by means of a boreal fluid. Fourier grew up in a France dominated by small producers, and though he believed in the beneficial calling of science, his ideal society is based on craftsmen and artisans rather than on engineers and factory workers. This aspect of Fourier's work made it attractive to the Brooks Farmers, who were mostly inclined toward literature and the arts. They never bothered much with the modern aspects of science and their implication, although they were outstanding in their knowledge of literature and the arts, and deeply interested in all social currents, including the ten-hour day. Not a single science has been connected with Brook Farm, although its influence has been felt in literature, in modern education, in social philosophy and in the cultivation of music.

In this respect the Brook Farmers lagged behind the Owen school of socialists, who practiced for a while at New Harmony on the Wabash River in what is now the state of Indiana. Some of America's outstanding leaders of science came in the twenties to this place in the wilderness, attracted by Robert Owen's ideals of the application of science to a modern industrial community for the benefit of all. When the colony was established, a "boatload of knowledge" sailed down the Ohio from Pittsburgh, and New Harmony became for a while a great center of learning in the wilderness. Here resided for a while Maclure, Troost, Lesueur, Say, Rafinesque, and the Owens, father and son. Robert Owen's influence extended far beyond New Harmony, and through William Maclure set its stamp on the early development of American geology. It is even reported that Owen's ideas influenced, of all people, some of the founders of the Lowell manufactures. A revealing sidelight is also thrown on the relation between the utopian socialism of the first part of the nineteenth century and modern industrialism by the curious evolution which attended the fate of the Hopedale community near Milford, Massachusetts. This colony, which under Adin Ballou had professed a Christian type of socialism, was taken over after its disintegration by some of the colonists and turned into a textile machinery center. The industry is still flourishing under the management of the very Draper family by which it was organized.

A typical New England combination of social protest and respect for science was the philosophy of Henry David Thoreau, another of the great Concord writers. His father was a small craftsman, a manufacturer of lead pencils. The son assisted him in his trade, and in time-honored Yankee fashion improved the product. Then, rather than specialize in this profession, he decided to follow his calling as an observer of nature. But even in tiny Concord the world pressed too closely upon the individual, and left him no vista. Besides, the proper wants of man were few:

> I see young men, my townsmen, whose misfortune it is to have inherited farms, houses, barns, cattle, and farming tools; for these are more easily acquired than got rid of. Better if they had been born in the open pasture and suckled by the wolf, that they might have seen with clearer eyes what field they were called to labour in. Who made them serf of the soil?

And so, to escape this civilized form of serfdom, Thoreau set up a cabin by the brink of Walden Pond, two miles from Concord, where he spent two years (1845–1847) in relative solitude. Later he settled in Concord and became a little more sociable, occasionally lecturing, and describing in books and essays his observations of New England nature. He died in 1862, only forty-five years of age.

Although he always remained the philosopher and poet rather than the man of science, Thoreau was a born naturalist. Some may claim that Emerson spoiled him as a man of science, others that Agassiz spoiled him as a poet. However, Thoreau was above all himself. In his description of the behavior of plants and animals, of rivers and forests, of all living aspects of his country, he stood without peer. His writings are a full expression of what we might call elementary naturalism, the pure love of life and of living things, from which all science of nature is derived, and to which it must constantly return. Thoreau liked to think of himself as "a mystic, a transcendentalist, and a natural philosopher to boot." It must be granted that, when he wanted to, he also could be a good scientific observer.

Indeed, Thoreau is not all just good "literature," but occasionally also good science. His *Walden,* as well as his *Civil Disobedience,* proclaim a philosophy which has influenced men as divergent in national background as Van Eeden, Tolstoy and Gandhi. As a case study of the natural history of a small body of water *Walden* is also a pioneer work in limnology, in the science of pond behavior. Walden Pond happens to be a good example of a seepage soft water lake of generally low biological productivity, and many limnologically important facts about such lakes were first studied by Thoreau. He observed the scanty crop of littoral vegetation, the absence of organic sediment except in the deepest water—which he correctly attributed to decomposing forest leaves and the relative purity of the water. "You think that they must be ascetic fish that find a subsistence here," he philosophized, and pointed out that Sandy Pond, very near to Walden, has ample aquatic vegetation, is "not remarkably pure" and "is more fertile in fish." Thoreau was also one of the first to study the thermal stratification of pond water, by means of crude measurements, though he, with a layman's logic, assumed it to be a linear function of the depth. "How much this varied temperature must have to do with the distribution of fishes in it! The few trout must oftenest go down below in summer." He also surveyed the contour of the pond and made a map of it; it shows a remarkable accuracy when it is compared to modern air photographs. Thoreau's books and his *Journals* are full of keen observations, which modern ecologists may find useful to study.

Walden Pond is now a favorite bathing place within easy reach of Boston, but on a quiet day it still looks very much as Thoreau must have known it. A cairn near the northern shore marks the spot where his cabin stood, and the visitor pays his respect to the memory of the remarkable man by adding his pebble to the mound.[1] Though his own contribution to science was small, he had a message which all men of science may do well to ponder. He knew that the great god Pan is not dead, and he was a constant worshiper at his shrine. But he also knew that he belonged to his own period and to its struggles. "I rejoice that I live in this age," he exclaimed, "that I am his contemporary." This was when John Brown of Osawatomie made his supreme sacrifice for the cause of Negro emancipation.

We cannot always follow Thoreau and the other Concord writers when they indulged in their transcendentalism, which had too often the flavor of old German *Naturphilosophie*. They read Carlyle and under his influence made more concessions to Schelling and other German idealist philosophers than we can accept at present. Like much of the utopian socialist doctrine, this aspect of American pre-Civil War thought belongs definitely to a bygone age. To the present generation Thoreau and his associates still speak through their vitality, their combination of love of nature and love of mankind, and their keen powers of observation.

9

Some of the idealism of the thirties and forties failed in its ultimate objectives. More fortunate in their immediate results were those humanitarians who selected a field of endeavor for which the social system was partly, or wholly, prepared. Great advances were made in the field of human welfare, often of considerable importance for the advancement of science. Medicine always had a direct humanitarian appeal. Now it was growing enormously in scope and understanding. With the advance in theoretical knowledge of the diseases of the human body came great work in hospital care, in the care for the deaf, the mute, the blind, the insane. This was the period in which Samuel Howe built the Perkins Institute for the Blind; in which Dorothea Dix traveled around caring for the mentally sick, and in which William

[1] Well meaning hero worship led, some years ago, to the construction of two dismal cemetery-like structures to mark the exact position of cabin and woodshed.

Morton applied the first ether as an anesthetic. With such outstanding achievements came others of perhaps lesser appeal to the heart, but of equal humanitarian and progressive importance. Lemuel Shattuck, in this period, accomplished great work in public health, and Loammi Baldwin, Daniel Treadwell and others in the specific field of improved water supply for the large centers of population. Even a Sylvester Graham, who went around preaching temperance, chastity and the life of nature, the "prophet of bran bread and pumpkins," as Emerson called him, may be counted as a pioneer in modern methods of nutrition. We honor his memory by eating Graham bread.

The transition from humanist to humanitarian, and from humanitarian to scientist, was marked by the philosopher of the breakfast table, the best scientist among the New England authors, and the best author among the New England scientists. Oliver Wendell Holmes was the son of a scholarly Cambridge minister who belonged to the American Academy of Arts and Sciences, and the grandson of a surgeon in the Revolutionary Army. He graduated from Harvard in 1829. Seven years later he received his M.D. after passing several years abroad visiting the hospitals of Paris and other European cities. He then taught for a while in Dartmouth College. When John Collins Warren resigned from the Harvard faculty in 1847, Holmes was elected to succeed him and became the incumbent of the chair of anatomy at the Medical School until 1882. He died in 1894, leaving a reputation as a beloved medical practitioner, a fine teacher, a poet, a novelist and a wit. His son and namesake became a Justice of the Supreme Court of the United States.

When, as a young man, Holmes decided to go to Paris for his medical studies, he followed the trend of the times. The fear of France as the center of dangerous thought had subsided; it was clearly recognized that Paris was more than ever the center of the learned, and in particular of the medical world. The French Revolution and the Napoleonic Wars had stimulated the development not only of natural science, but also of medicine, and the Restoration had not stopped its momentum. In Paris that beloved creed of the ancient practitioners, the old theory of humors or solids of the system, was definitely outmoded. The newer men, headed by Bichat, studied local pathology, devoting their energy to the study of special organs of the body, each with its own characteristics. This also influenced the ancient art of surgery. So long as a disease was vaguely conceived of in terms of body fluids, of muscular or nervous tension, no particular reason existed to operate on special parts of the body except as a last resort in emergencies. Once diseases were traced to lesions in specific organs, it became logical to remove the diseased part. And with the growth of hospitals and medical care in general came the introduction of statistics into case studies as a part of the development of statistics in many fields where mass phenomena were observed. With this the name of Pierre Louis was connected. British medicine followed the same trends, and in 1848 the first appendix operation occurred. American surgery also learned its lesson and soon became very efficient.

John Collins Warren, himself a surgeon of great fame, had studied medicine in Paris a generation before. He could only advise his younger colleagues to do likewise. Among the young Bostonian physicians who crossed the ocean for medical training in the same year as Holmes were Henry Ingersoll Bowditch, the son of the astronomer; James Jackson, Jr., son of the Boston practitioner; Charles Thomas Jackson and Jeffries Wyman. Letters of several of these young men have been published and give some understanding of the deep impression made upon the Americans by French medicine and scientific practice in general.

In his analysis of the contagiousness of puerperal fever, written after his return, Holmes showed that he had learned his lesson in France. This paper, the best of all his medical papers, is a study of childbed fever. It was furthermore a case study—as yet something unusual in America—and, above all, a plea for antiseptic treatment, a novel idea for the times. The physicians themselves, he charged, were the vehicles by which puerperal fever was transmitted. Holmes was taken to task by two of the country's leading obstetricians. The author, confident of the correctness of his conclusions, republished his paper with an introductory note in which he said: "There is no epithet in the vocabulary of slight and sarcasm that can reach my personal sensibilities in such a controversy." He never had reason to regret his early study. Not only were many mothers saved by his method, but the work of Lister and Pasteur proved decisively that he had been right from the beginning. Semmelweiss in Vienna, who had expressed ideas similar to those of Holmes, was even more acrimoniously assaulted. Lacking Holmes's social standing and literary poise, and being moreover a liberal in Metternich's Austria, he was dismissed from his hospital and ended his life in an insane asylum.

Other medical work of Holmes's dealt with auscultation and microscopy, and with homeopathy, his pet aversion. As a teacher of anatomy he seems to have made the human skeleton a subject of keen interest to even the dullest of his students. He was also a gifted biographer, and the sketches he presented in the later part of his life of some of the older physicians, of Waterhouse, Jeffries, Warren and Jackson—even of old Holyoke, whom he remembered vaguely—give a loving picture of Boston medical life in the first decades of the nineteenth century.

10

As a member of the staff of the Harvard Medical School, Holmes lent wisdom and grace to the instruction and in his own way contributed to the growth of this institution from a small provincial college to its present world prominence. Though the Medical School had started from very modest beginnings, it was never a mediocre institution, starting with such men as the Warrens, *père et fils,* old Jackson and Waterhouse on its staff. Both the comparative excellence of its instruction and its connection with Harvard had helped it overcome the competition from the small provincial schools, such as Castleton, Pittsfield, Woodstock; and to hold its own against the schools of more generously endowed Yale and Dartmouth. There were always good teachers and men of science on the staff. In Holmes's days there were Henry Ingersoll Bowditch and Henry J. Bigelow, both sons of well-known scientists, and Walter Channing, brother of the minister. The Massachusetts General Hospital was connected with the Harvard Medical School, if only in an indirect way. Here the professors of the Medical School and their assistants tended the patients and conducted their case studies. It was housed in a pleasant-looking structure built by Bulfinch and opened in 1821.

In 1846, a great event took place here, which spread the fame of the Harvard Medical School to the farthest ends of the world, and proved dramatically that America's scientific contributions could no longer be ignored. This was the discovery of ether as an anesthetic, an event as beneficial for the world at large as it was painful to the three men most immediately concerned. In retrospect this discovery loses much of its seemingly accidental character since it was the logical outcome of three developments—the growth of chemistry, the development of dentistry as a profession, and the new medical views on local pathology. To nonmedical contemporaries it came almost everywhere unexpectedly, and was variously considered as a boon sent from heaven or, by some old-time believers, as a new temptation offered by the Enemy.

The replacement of the old theory of humors and tensions by local pathology had a special importance for dentistry. This field of medicine was exclusively dedicated to one locality and one function of the body. Dentists, in fact, were a special group of surgical physicians, though the medical profession had some difficulty in recognizing it. Originally a craft in which ivory turners predominated, with perhaps a sprinkling of hot-poker specialists, dentistry evolved into a science during the earlier decades of the nineteenth century. One of the first serious practitioners was John Greenwood, a grandson of the first Harvard professor of mathematics and an able mechanic, who went to France in 1806 to study dental practices. He is reported to be the originator of the foot-power drill, of special springs to hold plates of artificial teeth in position, and the first dentist to use porcelain teeth. In his earlier days he was George Washington's dentist; his work, still in existence, has been praised by the experts. One of his pupils was Horace H. Hayden, a Connecticut man who in 1800 settled in Baltimore and is remembered as the man who established dentistry on a solid scientific basis in America. About 1825 he delivered lectures on dental physiology and pathology at the University of Maryland; in 1840 he was one of the founders of the Baltimore College for Dental Surgery. Hayden was also a geologist, after whom Parker Cleaveland has named a mineral.

It was to dentists rather than to surgeons that the possibility of painless operations occurred first, partly because the removal of a tooth is a small local treatment without the consequences which a major surgical operation may involve, and partly because the dentist was not affected, in his practice, by the belief in humors of the system which handicapped the growth of local surgery. The very fact that he was standing outside of the medical profession made him less likely to succumb to its general atmosphere. Though such a situation has certain obvious disadvantages, there do exist cases where an unorthodox approach may lead to the discovery of new ways and means. Some dentists witnessed with professional interest the "pleasure-producing" results of nitrous oxide, the so-called laughing gas discovered by Humphry Davy in 1800. The effects of this gas, which produced an intoxicating sensation with an "agreeable sense of well-being," as well as those of sulphuric ether, which could throw people "into a lethargic condition," as Faraday discovered in 1818, were a subject of popular interest. The "ether frolic" entertainers were able to interest Dr. Long in Georgia and Horace Wells in Hartford, Connecticut.

The general practitioner, Crawford W. Long, in 1841 performed some minor operations under ether but soon became discouraged. Horace Wells, a Hartford dentist with a considerable practice, was more persevering. After having seen a public performance of nitrous oxide in 1844 he declared his belief "that a man, by taking that gas, could have a tooth extracted or a limb amputated, and not feel the pain." He tried this theory out on his own person, and had a molar extracted without the slightest pain. He followed this by the successful administration of the gas in several cases of extraction, and other dentists of Hartford administered it with like success. In December 1844 he went to Boston to lay his discovery before the medical profession of that city. At the Medical School Dr. Warren, the veteran of the faculty, allowed him to perform before his medical class. Wells was not too well prepared and did not fully succeed in producing unconsciousness in his patient, a Harvard student who had volunteered. When he began to extract a tooth from the patient, he was rewarded by a yell. "Humbug, a swindle," shouted the onlookers and the unfortunate dentist was mobbed. Discouraged, Wells returned to his Hartford practice, but never recovered from the blow. The wound was further aggravated when Wells heard of Morton's success.

11

William Thomas Green Morton was born in 1819 in Charlton, Massachusetts, on a farm of Worcester County, like so many other men who have left their stamp on the development of America. He went to Baltimore in 1840 to study dentistry, later established himself in Boston as a dentist, and soon built up an extensive practice, for a while in partnership with Wells. Morton was an expert technician, who specialized in crowning teeth. He felt that his work could be more successful if he could devise means to lessen pain, but found that his limited scientific knowledge interfered with his ambitions. He entered his name as a medical student in 1844 and stayed for a while at the house of C. T. Jackson with its chemical laboratory; here Jackson told him of the properties of sulphuric ether. Morton now spent his time experimenting with ether, partly on himself and partly on animals, until, in 1846, he was able to extract a tooth painlessly from a patient after the administration of ether. This occurred on September 30, 1846. Morton now applied for a patent, together with Jackson, whose name he deemed necessary for the success of his discovery. After all, Jackson was a well-known scientist, and Morton only a technician. His success in the extraction of teeth led Morton to the serious consideration of the use of ether for major surgical operations. As had Wells two years before, Morton was asked by Dr. Warren to demonstrate his discovery, now on a man with a vascular tumor on the side of his neck. The demonstration took place October 16, 1846, and is one of the great dramatic moments in the history of science. It has often been recounted; we repeat it here in the words of Dr. A. A. Gould, who was present. Morton was delayed somewhat by a last-minute readjustment of the inhaling apparatus and arrived late in the operating room, where Warren and several of his colleagues and students were waiting.

Previous to the operation, Dr. Warren, having waited ten or fifteen minutes, again turned to those present, and said: "As Dr. Morton has not arrived, I presume he is otherwise engaged"; apparently conveying the idea that Dr. Morton did not intend to appear. The remark of Dr. Warren brought out a great laugh. Dr. Warren then sat down to his patient. Just as he raised his knife Dr. Morton appeared.

The first glance of Dr. Morton, on entering the side-door of the amphitheatre, was upon his patient, the next was upon the faces of the expectant crowd. But here was no pleasant or assuring picture; each betokened either common curiosity or plain incredulity.

From the confused state which his novel position had thrown him into, he was suddenly recalled by the clear, abrupt tones of Dr. Warren, who turning first to the patient and then to himself, said, "Well, sir! your patient is ready."

With a slight apology for having detained him, and a statement of his excuse that he had been compelled to wait for the completion of his instrument, Dr. Morton stepped to the bedside of the patient. Taking the man by the hand he spoke a few encouraging words to him, assuring him that he would partially relieve, if he did not entirely prevent, all pain during the operation, and pointing to Mr. Frost, told him there was a man who had taken it and could testify to its success. "Are you afraid?" he asked. "No!" replied the man: "I feel confident, and will do precisely as you tell me."

Having adjusted his apparatus, he commenced the administration, overwhelmed by a painful sense of responsibility and fear, lest a present failure should bring his long-cherished plans into contempt and perpetual disrepute. Under the first influence of the agent, the man became flushed and exhilarated, but soon its more powerful effects became manifest, and in four or five minutes he lay as quietly and soundly asleep as any child. . . .

The confused hum, or scraping of feet made by the audience in their endeavor to obtain a better view, had been succeeded by the most breathless silence, and as Dr. Morton turned to announce to Dr. Warren that his patient was ready, he noticed that the looks of incredulity and contempt had given place to an expression of astonishment and settled interest.

As Dr. Warren made the first incision through the skin, the patient made no sound nor moved one muscle of his body; as the operation progressed, all eyes were riveted on this novel scene in eager expectancy and amazement. At length the operation was finished, and the blood having been washed from his face, the patient was gradually allowed to come from his anesthetic state. When fully restored to consciousness and able to answer questions, he gave the triumphant and gratifying intelligence: "I have experienced no pain, but only a sensation like that of scraping the part with a blunt instrument." This arose from the fact that as the operation had taken longer than anticipated, Dr. Morton had several times removed the inhaling instrument from his mouth.

With the patient still lying like a log upon the table, Dr. Warren turned to the audience and said slowly and emphatically: "Gentlemen! this is no humbug."

Indeed it was no humbug. It is not easy to place ourselves in the period before Morton, when operations in the living flesh had to be performed with the patient fully conscious. Not easy—unless we think of the wars in China and other countries, where even nowadays operations have to be performed without anesthetic, sometimes for no deeper reason than because political machinations prevent the sending of medicine. But our ancestors all lived under the terrible shadow of unavoidable pain. The extraction of an infected tooth was a serious and grueling experience, decided upon by many people only as an ultimate resort; removal of a limb was, by our standards, a cruel piece of utter barbarism. Sometimes some liquor or opium was given, but often several strong men were needed to prevent the patient from spasmodic cringing when the surgeon applied the knife. The surgeon himself necessarily became hardened to the agonies he had to inflict, and repeated with Celsus, the Roman physician, that "resolved to heal the sufferer entrusted to his care, the surgeon must ignore cries and pleadings, and do his work regardless of complaints." To the physicians as well as to the laymen, Morton's discovery came as a gift from heaven, though at first not all were willing to recognize it.

Oliver Wendell Holmes, who like all members of the faculty of the Harvard Medical School was deeply impressed by Morton's discovery, told his class that "the fierce extremity of suffering has been steeped in the waters of forgetfulness, and the deepest furrow in the knotted brow of agony has been smoothed forever." In a letter to Morton he suggested for the new process the name of "anesthesia," which, like the term "technology" coined by his colleague Bigelow, has found general acceptance.

Mankind had indeed made a step forward, but the conquest of pain brought nothing but agony to its discoverers. The Mexican War of 1846 was the first armed conflict in which ether was used for surgical operations, but the great triumph of Morton's discovery also brought about the invalidation of his patent. Wells and Jackson had been aloof during the trying period of experimentation, which ended October 16, 1846; they now began to press for recognition, and the methods of Jackson were far from fair. Several Boston dentists organized a committee formally protesting against anesthesia, claiming that in many instances it had produced unfortunate effects. Another influential dentist tried to ridicule "Morton's sucking bottles." Here and there cries went up from the pulpit repeating the accusations which had greeted Boylston and Waterhouse in decades past, proclaiming that pain was ordained by Providence. And this was only the beginning of the trouble.

We cannot follow here the tortuous and painful path by which Morton sought recognition and some financial reward for his discovery. He was thwarted by Jackson, by Wells, by a public torn between recognition of the statements by a well-known academician and of a man who was "only" a dentist; by a legislature subjected to all kinds of political pressures and by a President of the republic, Franklin Pierce, who could never arrive at any clear conclusion. Morton's patent was invalidated, he was persecuted by creditors, his farm and furniture near Wellesley were sold by public auction, and he himself was hanged in effigy by his fellow townsmen. Capitalism has seldom been kind to the benefactors of man, but Morton was rewarded with even more than the average amount of sorrow. Through all these adversities the staff of the Harvard Medical School, Warren, Bigelow, stood by the great experimentor and supported his claims. His death came suddenly, in Central Park, New York, in 1860, after he had read a new attack upon his claims by Jackson.

The other claimants for the title of "benefactor of mankind" (a title bestowed upon Jackson and Morton in 1850 by the Paris Academy) came to an equally sad end. Wells experimented with chloroform, became an addict, got into a street brawl and ended his own life in a New York prison as early as 1848. Jackson continued to fight Morton's claims even after his adversary had perished, but at last he also paid the price. He died in the McLean Asylum, near Boston, in 1880. Even Crawford Long, the Georgia practitioner, did not entirely escape the doom bestowed upon the discoverers of anesthesia. He was ruined in the Civil War and died a disillusioned and discouraged man.

In the Boston Public Garden stands a simple monument "to commemorate the discovery that the inhaling of ether causes insensibility to pain, first proved to the world at the Massachusetts General Hospital in Boston, October A.D. MDCCCXLVI." In 1868, when Thomas Lee presented the monument to the city, the controversy was still raging and the donor found it wise to mention no names except the name of the Lord—the only safe one. Hartford has its Wells statue in Bushnell Park, and Georgia has honored Long in the National Capitol. The only places where Morton is remembered by name are his birthplace Charlton and Wellesley near Boston, where he lived for a while. The Wellesley memorial tablet stands, delicately enough, not far from the place where he was hanged in effigy.

12

The influence of Howe in the field of the care of the blind, and of Dorothea Dix in her campaign in behalf of the insane, backed by such men as George B. Emerson, Horace Mann, William Ellery Channing and the public opinion they represented, extended much farther than the scope of their immediate activities. They had connections in the medical world and the colleges, in newspaper offices, in the legislature, in the political parties and the labor movement. In this period of rapidly expanding economy, progressive work could find mass support against the powers of the past, and the revolution of the economic and technical system made many people aware of the necessity of changes in the field of human relations. A large amount of scientific-humanitarian work was accomplished in these years, and its influence extended into many directions of science and even literature. One of the fields in which this influence has been most felt was that of public health and vital statistics. The key figures here were Jarvis and Shattuck.

Dr. Edward Jarvis was born in Concord, Massachusetts, where he later taught school. He studied medicine with the kind and scholarly George Cheyne Shattuck, the Dartmouth friend of Nathan Smith, and a popular Boston physician. Dr. Shattuck was well-to-do, and spent his money well; he supported medical research and had a large practice among the poor in Boston's West End, where fees were rare. In 1830 Jarvis obtained his Harvard M.D. After having practiced in several places, including his native town, he established himself in Dorchester, near Boston, where he stayed until his death in 1884. He was primarily interested in the socially and mentally crippled. When Howe appealed for help in behalf of idiotic children—who could not be treated at the Perkins Institute—Jarvis was one of the men who established the Massachusetts School for Idiotic and Feeble Minded Children. His treatment of insane persons revealed sympathetic insight into the nature of their troubles and he took a great and active interest in the campaigns of Dorothea Dix. Bcause of his scientific training, Jarvis was better equipped than many other reformers to see the relations between the new discoveries in medical science, physiology, dietetics and the social requirements of his age. He embodied his ideas in many papers, but also in books; his *Practical physiology for the use of schools and families,* which appeared in 1847, was widely read; Jarvis also published shortly after a special edition for school children. It served to popularize the conception of "public" health, which Jarvis and his associates, above all Lemuel Shattuck, were trying to bring to the general American public.

Lemuel Shattuck—not a direct relative of Jarvis's teacher—began his career as a schoolteacher. From 1823 to 1833 he lived in Concord, where he made a living as a publisher and bookseller, still devoting time to teaching. He participated in the struggle to improve the public schools and in 1830 prepared a report on the Concord schools, the first annual school report in Massachusetts.

Subsequently, Shattuck moved to Boston, to continue his work as a publisher and bookseller. Here his interests turned more and more towards the improvement of sanitary conditions, and to the social importance of correct vital statistics. It is said that his understanding of the need of better vital records arose from his work on a history of Concord, published in 1835, and from his search for the genealogical records of the Shattuck family. He soon saw that what had started out as a personal hobby had much wider implications. He learned that Jarvis had begun to realize the importance of improved vital statistics during his fight against disease and intemperance. Both men understood that in a period of rapid city growth personal health problems became increasingly a matter of public concern, and that public health campaigns were impossible without accurate vital records. Shattuck and Jarvis, supported by other outstanding men in the Boston community such as John Collins Warren and George Cheyne Shattuck, began a campaign for the improvement of public health. It implied a campaign for better statistics.

The work of Jarvis and Shattuck established a connection between two entirely different fields, the humanitarian-scientific care of the crippled, and the analysis of statistical records. Yet the connection is clear. Whenever some campaign had to be organized for a noble cause, the question arose: How serious is the evil that has to be fought? The answer was especially urgent when legislative action was required, since in such cases unwilling solons, assisted by newspaper scribes, kept asserting that no problem existed at all. Vital statistics were necessary to find out how many persons in New England were blind, were deaf and mute, were insane. How many persons were victims of epidemics or other special sicknesses? What could be said, in a quantitative way, of the influence of alcoholism, of slum conditions, of factory work, on public health?

Statistical problems arose from the very nature of a rapidly industrialized society. Old cities were expanding, and many new ones were arising. With the advent of thousands upon thousands of immigrants the expansion took turbulent aspects. Many of these immigrants were poor and destitute, many were ignorant of city life. America was not kind to these people. True, it offered them employment and living—something they were often refused in their native country—but little more. These immigrants, segregated in factory towns and the poorest districts of the large towns, had to make the best of an extremely unfavorable situation. America witnessed the birth of slums as one of the results of modern capitalism, and with slums came disease. Disease had a way of spreading to the quarters of town with less poverty, or even to the districts of the prosperous. The new immigration, which brought millions of dollars to the industrialists, the railroad promoters and the real estate speculators, also brought in its wake grave problems of public sanitation. People became painfully aware of the fact that there is "public" as well as "private" health.

The new immigration only aggravated the problems which any industrialized society must meet sooner or later, and gave the reformers a better chance for a public hearing. It was during this period that the provision of Boston's drinking water was being solved. It was the period of improved hospitalization, of improved public schools, and the beginning of factory inspection. As a seaport Boston had always been subject to epidemics throughout the colonial periods and later there were large outbreaks of smallpox and yellow fever. And now, in 1831–1832, came the first epidemic of Asiatic cholera, followed by another in 1848. Shattuck found with alarm that the active causes of disease and death were increasing, and that the average duration of life was not as great as it had been forty or fifty years before.

13

In previous decades scant attention had been paid to vital statistics. Beginning in 1790, a federal census was held every ten years, so that some records had been produced from time to time, but these were not very informative. There were lists of local taxpayers, the parishes registered faithfully the births, marriages and deaths among the parishioners, and a beginning had been made of mortality tables in connection with insurance or disease. It was far from easy to find the vital information spread over these scattered and unrelated records, and many population data of interest to humanitarian campaigns were missing. Some records were decidedly erroneous; Jarvis found that in several towns the census of 1840 listed as insane more Negroes than there were local colored residents. Historians and natural scientists were also looking for statistical data. The statisticians found curious additional support in the genealogists, who began to appear through the efforts of old Yankee families to erect barriers between the older and the newer immigrants.

As in so many other cases, there were European examples, which could be studied and followed. Many states, municipalities and institutions had already established good statistical records. The mathematical theory of statistics had been and was being developed by Laplace, Poisson and Quetelet; handbooks of statistics were appearing; the Statistical Society of London was founded in 1834. Young physicians came back from the schools of Paris reporting on the new ways in which French medicine was using the analysis of case records in the investigation of diseases and their cures. And above all, from the point of view of the reformers, there was the work done in England by Edwin Chadwick, who started as Jeremy Bentham's assistant and as a member and secretary of the Poor Law Commission of 1832 carried through much important work on public health.

Before the days of Jarvis and Shattuck, vital statistics as well as economic statistics had not been entirely neglected in

America. Already in the eighteenth century, Benjamin Gale in Killingworth and Edward Wigglesworth at Harvard had worked on mortality tables. Timothy Dwight had been the moving force behind the attempts of the Connecticut Academy to obtain some statistical information on the different towns in the state. The only printed result of this action was Dwight's own account of New Haven, published in 1811. Another Connecticut man, however, had more time than Dwight to take up statistics seriously. This was Timothy Pitkin, a successful conservative politician from Farmington, active both in Congress and in the Connecticut legislature. While in Congress, he wrote *A statistical view of the commerce of the United States of America* a substantial book published at Hartford in 1816; it was for a long time an authoritative source book, thickly interspersed with tables on American commerce and statistics. A new edition appeared in 1835 with a handsome defense of Federalist and Whig economic policy, which is still of use to economic historians.

Some registration of vital records was effected by occasional communities; one of the oldest was that of Boston, which began in 1811. They were very incomplete; data on births in Boston between 1744 and 1849 are virtually lacking. A true understanding of the importance of systematic registration did not occur until the 1830's. Then, mainly through the activity of Jarvis and Shattuck, the general problem of statistics and the special ones of vital statistics began to be tackled more systematically. One of the results of this new interest was the foundation in 1833 of the American Statistical Association. George C. Shattuck, as well as Jarvis, was chairman for many years. Lemuel Shattuck now devoted his energies to the improvement of population statistics. In 1841 he published a study of the vital statistics of Boston, which was reprinted in 1893, in which he tried to bring some order in the remaining records of births, deaths and marriages. The bearing of such work on the struggle for social emancipation is shown by Shattuck's compilation of the mortality rates for Negroes, which was from two to three times higher than that of whites. Shattuck also extended his work to the vital statistics of the United States as a whole. As chairman of a legislative committee, appointed by the Massachusetts Legislature, he was the driving force for an improved registration system, which was accepted in 1842. In 1845 he organized a Boston city census with such care that he was called to Washington to assist in preparing the census of 1850. In 1860 Jarvis prepared the census report on mortality statistics, which was influential in establishing modern standards in this field.

In 1846 the first systematic study of the population trends of Massachusetts appeared. The author was Jesse Chickering. His book was based on the colonial census figures of 1764–1765 and the six United States census figures of 1790–1840; it listed comparative population statistics as well as data for agriculture and manufacture. Chickering tried to obtain the center of the population of Massachusetts by establishing a north-south and an east-west line of demarcation between two parts of the Commonwealth with equal population, measured according to towns. He found it to have been near Weston in 1790, near Waltham in 1810, near Watertown in 1830, and in 1840 in Brighton, only about two miles west of the Statehouse, where it had also been in 1765. Thus it was constantly moving towards Boston.

> Boston [he concluded from this fact and others] represents Massachusetts more fully than Paris does France, or London, England, and in a far more emphatic sense than New York City, Cincinnati or any other principal city in the Union, the states to which they respectively belong.

Chickering also studied the movement of the colored population extensively, and did not mince words on "the insulated and degraded position of the colored among the whites." He concluded that the increase of the colored population in the Commonwealth was not likely to keep pace with that of the whites.

It was in this way that under the influence of the humanitarian and democratic movements of the thirties and forties, vital statistics were modernized in New England.

14

Shattuck's main contribution to social progress was his

monumental report on public health. Boston had already established a Board of Health in 1799 to deal with such questions as burials and the spreading of contagious diseases, and some other New England towns had followed its example. But since that time, sections of the different public sanitation laws had been partly repealed, and this in a society in which the demand for some form of health control was becoming more and more urgent. The first improvement was effected in 1838 by a number of medical men, and from this time on the agitation for satisfactory public health regulations did not abate. In 1848 the Massachusetts Legislature passed an act establishing a committee which was to prepare a sanitary survey, with recommendations for action. The result was a report published in 1850 and written by Lemuel Shattuck. In breadth of scope, clarity of vision and wealth of recommendation it can be considered as a classic in its field.

Shattuck proclaimed himself the apostle of preventive medicine. He quoted from Jarvis:

> Our education has made our calling exclusively a curative, and not a conservative one, and the business of our responsible lives has confined us to it. . . . But with health, with fullness of unalloyed, unimpaired life, we, professionally, have nothing to do.

He proclaimed, therefore, that the old-fashioned medical outlook has to be replaced by a radically different type of thinking. His criticism of the average physician's point of view was as strong as was that of Holmes on the subject of puerperal fever, and his language was as outspoken as that of William Lloyd Garrison's appeal against slavery in the *Liberator:*

> We believe that the conditions of perfect health, either public or personal, are seldom or never attained, though attainable;—that the average length of human life may be very much extended, and its physical power greatly augmented—that in every year, within this Commonwealth, thousands of lives are lost which might have been saved; —that tens of thousand of cases of sickness occur, which

might have been prevented;—that a vast amount of unnecessarily impaired health, and physical debility exists among those not actually confined by sickness;—that these preventable evils require an enormous expenditure and loss of money, and inpose upon the people unnumbered and immeasurable calamities, pecuniary, social, physical, mental, and moral, which might be avoided;—that means exist, within our reach, for their mitigation or removal;—and that measures for prevention will effect infinitely more, than remedies for the cure of disease.

The report opened with a long historical introduction to public health regulations abroad and in America. It reviewed and analyzed the causes of the frequent epidemics and concluded with a list of specific recommendations. He stressed the need for better supervision of medical practitioners. "Anyone," he asserted, "male or female, learned or ignorant, honest person or knave, could assume the name of physician and establish a practice to cure or to kill, as either may happen, without accountability. It's a free country!"

Among Shattuck's recommendations were the foundation of a General Board of Health, assisted by local boards, with sufficient authority to enforce their regulations; furthermore, an improved system of laws regulating public registration of births, marriages and deaths. He made a strong plea for special sanitary surveys, and himself published such surveys of Lawrence, Attleboro, Plympton and Lynn; he advocated garden cemeteries, public bathing houses, and better methods for the collection of refuse and the disposal of sewage—for which he recommended utilization as fertilizer. He asked for protection against smoke, adulterated food and the uncontrolled sale of patent medicines. Hardly any aspect of public health escaped his attention; not only did he recommend the training of nurses and education in sanitary science, he even suggested libraries for spreading sanitary information. Shattuck, doubtless, was thinking of such books as Jarvis had just been writing.

Shattuck anticipated opposition to his sweeping condemnation of the "free enterprise" philosophy with regard to public health. His defense of his position is an eloquent and passionate part of the report. "The progress of the age," he said, demands this new form of health activity:

> The wonders of the steam engine, besides giving us a new and most important stationary mechanical power, have revolutionized our systems and habits of locomotion, by sea and by land. A journey from Boston to New York, which formerly required days for its performance, is now accomplished within a few hours. A voyage to England, once always of uncertain duration, and frequently requiring months for its performance, is now made in ten days. One month only, instead of six, is consumed in a voyage to the Pacific coast. Events which have taken place in the East Indies have been known here within a month afterwards!—Who would have imagined, a few years since, that a commercial order could be sent from Boston to New York, that order executed, and the answer received in Boston, and the whole occupying but *ten minutes!*

In such an age, Shattuck continued, public health is not only practical, since it increases vitality and usefulness, since it benefits the people and aids the physician; but also economical, since the proposed annual costs were only $3000, while $7,000,000 was the loss to be reckoned in unnecessary sickness alone. The proposed measures would be eminently moral, charitable and philanthropic. Since a dirty neighborhood turns even a clean and healthy countrywoman into a sickly slattern, it is the sacred obligation of the community to prevent such a willful neglect of human rights. "God Himself wants it!" he explained in the spirit of the true Crusader.

Shattuck appealed to the conscience of his fellow citizens: in Boston, from 1840 to 1845, 46.62 per cent of all the deaths were of persons under 5 years of age; in some classes of the population it rose to 62 per cent. The average age of all who died was only 21.43 years; of the Catholics who died, it was 13.43 years. The implication was clear, since most immigrants were Irish and Catholic, and the older settlers were Protestant. Statistical work could thus be used to convince the incredulous. "Every person," concluded Shattuck, "has a direct or indirect interest in every other person. We are social beings—bound together by indissoluble ties."

Shattuck's report, one of the most mature documents to emerge from this remarkable period in the history of New England, was accepted—and duly shelved. This plea for comprehensive government interference in what most people believed to be private affairs of the individual was too radical for Shattuck's generation. For twenty years the report gathered dust in the state archives. Shattuck's recommendations were not taken up until after the Civil War. By this time Shattuck was no longer alive. In 1869 Massachusetts at last established a Board of Public Health. Many of Shattuck's recommendations were only gradually put into practice and some even nowadays await their full application.

Among the great men of Concord Jarvis and Shattuck are all too often forgotten. Their memory deserves a better fate.

Chapter 9

Steam, Water, Electricity

Lorsque nous voulons citer un example d'application pratique des grandes découvertes utiles au progrès de l'industrie, et par conséquent à la prospérité des peuples, nous sommes obligés de tourner nos regards vers les Etats-Unis d' Amérique. La Providence semble y avoir placé la grande école pratique du monde pour faire fructifier toutes les découvertes modernes. C'est qu'il fallait pour cela une terre vierge, des hommes actifs, entreprenans et libres; l'Amérique du Nord présentait seule la réunion nécessaire de toutes ces conditions.

—Guillaume-Tell Poussin, 1836

1

In the late twenties the steam engine was still a novelty in New England. The Yankees were becoming familiar with it on the steamboats, which began to appear along the coast and sometimes on the rivers, but which by no means replaced the sailing vessels. The day of the clipper ships, with their magnificent display of sails, was yet to come. An occasional mill or shipyard may have boasted a steam engine, but such cases were rare. The rapidly expanding textile industry depended for power almost entirely on water. The smokestack did not yet dominate the skyline of the new industrial settlements at Manchester, Dover, Holyoke, Lowell and Pawtucket. When in 1820 Daniel Treadwell invented a power printing press, he could not find a single steam engine in Boston and had to work his press by horsepower. Samuel Slater began to experiment with a steam-driven textile mill in Providence in 1828, but this venture remained an exception for some time. The fact that until well into the forties mills operating with steam were specifically designated by the name of "steam mills" indicates that for many years the use of steam was exceptional rather than common. The textile concern incorporated in Salem in 1839 and closed in 1953 illustrated this attitude in its name—Naumkeag Steam Cotton Company.

Steam engines were in this period widely used in many sections of England and on the continent of Europe; they were not even uncommon in parts of America different from New England. The experiments of John Fitch and others with steamboats date back to the early days of the republic. Nicholas Roosevelt built steam engines of the old Newcomen type in his New Jersey foundry from 1793 on. Oliver Evans, the versatile mechanical genius of Wilmington and Philadelphia, who had been busy inventing all kinds of improvements in milling ever since the Revolution, had sent to England in 1787 and 1794–95 drawings of a steam engine which was even constructed on a high-pressure principle. But the six steam engines in this country, operating in New York and Philadelphia, with one in Boston "employed in some manufacturing," which were reported in 1803, were probably old-fashioned Newcomen engines. The lag in development between America and England is apparent from the fact that the Boulton and Watt engines, based on James Watt's inventions, dating from the years between 1770 and 1785, had been in extensive use for many years before they were imported to America. Robert Fulton was one of the first, if not the very first, to introduce them into America for use on his steamboat. By 1815 George Stephenson had locomotives running at the Killingworth colliery near Newcastle-on-Tyne, to pull freight cars, but for many years no American made an attempt to emulate him, notwithstanding certain proposals by Oliver Evans; and Evans died in 1819. Capital could perhaps have been diverted from turnpike and canal building for this purpose, but there was no adequate technical skill to cope with these problems. America was still a country of craftsmen, not of engineers. When the factory system was introduced water power was plentiful, and steam engineering remained for many years to many an esoteric art, despite Evans' "Abortion of the Young Steam Engineer's Guide" of 1805, even published in 1821 in a French translation. At the time of Evans' death, 50 steam engines were in use, including those at the Fairmount waterworks in Philadelphia built on the Evans high-pressure (200 pounds) principle.

2

In the early twenties, English and French engineering textbooks, such as the standard books of John Farey and of Thomas Tredgold, began to appear in a steady stream. There were men in New England, at West Point and in other parts of America, who bought and studied these books. Such a man was Loammi Baldwin, the canal and dry dock builder. In a corner of the library of the Massachusetts Institute of Technology we still can study books from his library, with careful descriptions of canals and steam engines, illustrated by fascinating plates. Some men went to Europe to inspect whatever machines and experiments they could see; in England they found that diemaker and engraver from Newburyport, James Perkins, now an honored expert on high-pressure steam engines. Among the New Englanders who went to England and the Continent was a Providence man, Zachariah Allen, who after his return incorporated some of his findings in a book, *The science of mechanics.*

Allen, the son of a pioneer in calico printing and a graduate of Brown in 1813, was still another of those jacks-of-all-trades that the Yankee soil bred in those days, a student of law and the sciences, a businessman and an inventor. By 1820 he was already a prominent man, deeply interested in the textile industry. After a tour through England and France to inspect their new inventions he wrote a book about these new achievements, which appeared in 1829 with a "doubleacting steam engine" as its frontispiece. He was not the first New Englander to write about steam engines, since Bigelow's *Elements of technology* appeared before *The science of mechanics* and contained a chapter on this subject, but where Bigelow was the academician, Allen's approach was primarily that of a practical man. "The steam engine," he wrote, "has become almost as important as the plough. Instead of indulging in the abstract and profitless investigations, once so common, philosophers seem now rather to delight in researches into nature for the purpose of making discoveries applicable to the useful arts."

The trouble with America, as Allen saw it, was that there were too few people who knew how to handle the new machine, despite the fact that "the high price of labor has had the tendency to direct the genius of the people to all kinds of mechanical inventions, from the simple apparatus for paring an apple, to the machinery for propelling a vessel of war." But the management of mills and machinery was often in the hands of people without the proper practical knowledge. Allen's description of the men of good will and poor training of the America of the 1820's again reminds us of the early years of the Soviet Union, a hundred years later:

> At some of the American mills, which have been erected only a few years, various kinds of machinery, abandoned after a course of unsuccessful experiments, may be found collected as rubbish, forming a sort of museum of injudicious and abortive contrivances.

Allen's book was eminently practical, and contained information the American public was very much in need of, notably on steam and hydraulics. He lived to see many of his ideas accepted, and played an important role in the development of Rhode Island industries. He is also remembered as an early promoter of mutual fire insurance.

3

It is sometimes forgotten that railroads were in existence and an object of wide discussion long before the steam locomotive was introduced. It had occurred to many people that the use of cars on rails facilitated traffic, especially when horses were used as motive power. Such tramways, as they were called, were long in use in English "collieries." The Bostonians used such a tramway perhaps as early as 1795, but certainly about 1804 to overcome the difficulties of reducing their three hills—Beacon, Pemberton and Vernon. The Mount Vernon Proprietors laid out an inclined road on which to slide dirt cars, which emptied their loads into the Back Bay at the present site of Charles Street. In the following decades other tramways were built in different parts of the country.

The most famous of these projects was the Quincy Railroad, south of Boston, built in 1826. It was the idea of Gridley Bryant, a master builder of Boston and a self-made engineer, who followed closely the technical developments in Great Britain. Bryant had granite quarries in the Blue Hills, and had undertaken to supply the construction of Bunker Hill Monument with material from these quarries. The tramway was three miles long and led from the works to the Neponset River. At first the cars were drawn by gravity—the loaded cars being used to pull up the empty ones—later horses and oxen were used. Considerable engineering thought was used in the construction.

> The road [explains a contemporary description] is constructed in the most substantial manner. It rests on a foundation of stone, laid so deep in the ground as to be beyond reach of the frost . . . the rails are laid on stones eight feet in length . . . at a distance six or eight feet from each other . . . the rails are of pine timber, on the top of which is placed a bar of iron. . . . The carriage wheels are of a size considerably larger than a common cart wheel. . . .

Thomas Handasyd Perkins, the Boston merchant after whom the Perkins Institution is named, paid the bills and took the stock—like Bryant he saw big things in the making.

The construction of this tramway coincided in time with two major events: the opening of the Manchester and Liverpool Railroad in England, where steam locomotives were used, and the opening of the Erie Canal in New York State, connecting the Hudson with the big lakes and opening the upstate territory for through water traffic to New York City. These great undertakings both intrigued and worried the businessmen of Boston. Was the canal going to divert all western commerce to New York? Or was the canal destined to be outmoded and replaced by the steam railroad? A sharp dilemma arose: should the Bostonians emulate that outstanding New Yorker DeWitt Clinton and build a canal to the west, or should they break with tradition and build a railway? The canal enthusiasts had one great difficulty: they had to produce sufficient capital and engineering skill to break through the wall of the Berkshire hills, a hundred and thirty miles west of Boston. Loammi Baldwin was consulted and made his survey of 1825. As we know, nothing came of these plans, because the railway promoters won the battle; we have already mentioned that the tunnel was later realized as part of the Boston railroad system.

The debate on canals versus railroads enlivened the Commonwealth for several years, and some of the arguments today make entertaining literature. "Is old Massachusetts in her palsied dotage?" asked one of the many Bostonians worried by the opening of the Erie Canal, "is her sun of prosperity . . . setting, to rise no more? This sun is irradiating the hills of Hudson and fertile vales of New York." Another Bostonian observed that in the years from 1790 to 1820 Massachusetts had increased its population by only 30 per cent, while the whole nation had increased by 150 per cent. This debate also had its angle of class conflict, since the railroad promoters came mainly from those groups representing merchant-industrialist interests, and their opponents consisted mostly of landed country proprietors who feared further expansion of the city capitalists, and were backed by turnpike, stagecoach and canal enterpreneurs; New England's growing isolation from the expanding United States did not mean too much to those people when their immediate interests were menaced. One of the chief promoters of the railroad schemes was the *Boston Daily Advertiser,* a Whig paper, whose editor, Nathan Hale, was one of the most enthusiastic and intelligent organizers of the pro-railroad propaganda.

Hale was the nephew of the man who regretted that he had only one life to give to his country. He was a Williams College graduate with a taste for mathematics, which he taught at Exeter Academy; he also had a taste for civil engineering. He had a lathe in his house, as well as toy locomotives, propelled by whalebone springs, and he studied the granite railway in Quincy with an engineer's love.

> I cannot remember the time when we did not have a model railway in the house [wrote his son Edward Everett

> Hale in his reminiscent *New England Boyhood*], in earlier years it was in the parlor, so that he might explain to visitors what was meant by a car running on rails. I can still see the sad, incredulous look, which I understood then as well as I should now, with which some intelligent person listened kindly, and only in manner implied that it was a pity that so intelligent a man as he should go crazy.

To such a man the famous Rainhill experiment of 1829 was what he had been looking and longing for, and the *Daily Advertiser* gave it wide and favorable publicity.

In this Rainhill experiment, held near Liverpool in England, Stephenson's *Rocket* averaged a speed of fourteen miles an hour, with a maximum speed of twenty-nine. The *Rocket* was a locomotive engine with a multitubular boiler and a blast pipe; a direct connection existed between the steam cylinders, one axle and one pair of wheels. Stephenson's triumph was also a triumph for the steam railroad, and the expansion of the railway system over the world dates from this experiment. In Boston, Hale was now no longer a strange visionary, and the Erie Canal began to lose some of its magic spell. Experiments with steam locomotives, which had already been undertaken outside of New England, now began to be taken seriously. The textile mill owners in Lowell were in the vanguard of the railroad promoters, notably Patrick T. Jackson; in 1830 they obtained a charter for the Boston and Lowell Railroad; in the same year two other charters were granted, one for a line to Providence, another for a line to Worcester, the first in the general direction of New York, the second toward the west. Opposition to horsepower died out, and an order was placed for a Stephenson locomotive. It arrived in 1832 and was subjected to a careful study in the Lowell textile machine shops.

The opening of railroads with steam locomotives in South Carolina, Baltimore and elsewhere in America gave further proof that these ventures were really practicable. All three New England lines were opened in 1834 and 1835 with due ceremonial. The public was impressed:

> Crowds of people assembled at the Tremont Street terminus of the Worcester railroad to witness the operation of the Locomotive Engine. It was the first time we ever saw one in motion, and we candidly confess that we cannot describe the singular sensation we experienced, except by comparing it with that which one feels when anticipation is fulfilled and hope realized. We noted it as marking the accomplishment of one of the mighty projects of the age, and the mind, casting its eye back upon the past, as it was borne irresistibly onward, lost itself in contemplation of the probable future. (*Boston Evening Transcript,* March 27, 1839).

After the opening of the three railroads radiating from Boston, the new mode of transportation began to be gradually accepted. A construction program was efficiently carried out with financial support from the legislature—which had not been granted to the earlier turnpike and canal builders. By 1842 the Boston and Worcester Railroad was continued through the Berkshire Hills as far as Albany on the Hudson—connecting at last the New England capital with the region served by the Erie Canal. By this time the Massachusetts Legislature had granted state aid to the extent of $5,555,000, a juicy stake for political intrigue in this era of unbridled economic expansion. In the building of the railroad system imported Irish laborers were employed almost without exception.[1] All these factors represented a profound change in the political and social structure of New England.

The public gradually came to accept the steam railroad. In 1830 the Proprietors of the Middlesex Canal had still

[1] They [the Irish] are, as a class, industrious and hard-working. Let the unprejudiced look to our public improvements, the rapidly expanding confines of our city, our new wharves, our bridges and railroads. Who built them? Yankee enterprise furnished the capital, but who supplied the labor, the indispensable muscular strength? Who dug down the hills and filled up the valleys? Who deepened the docks and extended the wharves? Who have done for us what we could not do for ourselves? Who laid the foundations of Central Wharf? Irishmen. The Western Avenue? Irishmen. Commercial Street? Irishmen. The Providence Railroad? Irishmen. The Worcester Railroad? Irishmen. The Lowell Railroad? Irishmen. (*Boston Evening Transcript,* March 20, 1835.)

protested against the construction of a railroad from Boston to Lowell with the argument that there never could be "a sufficient inducement to extend a railroad from Lowell westwardly and northwestwardly to the Connecticut." By 1840 such protests had lost their meaning. One of the main reasons for this triumph of the steam engine against misunderstanding and class feeling was the brilliant work done by American civil engineers.

4

Civil engineering was still in its infancy when at last the steam engine was introduced. There were no railroad engineers, and much pioneering had to be done in a landscape very different from England. It is astonishing to see how rapidly progress was made. The Lowell Railroad was constructed with great care and ingenuity, and was considered the best example of railroad engineering in the country despite many imperfections due to lack of engineering experience. The track was originally laid on ties of split granite and beneath each rail of the outward track a wall of stone about four feet in height was laid along the entire length of the road. The idea that granite was the correct material for tracks was soon abandoned, since the resulting firmness of the road was too much for the rolling stock, which rapidly pounded itself to pieces. In the words of a cockney coachman: "Hit ain't the 'eavy 'auling as 'urts the 'osses' 'oofs; hit's the 'ammer, 'ammer, 'ammer on the 'ard 'ighway."

When the *Chocktaw,* a 30-horsepower four-wheeled Stephenson locomotive, arrived at Boston in November 1832 from the Stephenson works at Newcastle, it was accompanied by complete railroad apparatus, including about 2000 bars of railroad iron. The technicians at the Lowell machine shop began to study the engine, which the public soon knew as the *John Bull.* A whole new industry was started. By 1835–1836 the shops already built their own locomotives, with 11-inch cylinders and 14-inch stroke, all burning wood. The early engineers on the locomotives were skilled mechanics employed by the Locks and Canal Works. American energy in railroad building aroused the admiration of foreigners who philosophized on causes and effects. Guillaume-Tell Poussin, *"ex-major du corps génie américain,"* who in 1836 published a book in Paris on American railroads, found the secret in the American motto: "Time is money." *"C'est qu'en Amérique on sait apprécier la valeur du temps: on l'employe; chez nous, on l'use."* [2]

The result was that with development of the railroads, civil engineering in the United States lost its dilettantist character, especially in New England. Despite all its turnpike, canal and dam construction, New England, in the beginning of the railroad period, could boast only one fully trained civil engineer, Loammi Baldwin, Jr., but Baldwin was fully occupied in building the great dry dock at Charlestown besides being otherwise variously engaged. Was the story of the early canal building to be repeated, and could talent come only from abroad? The fact that America was not caught off guard this time, but, on the contrary, had sufficient engineering talent not only to cope with the situation, but in a short period even to provide leadership abroad, was mainly due to the foresight of one man, a sturdy son of Braintree, Sylvanus Thayer.

Sylvanus Thayer is remembered as the grand old man of West Point. He was educated at Dartmouth and graduated from the United States Military Academy in 1808 as a second lieutenant of engineers. Under the watchful eye of Jefferson, West Point had from its beginning given some instruction in modern engineering. After the War of 1812, in which Thayer served with distinction, he was sent to Europe by President Madison to study the theory and practice of fortification. In Paris, during the eventful days of 1815 and 1816, he became acquainted with the École Polytechnique, the school famous for its great teachers, outstanding textbooks and equipment, and its many and world-renowned graduates. Of the teachers of Napoleonic days, Lagrange was dead and Monge dying, but Laplace, Poisson, Ampère, Hachette, Arago and Legendre were still active, as were many of Napoleon's engineers. Thayer understood the great importance of the work these men were doing and decided that

[2] "It is because in American people know how to appreciate the value of time: they put it to work; here, we use it."

the young and struggling republic across the Atlantic could profit by the experience of the French in engineering education. He had an opportunity to carry out his ideas after his return to America, when he was appointed superintendent of West Point, a position he held from 1817 and 1833.

Under Thayer's leadership the science and engineering departments at the Military Academy were strengthened. For many years they were among the best in the country. For several decades the engineering department had the field in America almost exclusively to itself. French educational influence was freely encouraged, not only in appointments and teaching, but throughout the whole disciplinary system. Crozet, one of Napoleon's engineers, was made head of the engineering department in 1816; another Frenchman taught drawing and French. French textbooks were freely used and partly translated. Jared Mansfield, assisted by his son-in-law Charles Davies, gave a thorough training in mathematics and mechanics, which included descriptive geometry, hydrodynamics, "pneumatics," and optics. Davies's mathematical textbooks, which appeared in regular succession for several decades, remained a fixture of American colleges for several generations; like those of Farrar they were partly or wholly based on the textbooks of the French. Another teacher was David Bates Douglass, whom we met as a canal builder, and who taught for many years civil and military engineering. And so we find that from this school, where Thayer was enthroned in "Sylvanic majesty," graduates of the army engineering corps were sent throughout the country wherever responsible engineering work was to be done. The government often granted army engineers permission to accept a supervising position in strictly civilian enterprises. The early railroad engineers were almost without exception West Point graduates.

5

The most distinguished of these West Point engineers operating in New England was undoubtedly Major Whistler. George Washington Whistler was born in Fort Wayne, Northwest Territory, then a wilderness outpost in command of his father. He graduated from West Point in 1819 and was appointed second lieutenant in the corps of artillery. After completing several assignments for the army, teaching descriptive geometry at West Point and topographical field work in the rough Middle West, he was granted permission to serve the newly founded Baltimore and Ohio Railroad Company. A trip to England in 1828–1829, in company with other army engineers, gave him the opportunity to study Stephenson's methods firsthand and even included a visit to the great man himself. After his return he did planning and construction work on the newly chartered Baltimore and Ohio, Baltimore and Susquehanna, and Patterson and Hudson River Railroads. In 1833 he came to New England as surveyor for the railroad from Providence to Stonington, planned as part of the projected Boston–New York line.

Working with Major Whistler on the Stonington Railroad was William Gibbs McNeill, another West Point graduate, who was two years his senior. Like Whistler, McNeill was an army engineer serving on some of the newly chartered railroad companies. Both men were friends and companions, they had been together on the trip to England, and McNeill's sister Anna became Whistler's second wife. Most of the engineering work in the early pioneering days of the New England railroads was accomplished under the leadership of these two men, supported by some other army engineers, such as William H. Swift. Moreover, Whistler was not only a road builder, he also became a machine constructor. Towards the end of 1833 he resigned his commission in the army and accepted the position of engineer to the Proprietors of Locks and Canals at Lowell, a company in search of men to supervise the analysis of the new Stephenson machinery which had arrived from England. It was under Whistler's directorship that the Lowell machine shops began to construct their own locomotives. Here the major remained for three years, from 1834 to 1837, after which he again found himself with McNeill and Swift engaged in railroad building, this time on the boldest piece of work yet undertaken in the United States. The Boston and Worcester Railroad at that time was busy extending its line across the Connecticut and through the Berkshires to Albany. There was little English precendent for such a railroad construction through the wild and mountainous country of western Massachusetts. It offered a host of new problems, which the major and his associates had to solve as they arose. The problems were quite varied, and included the topographical survey, the structure of the roadbed, the taking of the grades, the blasting of the rocks, the type of rails to be used on a curved incline and the construction of many types of new equipment, from locomotives to railroad tracks. One of Whistler's toughest battles was against repeated attempts by directors to reduce the cost of construction at the expense of its quality. These attempts became particularly insistent when the winding and rocky valley of the Westfield River had to be conquered. Whistler, however, knew how to handle his thrifty superiors. The story is told that, after a talk with the major, the president of the railroad declared that he would have nothing to do with a twopenny cowpath, and the line was built in the best possible way. For this purpose locomotives had to be built which could successfully and economically surmount a grade of over 80 feet to the mile, while pulling heavy trains; and they were built. American railroad engineering reached a first stage of maturity in this arduous work.

Between 1840 and 1842, while chief engineer of the railroad, the major lived at Springfield. While engaged on construction work, he was visited first by a nobleman, later by two Russian army engineers, emissaries of Czar Nicholas I. His Imperial Majesty had decided to connect St. Petersburg and Moscow by railroad; the story goes that the Emperor determined its course by the simple expedient of placing a ruler on the map between the two cities. The Russians were deeply impressed by Whistler's work, by his use of steep grades, sharp curves (600 feet radius), and of labor-saving machinery, particularly a steam excavating machine. They admired the American system of wooden bridge building, so well adapted to the needs of the country rich in timber resources. The result was that the Imperial Government invited Whistler to supervise the construction of the St. Petersburg-Moscow Railroad.

Whistler accepted the flattering invitation, which carried a salary of $12,000 a year and demonstrated dramatically the enormous strides which Yankee civil engineering had taken in less than fifty years. In 1842 he left for St. Petersburg, his family following him, and he spent the next seven years working for the Czar, supervising the construction of the railroad. The Baltimore firm of Ross Winans—who was one of his traveling companions to England—furnished the rolling stock. At Whistler's advice, the Russians adopted the wide 5-foot gauge instead of the standard continental and American 4 feet 8½ inch gauge, and this same wide gauge has remained standard in Russia and the present Soviet Union.

In Russia, Whistler preserved, in an autocratic atmosphere, the simple dignity of the republican. But the work was too much for him. The railroad was finished in 1850, but the major was one of the victims of the difficult project. The climate affected his health, and he died in 1849 at the age of forty-nine. He lies buried in his beloved Stonington, and a monument has been raised in his memory at Greenwood Cemetery, Brooklyn. He was not only a man of inventive genius and an exceptionally able engineer, but also an amateur musician and painter. James Abbott McNeill Whistler, one of his sons by his second wife, who was born in Lowell, attained fame in a field that was only an avocation of his father's. Some of his paintings have been found to reflect the treeless atmosphere of the streets of the St. Petersburg of the forties, where he spent part of his youth. Perhaps it would be juster to look at Whistler's "Mother" not as an expression of saccharine sentimentality, but as a portrait of the wife and sister of two great builders of modern America.

6

Much of the new work in railroad engineering was accomplished in the new town of Lowell, on the Merrimack River. It was not the only type of civil engineering in which this recently founded city excelled. With its mills, waterworks, machine shops and railroad, it was a general center of scientific and engineering activity. In 1796, long before the city was founded, Newburyport merchants had built on its future site the first canal across the Pawtucket falls. This canal was the property of the Proprietors of the Locks and Canals on Merrimack River, founded in 1792 and existing up to the present day. After the success of the Waltham Manufacturing

Corporation, operating with its new power looms, the city was founded in 1820 as an experiment in large-scale cotton manufacture by several Boston merchants. They bought up the land around the Merrimack falls as well as the majority of the stock of the Proprietors of the Locks and Canals, and thus secured the water power they needed. The drop of the falls was about 35 feet; at their head a dam of about 95 feet was built, which deadened the current of the river for about 18 miles. In this way a reservoir of more than 1000 acres was formed, from which the different mills could draw their power. This power was transmitted by means of water wheels, principally made of wood and often of large dimensions, up to 30 feet in diameter, with buckets 12 feet long. The manufacturing venture was successful from the beginning, and soon became famous both for the scale of its operations and for its paternalistc treatment of the factory workers. We have already mentioned that the early operators were mostly young women from New England farms, lodged in strictly supervised boardinghouses. The difficulty of obtaining a sufficient supply of labor made this type of paternalism almost imperative; it included not only material supervision but also cultural programs. The neatness of the girls and their cultural efforts were widely advertised, and many a foreign visitor was brought to Lowell to admire the way the Boston industrialists solved the human problem in modern capitalism. The Lowell system prevented some of the worst excesses of the early factory system as it existed in England. Its very strictness, however, prevented the public from being well informed on much of the misery, the exploitation and the enslavement it involved. When English and Irish workers came to the factories and began to replace the Yankee farm girls, some of the paternalism slackened, but the exploitation remained. Unions were and remained bitterly opposed by the corporations; black lists and other persecution of labor organizers were characteristic features of the system. The first "yellow dog" contract in this country that we know of dates from one of the Dover mills of the Boston corporation; it typified the current manner of dealing with workers. Lowell became the model of industrial feudalism, based on absentee ownership.

Much of the European science was cultivated under the benevolent patronage of a semifeudal monarch. Something similar occurred on American soil under the paternalistic system of the Boston manufacturers at Lowell. It tried to stimulate textile manufacture by means of the best that science and technology could offer. Talented and energetic people were invited to the new city. The result was that the town of Lowell struck visitors as a town of youth and progress. Few residents were more than thirty years old. "It was in truth a boys' town that rose on the streets carefully laid out over the old Tyler and Fletcher farms," writes Frederick W. Coburn, Lowell's historian. It is true that those who could not stand the pace disappeared, died young or returned to their more sedate inland farms. Those who remained formed a remarkable group of men and women, which included a large number of mechanics, inventors, engineers and scientists. America, as well as Europe, watched Lowell with amazement.

A typical Lowellite was Paul Moody, a weaver from Newbury; he had built his own nail-cutting machine, helped Francis C. Lowell in his work on the power loom and on many other technical improvements necessary to make the Waltham enterprise a running concern. He was appointed superintendent of the Waltham factory. The founders of Lowell took his advice in selecting the site for their new industrial Utopia, and shortly after the chartering of the Merrimack Manufacturing Company, Moody became its superintendent. In this position he continued to devote his remarkable talent to the solution of the many mechanical problems created by the growing textile industry. He died in 1831, in the full force of his life. The name of a main street both in Waltham and in Lowell keeps his memory alive.

Whatever may have been the faults of the paternalistic employers, they recognized and knew how to use technical talent. One of them, Kirk Boott, particularly distinguished himself as an autocratic superintendent, a driver of men and women and a keen discoverer of technical ability. Boott himself—he died at forty-seven years of age—had a keen understanding of practical hydraulics and of the problems of the machine shop. He came from a distinguished English family; his brother Francis collaborated with Jacob Bigelow of Boston in the study of the White Mountain flora and later became a leading London physician. Boott brought many mechanics and inventors to Lowell, if necessary employed at high salaries, some imported from near-by textile towns and from other places in New England, others from as far away as Lancashire. Among them for a while was Elias Howe, the nephew of the inventor of the Howe truss; we pointed out that the mechanical training he received at Lowell might have helped him in his later work on the sewing machine. Boott brought Whistler to Lowell. We may see his hand in the invitation extended to Warren Colburn, the educational reformer, to come to Lowell as a superintendent; Colburn accepted and until his death—he also died young—devoted much of his time to lectures on astronomy, mechanics and natural history.

The Lowell influence was felt throughout the country. After the introduction of the railroads, the Lowell Machine Shop pioneered in the construction of locomotives, and many of the early locomotive engineers were recruited from the mechanics associated with the shop. In 1834 James Tyler Ames and his brother Nathan, who were mechanics in their father's shop in Lowell, founded the big Ames Manufacturing Company in Chicopee on the Connecticut River. Samuel Batchelder, manager and later president of the Hamilton Company in Lowell, was a prolific inventor of textile machinery, whose dynamometer for the registration of power of belt-driven machinery was internationally known. Even Lowell's first mayor, Elisha Bartlett, was a man of science, who had taught medicine at the Pittsfield Medical School and in 1842 published a study of fevers. The experience of Lowell encouraged the foundation of other factory towns, notably of Lawrence on the Merrimack and of Holyoke on the Connecticut.

Early Lowell was particularly distinguished in two technical fields, industrial research in chemistry and in hydraulics. Here two figures stood out, whose work had a fundamental influence upon the entire development of the American industrial system. One was Samuel Luther Dana, the first in the country to practice industrial chemistry as a profession;

the other was James Bicheno Francis, the first to construct turbines.

Samuel Luther Dana, a Harvard graduate of the class of 1812 and a Harvard M.D. of 1818, began his professional career in Waltham. In his early years he and his brother had made a hobby of the geology of Boston, which has been discussed in a previous chapter. Through his practice as a physician in Waltham he was in regular contact with the managers of the Boston Manufacturing Company. This gave him the idea of applying chemistry to industrial research, and led him to the establishment of a chemical laboratory for the manufacture of oil of vitriol and bleaching salts. In 1826 he accepted an appointment as chemist of the Merrimack Manufacturing Company in Lowell, where he devoted his time to many questions of industrial chemistry, especially to the printing and bleaching of cotton. One of his methods, developed in 1836–1837, became known as the "American system of bleaching," and was widely used by this company. While studying the use of cow dung in calico printing, he found that its effect was due largely to the sodium phosphate it contained; after this discovery sodium phosphate was obtained from bones. This work in phosphates brought Dana to the study of manures and to the application of chemistry to agriculture—at that time a novelty in America. When Dana's *Muck Manual for Farmers* appeared in 1842, it became immediately popular. The book passed through many editions, even after better knowledge of Liebig's work in Germany had made Dana's advice somewhat antiquated. Dana's *Essay on Manures* of 1850 received a prize from the Massachusetts Agricultural Society. With these publications agricultural chemistry began to be more widely known in New England.

Dana's work extended into many other fields. He contributed to the agricultural and geological surveys of the state and to the study of lead poisoning. He found that the bright inner surface of the lead service pipes of the Lowell water system was due to corrosion resulting from the chemical action of gases in the water. In 1848 he published a paper on lead diseases; in 1851, a study of the manufacture of oil from rosin. Dana continued to work in the constant employ of the

Merrimack Manufacturing Company until his death in 1868. He was an early example of an industrial scientist, and deserves far more attention that he has yet received.

7

In 1833 James Bicheno Francis, then a lad of eighteen, had come from Oxfordshire in England to work under Whistler and McNeill on the New York to Boston railroad. He was made an assistant engineer because of some previous experience in harbor and canal construction. He followed Major Whistler to Lowell and was set to work on the Stephenson locomotives. After Whistler left, in 1837, he became the chief engineer of the Proprietors of the Locks and Canals. He was then twenty-two years of age; Lowell was not afraid of young men in responsible positions, and tried to keep them if they did well. Francis held his position for forty-eight years. Methodical and conscientious, he was for many years identified with the construction work of the company, while at the same time he found opportunity for independent research on strength of materials and on the flow of water. The results of this painstaking and sustained work were published in his *Lowell Hydraulic Experiments,* a large and handsome volume, published in 1857. It was one of the first original scientific texts written in New England. Republished and enlarged in 1868, 1871 and 1883, it remains even now a standard work of reference, and truly marks, in the words of his biographer, "an era in the engineering literature of America as well as in the growth of the profession." The profession had grown indeed. It is strikingly revealed by comparing Whistler's trip as a pupil to England in 1825 with his trip as a master to Russia in 1841, or by comparing the modest book on hydraulics, written in 1833 by Storrow, with Francis's stately tome of 1857.

The explanation of this rapid development lies in the enormous growth of America's industrial power between 1825 and 1850. Francis, in his introduction, stressed the extent to which the Merrimack industry had developed. It was estimated not long since, he wrote, that the total useful effect derived from water power in France was about 20,000 horsepower. An amount of power far exceeding this was now already derived from the Merrimack River and its branches in Massachusetts and in New Hampshire alone. What, he exclaimed, must be the amount of the population and the wealth of the Northern states, when the other rivers that water them are equally improved?

The supply of water power at Lowell had been generous to start with, and the water wheels, usually of the so-called breast wheel type, in which water was let into the bucket near the top and ejected near the bottom, were quite impressive. A vast amount of ingenuity had been spent by mechanics to improve these wheels, so that in the first thirty years no less than three hundred patents had been granted. This culminated in the perfection of water turbines, with which the name of Francis and his friend and colleague Boyden are permanently connected.

Most rules in hydraulics had to be taken originally from texts published abroad, based on experiments taken with apparatus and on a scale not comparable to standards required for the large operations at Lowell. This induced the Proprietors of the Locks and Canals to lay out funds to determine anew some of the most important rules for gauging the flow of large streams of water. It was one of the first instances on record in which a private American company appropriated a considerable sum of money for industrial research—a grant rewarded not only by the increased efficiency of the mills on the Merrimack, but also through the publication of Francis's outstanding work.

The book was divided into two parts, the first dealing with hydraulic motors, the second of more varied content. The invention of the water turbine is usually ascribed to the engineer Benoît Fourneyron, who in 1834 proposed to increase the efficiency of the ancient water wheels by letting the water enter around the entire circumference instead of upon one portion only. His way of obtaining such result was to place the axis vertically instead of horizontally, so that all the moving vanes were simultaneously agitated by the dynamic pressure of the water as it changed its direction and velocity. The Fourneyron turbine received all the interest it deserved at Lowell, and Francis described 92 experiments on improved turbine water wheels of Fourneyron's type, as well as many other experiments on other types of turbines. A special new type of turbine was eventually developed from these tests; it was used originally in the Tremont Mills at Lowell and is still named the Francis turbine after its inventor. It is an inward-flow turbine, and as such differs from the outward-flow type of Fourneyron. This means that in the Francis turbine the water enters the wheels around the entire outer circumference and passes out around the entire inner circumference; in the Fourneyron type the motion is reversed. It was soon improved, also in Lowell, by Francis's collaborator, Uriah Atherton Boyden, after whom the Boyden turbine in named.

In the second part of Francis's book we find a description of more than 160 experiments made for the purpose of determining formulas for the flow of water over weirs, with several tables for comparison with the results of other experiments, as well as experiments on the effect of backwater on the flow over weirs, on the flow over a dam in which the crest was of the same form as the Lawrence dam, on the effect of observing the depths of water on a weir at different distances from a weir, and a series of tests to determine the rule for gauging streams of water in open canals. The book ends with 101 experiments on the discharge of water through submerged orifices and diverging tubes. In later editions many more results were added.

These experiments were only part of the extensive engineering and research work which Francis accomplished. He was an authority on strength of materials and published works on such varied subjects as the preservation of timber, the deflection of beams and the evaporation of steam boilers.

He lived to be seventy-seven, and remained until the end a resident of Lowell. His work as a consulting engineer carried him into many states and into Canada; he even served a year in the state legislature. After the Civil War, he became a member of the corporation of the Massachusetts Institute of Technology. At the death of his friend Boyden, Francis was one of the trustees of a fund of $230,000 to establish an observatory "at such an elevation as to be free, so far as practicable, from the impediments to accurate observations which occur in the observatories now existing, owing to atmospheric conditions." The observatory was built by Harvard College, and stands at an altitude of more than 8000 feet near Arequipa in Peru.

8

To early Lowell belongs still another notable technical figure, Charles Storer Storrow—though he achieved fame mostly because of his work at Lawrence, the younger sister town of Lowell. Storrow had a Harvard education, graduated in 1829 and learned engineering under Loammi Baldwin. He later went abroad, studied at the École Polytechnique and at other places in France and in Great Britain, and thus became one of the first foreign-trained engineers in this country. After his return to Boston in 1832 he took employment with the Boston and Lowell Railroad, where he directed the running of the first train from Boston to Lowell and back, on that memorable day in May, 1835. He was manager of the railroad until 1845, when he joined the Boston capitalists who were planning to repeat the Lowell experiment on a promising site ten miles lower on the Merrimack. The new vassal town was to be laid out fully utilizing the experiences of Lowell. In three years, hills were leveled, valleys filled in, streets projected, buildings established and two canals and a reservoir as well as machine shops constructed. The Essex Company chartered for this purpose found in Storrow a faithful treasurer, agent and engineer. The masonry dam across the river was built under his supervision, 100 feet across, 35 feet wide at the base, and 12 feet at the top, and was bolted to a solid ledge at the river bottom. It was a respectable piece of engineering in its day, and still commands the attention of all who pass over the bridge into the city and look over the railing. The town was a successful textile center from the start, and was named Lawrence after two of the main investors. After its incorporation as a town in 1853, Storrow became its first mayor.

Storrow and Baldwin were members of the Boston ruling caste and therefore were well qualified to lift from engineering as a profession the odium still attached to it by their circle. As a matter of fact, the prejudice embraced not only engineering, but even natural science, not so much when cultivated as a means of gentle diversion as when considered as a fully acceptable university subject. We shall meet Storrow again in a later chapter as one of the moving spirits behind the foundation of the scientific school in Harvard College.

Storrow was above all a hydraulic engineer, though he also worked in the machine shop. Such a man was naturally interested in another of the major engineering works of that period, of paramount interest to all Bostonians. How could the rapidly growing capital be provided with good drinking water? The time when every family had its own well was long since past. An "Aqueduct Corporation," chartered in 1796 in the turnpike and canal company days, supplied fifteen hundred houses with water from Jamaica Pond, a few miles south of Boston; the water was brought in through pipes of pitch-pine logs. Agitation for improvement had led to several committees, set up partly by the city government, partly by private persons. This was one of the first technical undertakings of which the scope transcended the interests of any special group, privileged or nonprivileged; it was truly a community project. Should it be in private hands, or should such an enterprise be publicly controlled? The issue, with its modern ring, was widely discussed. The first report was made by Daniel Treadwell in 1825, who suggested tapping Charles River above the falls at Watertown, and Spot Pond in Stoneham, both only a few miles away from Boston. Dr. John C. Warren supported him by pointing out "that many complaints of an obscure origin owe their existence to the qualities of the common spring water of Boston," and also made the point that "if there be any privilege which a city ought to reserve exclusively in its own hands, and under its own control, it is that of supplying itself with water." The fight between "free enterprise" and public control remained undecided at that time, because the city fathers were reluctant to spend money for such unusual undertakings. However, the issue became more urgent with every passing year.

In 1834 the City Council set up another committee, which appointed Loammi Baldwin to make an examination. The dean of New England's engineers presented a report as substantial as his earlier report on the proposed canal through the Berkshires to Albany. His choice was Long Pond in

Natick and two smaller ponds near by, though they were from 24 to 27 miles away. Since they were 128 to 155 feet above marsh level, water could be brought to town through an aqueduct by the force of gravity alone. No pipes would be necessary until a nearer point of sufficient height was reached, from which the water could be carried by means of cast-iron pipes into a reservoir placed high inside the city. Baldwin thought of the aqueduct very much in terms of a canal builder. He did not believe in pumping the water up, since this would involve steam engines, placing the city under necessity "of maintaining the fires, which must never go out, by sea-borne coal, a supply of which may be interrupted or entirely cut off by the act of our own government, or the interference of foreign powers." It seems that Baldwin, by now a man of fifty-four, did not place complete confidence in the new steam-driven contraptions, with which he never became quite familiar. His proposal to select Long Pond as the source of water supply was eventually adopted, but not before 1846, and after a wild controversy and veritable free-for-all had taken place in which all kinds of alternatives were proposed, including the use of the old Middlesex Canal, now in decline because of the railroad competition, as an aqueduct.

Baldwin's pupil, Storrow, just returned from Europe, contributed to the solution of the problem by publishing a neat *Treatise on water works* (1835), of some interest as one of the first engineering texts written by an American. "An abundant supply of pure and wholesome water is everywhere one of the first requisites for health, cleanliness and comfort," was Storrow's opening statement. The booklet reported on the achievements of continental authors in the fields of hydraulics, especially the French. Following the customary pattern of American textbook writers of his day, he did not try to introduce any new idea of his own. His sole new contribution was the conversion of the tables from French into English measuring units. The ten chapters deal with the motion of water in canals and in pipes, and conclude with a report on artesian wells. He used some continental calculus, something of a novelty in an American book of this type.

A decisive factor in the final solution of Boston's water supply problem was the successful completion of the water system of New York. Then, in 1846, Boston followed the New York example. The aqueduct finally decided upon was a covered conduit, made of brick and oviform in shape, with the broad end down, 5 feet wide and 6.33 feet high. The length was more than 14 miles, and its capacity of delivery 16 million gallons daily—as compared to the 1.5 millions daily proposed in the first Treadwell report.

The final action took place in October 1848, when the water from Long Pond—now renamed Lake Cochituate—was officially introduced into the city. It was a great public celebration, with a speech by the mayor, Josiah Quincy, an ode by James Russell Lowell, a hymn by George Russell, sung by the Händel and Haydn Society, and a procession, in which fire companies, military men, civil organizations, waterworks employees and school children participated. The procession reached the Common, and listened to the mayor. "At the conclusion of the Mayor's address, he asked the Assembly if it were their pleasure that the water should now be introduced. An immense number of voices responded 'Aye!' Whereupon the gate was gradually opened, and the water began to rise in a strong column, six inches in diameter, increasing rapidly in height, until it reached an elevation of eighty feet."

> After a moment of silence, shouts rent the air, the bells began to ring, cannon were fired, and rockets streamed across the sky. The scene was one of intense excitement, which it is impossible to describe, but which no one can forget. In the evening, there was a grand display of fireworks, and all the public buildings and many of the private houses were brilliantly illuminated.

It is interesting to see how many outstanding scientific men of New England were at one time or another connected with the Boston water supply discussions. In addition to Treadwell, Baldwin, Warren and Storrow, already mentioned, we find Webster, Horsford, Silliman, Walter Channing and Charles

T. Jackson, engineers, inventors, chemists and physicians. Science and engineering began to find a place in the new community undertakings.

After a hundred years, and a considerable extension of the system of water supply, the aqueduct is still in place and in use, winding through pleasant country, passing by landscaped reservoirs and stations built in the "severe and simple" taste praised by Nathan Hale. It offers excellent opportunities for an arduous hike as well as an easy stroll.

9

The technical development of the remarkable and turbulent decades preceding the Civil War was not, of course, characterized solely by the introduction of the steam engine and the widespread use of water power. Technical improvements were made in a large number of industries, in the iron tool factories of Rhode Island and Worcester, in the shoe industry of Lynn and Randolph, in the many textile mills spread over the whole country. The business of the Patent Office in Washington increased regularly; in the years from 1840 to 1849 the average of applications for patents was 1000, and the average of patents granted was 550; in 1858 the number of patents granted had risen to 3710, and in 1860 to 4819. A large number of these patents were taken by men from New England, where industries were rapidly expanding. The inventive genius of the Yankee had at last found an unlimited field of opportunity. Added to the chance of making and improving a tool or a material, there was always the chance of making money or even a fortune. Inventors, however, were poorly protected and the list of talented men whose lives were poisoned by patent litigation is distressingly long. The northern United States of a hundred years ago was a free-for-all country, with a generous opportunity for all—to make a fortune, or to sink into poverty; to help each other, or to spread calamity.

New England slowly became familiar with the steam engine, primarily through its use on the railroads and the steamboats. By 1845 smokestacks became a more common sight in the industrial settlements. With the wider use of steam as motive power the manufacture of steam engines began also to develop. In 1844 the firm of Fairbanks, Bancroft and Company in Providence, machine and steam engine builders, hired George Henry Corliss, a young storekeeper from Greenwich, Connecticut, as a draftsman. Corliss, the son of a physician in New York State, was an inventor who had received a patent on a machine for sewing boots. In the Providence plant the draftsman dropped his ideas about sewing machines and became interested in improving steam engines. The results was the famous Corliss engine, one of the most important improvements made in steam engines since Watt.

In practically all engines built previous to 1849, when Corliss took out his first patent, steam was controlled by a single valve. Corliss used four valves. Two of these admitted steam to the cylinders, one at each end, while the other two controlled the exhaust. A system of levers controlled the valves and the admission of steam to the cylinder. This invention allowed a more satisfactory control over the engine, since one of the valves might be altered without changing the others in any way. The admission ("cut-off") for one end of the cylinder could be accurately adjusted without impairing the functioning of other parts of the machine. The amount of condensation was reduced, because the entering steam was not cooled by the exhaust steam. This Corliss "valve-gear and drop cut-off" resulted in a great saving of fuel, and has been universally used.

In 1848 Corliss organized his own compay, Corliss, Nightingale & Company, in Providence, to build his new engines; later he founded the Corliss Engine Company. He constantly improved his machines, and built up his company until it employed 1000 men, not without being involved in a number of copyright struggles. One of his most spectacular performances was the building of an enormous engine which provided the motive power of the 8000 different machines at the Centennial Exhibition in Philadelphia in 1876. The Corliss engine is at present still manufactured in Providence.

Two famous inventions of this period belong to New England: Elias Howe's invention of the sewing machine, on which a patent was issued in 1846, and Charles Goodyear's vulcanization of rubber, which was discovered in 1839 and on which the first patent—a French one—was taken in 1844. Howe worked on his machine in Cambridge, Goodyear on his process in Woburn, both in Massachusetts. We need not describe the inventions, nor the sufferings of the inventors, in detail, since they are a well-known part of American history, and their influence on the development of science and technology in New England before the Civil War was small. For us they are important principally as examples of that Yankee inventiveness which led to so many radical changes in our mode of living, whose origins go back to the thirties and forties of the century—those days when the creative energy of New England, which had been slumbering for two generations after the Revolution, at last found full opportunity to develop.

10

Goodyear's invention did not find full application until the years after the Civil War. A somewhat similar situation existed in the field of electricity. Only one electrical invention, the telegraph, found wide-spread application before the Civil War. The first machine in America moved by electromagnetism was invented by Joseph Henry in 1831, and the first actual electric motor was invented a few years later by Thomas Davenport. However, the revolutionary meaning of these inventions was not understood until much later. "Mechanical movements of considerable power through the agency of electromagnetism have by some been considered as a rival of steam power," wrote the *New American Cyclopedia* in a skeptical vein as late as 1859. But Morse's invention, the telegraph, was already widespread when the notion of electricity as a source of power was still a subject of speculation.

The great figure who influenced and inspired all early American work on electricity was Joseph Henry. Henry began his career as a protégé of Stephen Van Rensselaer in Albany, and did his fundamental research on electromagnetism at Albany Academy. He began his studies on electricity not much later than Faraday and under the same influence: the discoveries of Oersted and Ampère concerning the action of an electric current on a magnetic needle. In 1827 he read his first paper on modifications of the Ampère experiments; in 1829 he discovered electrical self-induction (sooner than Faraday), at the same time building strong electromagnets. In 1831 he caused a steel bar, suspended between the poles of an electromagnet, to swing and give signals by striking a bell. He operated this apparatus through more than a mile of wire; it can be considered the first magnetic and acoustic telegraph.

When the Smithsonian Institution in Washington was organized in 1846, Henry was elected secretary. From this year until his death in 1878 he was truly a national scientific figure—the first American rather than local scientist. Under his remarkable management the Smithsonian Institution became a great power for the organization of science in the country, especially in the field of anthropology and meteorology.

Henry's research in electricity inspired all great American achievements in this field. Two revolutionary conclusions could be drawn from his astonishing experiments. It was possible to utilize electricity for long-distance communication, and it could be used as a source of power. The first to think of utilizing Henry's discoveries to drive a machine was Thomas Davenport, a prosperous blacksmith in the rural settlement of Brandon, Vermont. The Penfield Iron Works at near-by Crown Point, New York, had installed one of Henry's electromagnets. Davenport inspected it and was fascinated. He realized its possibilities as a source of power, and in 1834, after having obtained four electromagnets, he built a little motor. However, public appreciation of Davenport's invention never progressed beyond a mild curiosity towards a new form of entertainment. He continued to make improvements and even constructed a prototype of the electric trolley car. The patent for his motor dates from 1837. But he never found a market for his inventions, though he established a shop and laboratory in New York. His health broke down and he returned to Vermont, where he died in 1851 without ever having met with the recognition he deserved. Electric power could not yet compete with steam or water.

The men who tried to apply Henry's ideas to communication were more fortunate. In 1833 Henry demonstrated his project of the electromagnetic telegraph, explaining how mechanic effects might be produced at a distance by means of an electromagnet. At this time Samuel Finley Breese Morse was painstakingly working in New York at the invention of a practical telegraph. We have seen how Morse, then a struggling painter, had received the idea of an electric telegraph from Charles T. Jackson, while returning from Europe on the same boat. Through a friend, Morse became acquainted with Henry's ideas; after many experiments he was able to incorporate them in a practical apparatus. By 1840 he had the "Morse" code. The public test came in 1844, when Morse, sitting in the Capitol in Washington, sent a message by wire to his collaborator Vail in Baltimore. This message, "What hath God wrought," opened the era of the telegraph.

In Dover, New Hampshire, Morse's work was closely followed by Moses Gerrish Farmer, the principal of a school for girls. Farmer had already made several inventions, among them a machine to print paper curtain shades. Farmer became deeply interested in electricity, at first seeking to apply it for power generation. With his brother's help and with money earned from his curtain manufacture, he constructed a miniature electric train, consisting of two cars. On one of the cars the batteries and a motor were mounted; the other car was for passengers. The brothers exhibited the machine during 1847 in New Hampshire and Maine, allowing children to ride in the passenger car. Like Davenport's machines, Farmer's inventions remained mere toys. Unlike Davenport, however, Farmer decided to discontinue the search for an efficient motor and to concentrate on Morse's ideas. In 1848 Farmer became an operator of the telegraph office in Salem, and subsequently obtained several technical jobs in the new telegraph system. He invented many electrical devices, such as a fire alarm system for Boston, and in 1858–1859 invented an early type of incandescent lamp. In 1868 he lighted a residence in Cambridge with forty such lamps, using a dynamo of his own invention as a substitute for a galvanic battery. Farmer later became a consulting electrician, retiring eventually to Eliot, Maine, where he lived until 1893 in moderate prosperity.

The glorious apotheosis of the telegraph came in 1858, when Queen Victoria sent a message to President Buchanan by means of a trans-atlantic cable. New York was excited, and held a "cable carnival" of two days, in "glorious recognition of the most glorious work of the age." Unfortunately the cable stopped working at about the same time. For the promoter of the cable connection, Cyrus West Field, this contretemps occasioned an abrupt transition from exalted glory to miserable defeat. He persevered, however, and in 1866 the cable was reinstalled. The telegraphic connection between the two continents now remained permanent.

Cyrus West Field, son of a Congregationalist minister in Stockbridge, Massachusetts, a successful businessman and brother of the famous jurist David Dudley Field, deserves the full credit for the success of this remarkable enterprise, for which he fought through ten years of disheartening failures. "Suppose you don't succeed? Suppose you make the attempt and fail—your cable is lost in the sea—then what will you do?" he was asked by a British nobleman at the beginniing of his labor.

"Charge it to profit and loss, and go to work to lay another," said Field.

Chapter 10

The Great Naturalists

> Here we are impressed with the inexhaustible riches of nature. The universe is a more amazing puzzle than ever, as you glance along this bewildering series of animated forms,—the hazy butterflies, the carved shells, the birds, beasts, fishes, insects, snakes, and the upheaving principle of life everywhere incipient, in the very rock aping organized forms. Not a form so grotesque, so savage, nor so beautiful but is an expression of some property inherent in man the observer,—an occult relation between the very scorpions and man. I feel the centipede in me—cayman, carp, eagle and fox. I am moved by strange sympathies; I say continually "I will be a naturalist."
>
> —RALPH WALDO EMERSON, 1833

1

THE PERIOD OF THE scientific leadership by gentlemen dilettantes had come to an end; their place began to be taken by the professional. The study of plants, animals and minerals was no longer an avocation of Manasseh Cutler, Benjamin Waterhouse, Jacob Bigelow or James Freeman Dana. The new generation was not satisfied with men of lesser stature than Louis Agassiz, Asa Gray, James Dwight Dana and William Barton Rogers. These men were specialists; they were mineralogists, geologists, botanists, ichthyologists, paleontologists—but they carried the name of naturalists, an appropriate name, indicating their broad grasp of nature despite their necessary specialization. The most outstanding of these naturalists were usually professors attached to well-established institutions, with considerable professional duties connected with academic life, the teaching of students, the organization of departmental work and the administration of finances allotted for research. They were helped, encouraged and admired by a large group of semiprofessionals and amateurs, as well as by businessmen and other men in public life. The tempo of scientific work as well as its content was changing. The popular interest in science and education gradually eliminated the solitary pioneer, at any rate in the accepted academic fields of natural science. Respect for the study of nature for its own sake began to be the accepted attitude.

The men who opened the study of natural history in New England had to defend their activities by a direct appeal to utilitarianism. They stressed perforce the medicinal value of certain plants, the nuisance value of certain insects, the economic value of certain minerals. Their task consisted in the most primitive type of classification, necessarily concetrating on the common specimens. In 1818 Benjamin Silliman, one of the first men in America to understand that the study of nature had to be undertaken by professionals in a systematic way, declared that in the beginning of the century it was extremely difficult to obtain the names of even the most common minerals, such as quartz, feldspar or hornblende, or granite, porphyry and trap among the rocks. In 1805 the whole Yale collection of minerals was carried in a single box to Philadelphia for classification. The gradual change of the country from an agrarian and mercantile economy to an industrial society gave encouragement to the work of popularization carried out by Silliman, Eaton and similar men of science. Bit by bit people began to look with something more than amusement at the "posy-seekers and bug-hunters" whom they met searching their fields and back yards.

The love for classification had led, at an early date, to valuable cabinets of minerals, birds, plants and other curiosa. Many a ship captain or supercargo came home with remarkable and precious mementos from foreign lands. Such objects were usually an exclusive exhibition in some wealthy home or in the libraries and meeting halls of learned societies, colleges or marine societies. In 1816 Josiah Quincy, then a Massachusetts State Senator, drew up a report proposing a building in Boston to house the libraries and cabinets already in existence. He proposed that the officers of the American Academy, the Massachusetts Historical Society, the Agricultural Society, the Medical Society, the Athenaeum and the

Linnaean Society should "form a body politick and corporate, by the name of the Massachusetts Institution, then the State should authorize a loan of 50,000 dollars at 5 per cent to erect a building," and offered it as a bill. The proposal was killed in the House with a vote of 119 to 12. There was no understanding as yet of the importance of such a move, and the state of the treasury after the War of 1812 served as a good excuse. For several more years naturalists had to work on their own with little public encouragement.

2

Such a solitary naturalist lived for many years in the house which William Dandridge Peck had built in the Harvard Botanical Garden. Thomas Nuttall was born in Yorkshire, England, and started as a printer's apprentice; he sailed to America in 1807 when twenty years of age. He soon showed keen interest in plants, which was encouraged by Benjamin Barton, the Philadelphia botanist. Barton later described his relation with young Nuttall as follows: "Observing in him an ardent attachment to and some knowledge of botany, I omitted no opportunity of fostering his zeal and of endeavouring to extend his knowledge. He had constant access to my house, and the benefit of my botanical books." Barton, who was a man of means, suggested to Nuttall that he undertake a scientific expedition into the interior of the continent. Nuttall eagerly consented and so began his extensive and solitary travels into the wilderness in search of plants, seeds, minerals and birds. During 1809–1811 he traveled for many months in the regions of the Kansas, Platte and Missouri Rivers; for two years, from 1818 to 1820, he explored the region of the Arkansas River. Neither fatigue, hunger nor native people could restrain his perseverance. On the first trip Nuttall was overcome by starvation and exhaustion, and only rescued from death by a friendly Indian. During his second trip he was robbed by Indians, who threatened his life. He finally reached Fort Bellepoint, where he lay ill for many months. Between the two journeys he stayed at Philadelphia; his summers he spent in excursions along regions east of the Mississippi, from the Great Lakes to Flordia; his winters he spent studying his collections. In 1822 he accepted the position of Curator of the Botanical Garden at Harvard College, where he had some light duties of instruction, but did not receive a professor's salary. He kept this position till 1834, when he resumed his travels.

Nuttall, in Cambridge, made a name as the author of the *Manual of Ornithology* (1832). No one, it has been said, has portrayed so tenderly the song of the birds, nor so charmingly their domestic life. He is also remembered as a man of marked peculiarities. He had lived too long with the birds, the plants and the Indians of the wilderness to feel at home among civilized people. He did not care much for his position, in which he could not work for science as he was wont to do, but used to say that he was, like his own plants, only vegetating. Mrs. Asa Gray, who lived afterwards in Nuttall's house, has left us some stories about this curious bachelor.

> He was very shy of intercourse with his fellows, and having for his study the southeast room, and the one above for his bedroom put in a trapdoor in the floor of an upper connecting closet, and so by a ladder could pass between his rooms without the chance of being met in the passage or on the stairs. A flap hinged and buttoned in the door between the lower closet and the kitchen allowed his meals to be set in on a tray without the chance of his being seen. A window he cut down into an outer door, and with a small gate in the board fence surrounding the garden, of which he alone had the key, he could pass in and out safe from encountering any human being.

Nuttall's travels carried him in 1834 with the Wyeth expedition across the Rocky Mountains to the Columbia River and in 1835 to the tropical vegetation of the Sandwich Islands. From there he returned to California, then a desolate part of Mexico, and returned home on a Boston ship by way of Cape Horn. This ship happened to be the good ship *Alert* on which Richard Henry Dana was serving his "two years before the mast." Nuttall was the last man Dana expected to see in that out-of-the-way place. "I had left him quietly seated in the chair of Botany and Ornithology in Harvard

University, and the next I saw of him, he was strolling about San Diego beach, in a sailor's pea-jacket, with a wide straw hat and barefoot, with his trousers rolled up to his knees, picking up stones and shells." Later when the *Alert* was rounding Cape Horn, Nuttall's passion for plants was aroused by the sight of land. "In the general joy, Mr. Nuttall said he should like to go ashore upon the island and examine a spot which probably no human being had ever set foot upon; but the captain intimated that he would see the island, specimens and all, in—another place, before he would get out a boat or delay the ship one moment for him."

Nuttall continued to work on his collections in Philadelphia until he returned to England in 1841, to take possession of Nutgrove, an estate near Liverpool bequeathed him by an uncle. A condition attached to the bequest was that Nuttall should reside in England at least nine months of the year. At Nutgrove he lived a simple farmer's life for the remaining seventeen years of his life, cultivating rare plants, especially rhododendrons from the mountains of Assam and Bhutan. He returned to America only once; he stayed for six consecutive months, combining the three last months of 1847 with the first three of 1848. He died in 1859 at the age of seventy-three.

Nuttall's publications are almost all in the field of descriptive botany and ornithology, where he classified a large number of new genera and species. Though he was not prominently identified with Cambridge, his cultivation of the Botanic Garden and his book on birds contributed to the growing appreciation of natural science in New England during his stay at Harvard. His influence was rather through his scientific zeal and firmness of purpose than through his organizational activity, though he joined the new Boston Society of Natural History. There are several good pictures of this curious bachelor at the Harvard Herbarium and also a life mask, discovered in a storage warehouse twenty-five years ago. The story of this discovery adds another pleasant touch to the Nuttall legend.

3

Despite his official position at Harvard, Nuttall always

remained the distinguished foreign guest. He was not the only one. The time was past when foreign scientists either avoided New England completely or traveled through it without finding much to draw their attention. Now, when a man of science came to New England, he found both things to arrest his interest and people eager to extend a hand of welcome. Boston became host to many distinguished naturalists, who helped stimulate the awakened love for natural science.

To Boston came Constantine Samuel Rafinesque, engaged on one of his interminable journeys. He was a Sicilian even more eccentric than Nuttall, who with all his vagaries always remained the solid Yorkshire printer. Rafinesque knew the botany of the United States as well as any born American, though his superficiality was criticized. After all, this could be expected of a restless traveler who also dabbled in antiquities, civil history, philology, political economy, philosophy and poetry. He could write in a fascinating way, describing the flora and fauna of unexplored America with the vivid color of a man who had been trekking through woods and swamps, over mountains and across rivers, along trails known only to trappers and Indians:

> The mere fatigue of a pedestrian journey is nothing compared to the gloom of solitary forests, when not a human being is met for many miles, and if met may be mistrusted; when the food and collections must be carried in your pocket or knapsack from day to day; when the fare is not only scanty but sometimes worse; when you must live on corn bread and salt pork, be burned and steamed by a hot sun at noon, or drenched by rain, even with an umbrella in hand, as I always had. Mosquitoes and flies will often annoy you or suck your blood if you stop or leave a hurried step. Gnats dance before the eyes, and often fall in unless you shut them; insects creep on you and into your ears. Ants crawl on you whenever you rest on the ground; wasps will assail you like furies if you touch their nests. But ticks, the worst of all, are unavoidable whenever you go among bushes, and stick to you in crowds, filling your skin with pimples and sores. Spiders, gallineps, horseflies and other obnoxious insects will often

beset you, or sorely hurt you. Hateful snakes are met, and if poisonous are very dangerous; some do not warn you off like the Rattlesnakes. You meet rough or muddy roads to vex you, and blind paths to perplex you, rocks, mountains and steep ascents. You may often lose your way, and must always have a compass with you as I had. You may be lamed in climbing rocks for plants, or break your limbs by a fall. You must cross and wade through brooks, creeks, rivers and swamps. In deep fords or in swift streams you may lose your footing and be drowned.

This was his experience in the Alleghenies, but the White Mountains and the Maine woods were not much better in those days, and even now the description fits in many details. But there was another side to the picture:

> The pleasures of a botanical exploration fully compensate for these miseries and dangers. . . . Many fair days and fair roads are met with, a clear sky or a bracing breeze inspires delight and ease, you breathe the pure air of the country, every rill and brook offers a draught of limpid fluid. . . . Landscapes and plants jointly meet in your sight. Here is an old acquaintance seen again; there a novelty, a rare plant, perhaps a new one, greets your view; you hasten to pluck it, examine it, admire, and put in your book. . . . You feel an exultation, you are a conqueror, you have made a conquest over Nature, you are going to add a new object or a page to science.

Rafinesque came to Harvard commencement in the summer of 1827, and stayed a week in the neighborhood. He browsed through the books of the Athenaeum and of the college, saw Jacob Bigelow, T. W. Harris "and other savans." On his return to Troy he stopped at Worcester to visit the library of the Antiquarian Society, of which he was a member, and herborized around Northampton and Pittsfield. "Crossing the Tacomick Mts. I went to New Lebanon to visit the mineral waters, and the village of the Shakers. This is the central settlement of these modern Essenians. I became acquainted with their botanist and gardener Lawrence."

Rafinesque does not tell us if he liked the Bostonians he met, something we should like to know in view of a previous experience of his which it can best be related in his own words:

> In 1825 I had sent two memoirs for prizes offered. One to Washington for a prize of $1,000 for the best means to clear the R. Ohio of snags and trees. The prize was awarded thro' political influence to a Contractor who has not cleared the river! but my memoir was returned with all the plates. The other was sent to Boston for a prize of $1000 offered by the Academy of Sciences for the best account of the materials existing on the history of the native tribes of America: altho' my memoir was the best, as appears by the report of a Committee communicated to me by Mr. Everett, yet the prize was not awarded, because my memoir was too long, etc., if it had been shorter, it would have been too short! but the worse was that my memoir was never returned; but stolen or mislaid by Mr. Holmes, the writer of pretended annals of N. America. It is thus that learned men are often served here. Prizes are sometimes offered merely to help a favorite to fulfil his part. I have since again written a memoir for another prize, which has been postponed from year to year. I do not state names as the subject is not yet decided; but if I am served there as I was in Boston, I never mean to write again for prizes offered by public Societies thro' the doubtful motives of inducing learned men to labor for nothing.

The villain of this plot is easily recognized as the Reverend Abiel Holmes, the father of Oliver Wendell Holmes. The neglect of Rafinesque's paper may have been due to a certain contempt for his idiosyncrasies among Boston's notables, who could not then appreciate the positive side of his work. He even played with the idea of evolution, which he called "the great universal law of Perpetual Mutability in everything," an idea, published in 1832, which was considered in most circles to be at least heretic. Yet Asa Gray, one of Rafinesque's severest critics, in reviewing his botanical work in 1841, a year after his death, had to state that it was indeed a subject of regret

> that the courtesy that prevails among botanists of the present day (who are careful to adopt the names proposed by those who even suggest a new genus) was not more usual with us some twenty years ago. Many of Rafinesque's names should have been adopted; some as a matter of courtesy; and others in accordance with strict rules.

Rafinesque's work was also instrumental in focusing the attention of European naturalists on America's natural treasures. It was one of the sources of Agassiz's interest in the United States.

4

In 1831 Boston had another distinguished visitor, with more social graces than Rafinesque, but with an equally adventurous life. John James Audubon landed in Boston from England on his return to America. He was already famous as the painter of American birds, whose work was admired on two continents. Never had birds been painted in such a wonderful way—"true to Nature as if the creatures were in their native haunts in the forests, or on the seashores. Not stiff and staring like stuffed specimens, but in every imaginable characteristic attitude, perched, wading, or a-wing—not a feather, smooth or ruffled, out of its place—every song, chirp, chatter, or cry made audible by the power of genius." This was the feeling of an Edinburgh critic, whose opinions still adequately express the admiration which Audubon's paintings inspire. When he arrived in Boston, the first volume of his *Birds of America* had appeared in London. Audubon's main interest was to obtain subscriptions to his work, and he seems to have been satified with the interest of Boston's wealthier friends of nature. He left Boston to visit the coast of Maine, accompanied by his family, to study not only birds, but also lumbering, as Michaux had done before.

On his return to Massachusetts he was taken ill in Boston, where the cholera prevailed, but recovered through the good efforts of his medical friends, the doctors Parkman, Shattuck and Warren. Audubon was moved by the kindness and appreciation he met in Boston: "Although I have been happy in forming many valuable friendships in various parts of the world, all dearly cherished by me, the outpouring of kindness which I experienced in Boston far exceeded all that I have ever met with." Audubon's Creole wit, his handsome appearance, his fine enthusiasm and intelligence, were in sharp contrast to the social eccentricities of a Nuttall or a Rafinesque, and Boston's leading circles recognized in the artist the son of a Southern planter, the gentleman and the scholar. Parkman's financial support enabled Audubon to leave the following year for Labrador on the schooner *Ripley*. His youngest son and four young Bostonians of good families went with him. They explored the forests of New Brunswick and the shores of the Bay of Fundy, the Gulf of St. Lawrence, the Magdalen Islands, the coast of Labrador, and returned by way of Newfoundland. The expedition lasted from June till the end of August. After its homecoming Audubon left immediately for the South.

We find Audubon back in Boston in 1836, arriving now in the modern way, by the steamboat *Massachusetts* from New York to Providence and thence by newly opened railroad to Boston at the rate of fifteen miles an hour. He was, as always, eager to sell subscriptions to his monumental work. The artist was as graciously received as before, stayed with Dr. Shattuck, met Everett, Bowditch, Gould and the newly wed Mrs. John Lothrop Motley, whom he described as "as handsome as ever, and her husband not far short of seven feet high." He liked to frequent the Faneuil Hall market to spend a few cents on a pigeon hawk or a blue teal, and have a chat with David Eckley the salmon fisher. More than once he went to Roxbury, to see the collection of skins and eggs which young Thomas Mayo Brewer, the budding ornithologist, had assembled. It happened that during his stay in Boston the *Alert* returned from California, bringing back Richard Dana and Thomas Nuttall. Audubon enjoyed meeting his colleague: "This has been a day of days with me, Nuttall breakfasted with us, and related much of his journey on the Pacific, and presented me with five new species of birds obtained by himself, and which are named after him." Audubon's further entry in his diary that day—September 22, 1846—is characteristic:

> One of Dr. Shattuck's students drove me in the doctor's gig to call on Governor Everett, who received me as kindly as ever; and then to the house of President Quincy[1] of Harvard University, where I saw his family; and then to Judge Story's. Then crossing the country, we drove to Col. J. H. Perkins', and on the way I bought a fine male white-headed eagle for five dollars. On my return I learned that at a meeting of the Natural History Society yesterday a resolution was passed to subscribe for my work. Dr. Bowditch advised me to go to Salem, and with his usual anxiety to promote the welfare of everyone, gave me letters to Messrs. Peabody and Cleveland of that place, requesting them to interest themselves to get the Athenaeum to subscribe for my work.

Audubon went to Salem next day be stage and visited several persons to obtain subscriptions for his book. He describes in detail his visit to the wealthy Miss Sitsby, "a beautiful blue":

> Although she has the eyes of a gazelle, and capital teeth, I soon discovered that she would be no help to me: when I mentioned subscription, it seemed to fall on her ears, not as the cadence of the woodthrush or mocking bird does in mine, but as a showerbath in cold January.

Audubon stayed several days more in Boston, met Daniel Webster, received from the great man an introduction useful for his coming journey to the West, and left for New York. It was Audubon's last visit.

5

Charles Lyell, the British geologist, the author of the *Principles of Geology,* was made to feel equally welcome. The publication of this book, in the years 1830-1832, had not

[1] The printed diary spells the name "Tinnay."

only deeply influenced geological thought, but had made a deep impression upon the public mind. Studying the theoretical errors which had retarded the progress of geology, Lyell had attacked the widely cherished belief that the changes in the earth's surface were due to causes in kind and degree distinct from those now constituting "the economy of nature." This doctrine, which was in harmony with orthodox teachings based on the Bible, seemed also to be in harmony with the enormous geological changes which had taken place in the course of the earth's history, and was carefully scrutinized in the *Principles of Geology*. Lyell's conclusion was that all former changes of the organic and inorganic creation are referable to one interrupted succession of physical events, governed by laws of nature in operation at the present time. All so-called catastrophes were banished in this theory. Even mountain ranges and oceans were not exempt from this development. Rivers, springs, tides, volcanoes, earthquakes have always acted as they now act, land has been elevated, and has subsided in former eons as it does at present. The cumulative effect of long-continued processes was fully grasped. Any special "catastrophic" forces active in bygone periods but which are no longer in existence were carefully rejected. This was geological dynamics, and thoroughly distasteful to many persons, scientists and laymen alike, who thought either in static terms or in terms of the Mosaic story. In Massachusetts, Lyell's theory was opposed by Edward Hitchcock, the leading geologist, in his reports on the geology of the Commonwealth. It was indeed a revolutionary theory, but the quiet and demonstrative way in which Lyell applied it to the explanation of geological phenomena brought his point of view to an early victory.

Boston was willing to listen to the new theory, when Lyell arrived in 1841. Lyell lectured at the Lowell Institute before large audiences, and was pleased. The Bostonians met in him one of the first bold scientific thinkers who ever came to their city. They found in him a brilliant example of a scientist who does not stop with classification, as most Americans did, but who asked for general theories to account for known facts. This lesson was not lost.

Lyell traveled over America as far north as Canada and

Nova Scotia, as far south as Kentucky. At Boston he was a welcome guest at the home of George Ticknor, once professor of French and Spanish in Harvard College. At that time he was devoting his time mainly to the history of Spanish literature. His house with its famous library—later to become the nucleus of the Boston Public Library—was open to his friends, among whom the British geologist, an Oxford graduate and a man of means, soon found himself at home. After his return to England Lyell dedicated his *Travels in North America* to Ticknor, "in remembrance of the many happy days spent in your society, and in that of your family and literary friends in Boston." During this visit Lyell attended the congress of geologists of the United States in Boston, one of the first scientific meetings of this kind held in the United States. He returned in 1845-1846, to travel as far west as the Mississippi, where he studied the influence of this mighty stream as a geological agent. Lyell's *Second Visit to the United States,* published in 1849, described his experiences on this second trip.

Lyell rewarded the hospitality which America had extended to him not only by writing his travelogues. One of the greatest services he rendered to American science was his recommendation to the Lowell Institute to invite Agassiz for a series of lectures. This marked the beginning of Agassiz's teaching in the United States.

6

When Jean Louis Rodolphe Agassiz came to the United States, he was already, at thirty-nine years of age, a scientist known and respected throughout the world. He was born in French-speaking Switzerland in 1807, the son of a country clergyman who was himself the sixth clergyman in direct descent from a line of French Huguenots. Agassiz's first ambition was to become a physician, but his interest in medicine disappeared gradually before an overwhelming desire to become a naturalist. At the University of Munich, which he entered in 1827, he wrote his book on Brazilian fishes, a folio with ninety plates, and written in Latin—*"digesset, descripsit et observationibus illustravit Dr. L. Agassiz"*—which made Agassiz's name well-known in the professional world. It was dedicated to Cuvier, the leading zoologist of his day, who was then engaged on his monumental *Histoire des Poissons.* The success of this work decided his future. Full of confidence he wrote to his father:

> I wish it may be said of Louis Agassiz that he was the first naturalist of his time, a good citizen, and a good son, beloved of those who knew him. I feel within myself the strength of a whole generation to work towards this end, and I will reach it if the means are not wanting.

It must be said that the young man knew his mind and recognized his own abilities. He now decided upon a more ambitious plan, the study of fossil fishes, a subject which had baffled paleontologists because of osteologic difficulties and the technical difficulty of finding persons able to make correct drawings. Although he had few financial resources, Agassiz, at this stage of his career, already employed at least one artist to draw his figures.

In 1831 he traveled to Paris, where he met Cuvier himself, then a man of sixty-two and the great master of what was known as comparative zoology. Cuvier had formed a system based on the characters of anatomical structure rather than upon external resemblances. Beginning in 1800 with his first memoir on fossil elephants, he had made the study of extinct animals as fundamental a part of his work as that of living creatures. He amazed the world by his ability to perform the restoration of a fossil animal from the structure of a single bone or a portion of one, by using his principle of the unchangeable relations of organs. Agassiz's work on living and on fossil fishes fitted exactly into Cuvier's great scheme of natural classification. No wonder that Cuvier, despite his natural reservation, received the brilliant young man with great friendship and placed all his material on fossil fishes at Agassiz's disposal.

Agassiz was only a short time under Cuvier's personal influence, but he considered himself Cuvier's pupil during his entire life. He recognized in the French zoologist his master, the only man he ever met whom he considered superior to himself. His greatest ambition always was to be a second

Cuvier, to work like Cuvier and to think like Cuvier. It set its stamp on Agassiz's lifework, and because of this, on much of America's early venture into modern zoology and geology. It also determined Agassiz's outlook upon the origin of species, which later played such a role in his opposition to Darwin, but which was established as early as his work on fossil fishes.

A year before Agassiz came to Paris, the celebrated debate had taken place before the French Academy of Science between Cuvier and Geoffroy Saint-Hilaire, on what we now call the principle of evolution. It was this debate which the aged Goethe in Weimar considered more important than the political revolution which took place at the same time. Geoffroy Saint-Hilaire defended the essential unity of composition in the organs of different animals, and even admitted the possibility of change of species because of the influence of environment. Cuvier bitterly opposed this point of view. He defended the doctrine of the invariability of species, and introduced his theory of catastrophes in order to account for the gradual secession of the different groups of animals in different geological periods. This theory, which we have already met in the discussion of Lyell's work, postulated the occurrence on earth of a series of formidable geological revolutions, changing the surface of the earth and destroying the living creatures. Constant interference of the Creator was necessary to repopulate the earth with plants and animals. The gradual progress in the structure of these living creatures was a token of the unfolding of God's plans of creation. The debate did not solve any problem and made the two antagonists even more determined to pursue their own contentions. The controversy continued, and became the subject of popular discourse, mainly because Cuvier's position seemed in close accordance with the traditional Bible story of creation. Agassiz, bred to a family of Protestant orthodox clergymen, had no difficulty in adapting himself to the ideas of his teacher, Cuvier.

In 1842 Agassiz, now teaching at Neuchâtel, began the publication of his *Nomenclator Zoologicus,* an alphabetical list of every zoological genus, describing upward of 17,000 names. In addition to this, he developed his theory of glaciers.

In 1837 he placed his conclusions before the Swiss Society of Natural History. His address, presented at a meeting in Neuchâtel, marks the beginning of all that has been written on the so-called ice age:

> *During the age which preceded the rise of the Alps and the appearance of living beings as we know them today, there must have been a drop in temperature well below that of our own day; and it is this drop in temperature that caused the formation of the huge masses of ice which covered the earth again, especially where one finds the wandering blocks with the boulders polished by the ice friction such as those we know. Without doubt it is this great cold which trapped the mammoths of Siberia, which froze our lakes and piled up ice even to the peaks of our own Jura which existed before the rise of the Alps.*

Several years of further observations of glaciers in the summer months led in 1840 to the *Études sur les glaciers,* where the thesis of the Neuchâtel address was elaborated in greater detail. Previous to the elevation of the Alps, Agassiz declared, there had been a great reduction in temperature, and from the North Pole to the Mediterranean the globe had been covered with an immense sheet of ice. This ice had disappeared under the influence of a gradual elevation of temperature. While retiring, it had left the peculiar remnants, erratic blocks, dikes, polished rocks, moraines and scratches, which now allow us to read the story of the glacial age.

In 1847 this book was followed by a second book on glaciers, the *Système Glacière.*

7

By the time this second book appeared, Agassiz had entered upon the American period of his life. It was the result of an invitation from the new Lowell Institute to deliver a series of lectures on the solemn subject of "The Plan of the Creation, especially in the Animal Kingdom." Cuvier himself could not have inspired a better subject. Agassiz accepted eagerly.

One fine morning in the first week of October 1846

[writes Jules Marcou] a stranger recently disembarked was seen in the streets of Boston looking to the right and left, in some astonishment, but steadily making his way to Pemberton Square, a rectangle with a garden in the centre, and surrounded by fine three-storied brick houses, at that time a very aristocratic part of the city, recalling many squares and circles of the London West End. After looking at the numbers of several houses, the foreigner pulled the bell at the door of Mr. John A. Lowell, who, on opening the door, was surprised to have a stranger, with a strong foreign accent, ask if Mr. Lowell was at home. The astonishment was quickly changed into undisguised satisfaction when the stranger added: "I—a-m P-r-o-f-es-s-o-r A-g-a-a-ss-i-z," with the drawling pronunciation so characteristic of Roman or French Switzerland, & more specially of Neuchâtel. Mr. Lowell very cordially extended both hands and congratulated him on his safe arrival; and, in this auspicious manner, Agassiz made his entry into American life, and was launched into American society.

Agassiz began his career in the United States by traveling around, visiting his colleagues in Boston, New York, Princeton and Washington, vitalizing with his characteristic impetuosity all aspects of American scientific life with which he came in contact. The sedate American scientists, reared in the school of Timothy Dwight and Nathaniel Bowditch, had never before set eyes upon such a man. They found him a fascinating teacher, a serious scholar, an inspired man of science, a tireless organizer, a prodigious author, the friend and honored colleague of the most brilliant scientists in Europe and a rollicking good fellow with a touch of Bohemianism to boot. Agassiz was, besides, one of those rare men who could make even the most benighted layman understand the fascination of science. Supported by Bache and Henry, he even succeeded in improving considerably the way in which scientific publications of the United States Government were published. There was a decided change for the better after 1847.

Agassiz's boundless energy and his full understanding of the standards of the modern science of his day affected American scientific life from the time of his arrival in this country until long after his death. In Boston and Cambridge he fascinated both his colleagues and the general public by his delivery of the Lowell Lectures. His scientific enthusiasm and magnetic personality fully overcame his difficulties with the English tongue and his lack of experience with the tastes of the American public. The appreciation was mutual. Agassiz until then had never seen a scientific lecture delivered before so many people. Even Cuvier's more popular lectures at the Collège de France and the Jardin des Plantes were held before three or four hundred people at most. At the Lowell Institute Agassiz had a regular audience of fifteen hundred. Like Lyell before him he was impressed by the seriousness of his listeners, their desire to learn despite lack of preparation, and even by their good-natured acceptance of his thick French accent. Here was a democracy on the march and Agassiz, with his Swiss background, could easily grasp it and feel his way to the heart of the public. This mutual willingness of Agassiz and his public to understand each other was one of the factors which made Agassiz decide to remain permanently in America.

What attracted Agassiz to the Americans above all was that "their look is wholly turned towards the future"—again a curious indication of the similarity between the way the United States affected outsiders in the forties of the last century and the way the Soviet Union affects them in the present. Agassiz also hoped to free the American savants from their dependence upon Europe, from the subservient role in which they had voluntarily placed themselves. Somewhat condescendingly he exclaimed in a letter: "But since these men are so worthy to soar on their own wings, why not help them to take flight? They need only confidence, and some special recognition from Europe would tend to give them this."

The Boston aristocracy was well pleased with Agassiz, whose knowledge, manners, scientific ardor and religious outlook it appreciated. Here was a man who could help modernize American science, who was sufficiently democratic to appeal to the general public and sufficiently conservative to please the industrialists and most of the clergy. The new Lawrence Scientific School was being organized, and the chair of natural history was offered to Agassiz. Agassiz accepted and from this moment on was no longer a guest of the United

States, but became a recognized leader of American science.

Agassiz lived at first in East Boston and later settled permanently in Cambridge, in a house on Oxford Street, where he gathered around him a small and intimate group of naturalists and artists, with whom he had collaborated in Europe and who preferred to follow him to the new world. Among them were the Count Louis de Pourtalès, Arnold Henry Guyot and Jules Marcou, who became important American men of science. They usually stayed, if only for a while, in Agassiz's house, which used to look like a caravanserai. Marcou has described the scene:

> Mattresses were laid on the floors of different rooms, even in the parlour; the only unoccupied room being the dining room, where the table was always abundantly furnished. It was a second "Hôtel des Neuchâtelois" transferred from the glaciers of the Aar to Cambridge. In all, there were twenty-three persons, twenty-two of them came from Neuchâtel, town or canton. In fact, it was *la Maison du Bon Dieu,* as the French call it, everyone entering any room, and Agassiz receiving with a smile every new arrival.

All sorts of zoological specimens in alcohol, or even alive, were constantly coming by express from all parts of the country and occasionally added to the merriment. At one time a big black bear was sent from Maine by an admirer of the Lowell Lectures, at another time an enormous sea turtle arrived. Such presents eventually ended up in the dissecting rooms.

When he sailed to his new destination Agassiz had left his wife and children in Europe. Mrs. Agassiz died shortly afterwards. In America Agassiz married again, and his second wife, Elizabeth Cabot Gray, bound him in his personal as well as his professional life to America and to the leading circles of Boston. After his marriage some order was established in his household, where he was now joined by his daughters and his son Alexander from Switzerland.

In 1848, after Agassiz had delivered his first course of lectures as a Harvard professor, he set out for an excursion to Lake Superior in the company of ten students. He lectured to them on the way about the landscape through which they were passing, aided by black canvas and chalk. On the first day, traveling through New England, he made this interesting remark:

> All the plants growing on the roadsides are exotics, as are also all the cultivated plants and grasses. Everywhere in the track of the white man we find European plants; the native weeds have disappeared before him like the Indian. Even along the railroads we find few indigenous species. For example, on the railroad between Boston and Salem, although the ground is uncultivated, all the plants along the track and in the ditches are foreign.

Agassiz was especially impressed by the glaciated remains of the landscape. Here was the chance to test and extend his theories. The results of the excursion were published at Boston in 1850 under the title of *Lake Superior,* with a narrative of the tour by J. Elliot Cabot, one of the participants, and was "elegantly illustrated," to show that Agassiz insisted on the best possible execution of his works. An important scientific result was the extension of Agassiz's glacial theory to a large section of the North American continent:

> The action of this cause [which transported the erratics over northern Europe and northern America] must have been such, and I insist strongly upon this point, as a fundamental one, the momentum with which it acted must have been such, that after being set in motion in the north, with a power sufficient to carry the large boulders which are found everywhere over this vast extent of land, it vanished or was stopped after reaching the 35th degree of northern latitude.

This publication was influential in winning most American geologists over to Agassiz's theory of an ice age. Before he came to America his theory was little known and even those who had taken notice of it were skeptical. Agassiz's theory, however, had begun to act as a leaven in the works of the geologists even before he came to the United States. Hitchcock, in 1842, speaking before the Boston meeting of the American geologists, yielded to the idea that the phenomena of the drift were the result of joint and alternate action of ice and water. He even coined a name for it: aqueoglacial, and this theory of aqueoglacial action found many defenders. Henry Rogers, in 1844, played with it, and so did Ebenezer Emmons. The objection of Hitchcock to Agassiz's theory, as he explained in his address of 1842, was its assumption of an immense accumulation of ice and snow around the poles during the glacier period and its consequent emission of enormous glaciers in a southern direction, followed by floods of water and icebergs upon the return of a warmer period. Neither, indeed, could he understand how such causes could operate when the land was rising from the ocean and the water consequently retreating, water which, loaded with ice and detritus, floated at least for centuries over a large portion of the earth's surface.

Agassiz's book of 1850 took all these arguments into account. Material, transported by water, could not have caused straight furrows and scratches, and not even drifting icebergs could have caused the drift:

> Water may polish the rocks, but it nowhere leaves straight scratches upon their surface; it may furrow them, but these furrows are sinuous, whilst glaciers smooth and level uniformly, the hardest parts equally with the softest, and like a hard file, rub to uniform continuous surfaces the rocks upon which they move.

He pointed out that the northern erratic blocks were rounded and widespread; that the highest hills were scratched and polished to their summits, while to the south the mountaintops had protruded above the ice sheet and supplied the glaciers with their load of angular boulders. He also called attention to the absence of marine or fresh water shells from the ground moraines, to show that this was not the result of water action.

This set American geologists to a new appraisal of the phenomena of glaciation, but they were not willing to accept Agassiz's ideas without further investigation. In 1849 Henry

Rogers (before Agassiz's book was published, but after his results were made known) still adhered to his theory of violent and interrupted water action, and as late as 1855 Dana suspended definite opinion "until the courses of the stones and scratches about the mountain ridges and valleys shall have been exactly ascertained." There was certainly a cold or glacial epoch and the increase of cold was probably due to an increase in the extent and elevation of the northern lands. Further than this he was not inclined to go. Only after the Civil War period do we see a gradual acceptance of the glacial hypothesis. By 1856 Hitchcock was so far convinced that he remarked that there might have been more than one glacial period. Even now, when the glacial theory is universally accepted, it is still problematical what causes have led to the glaciation.

8

To Agassiz's first years in the United States also belongs an excursion to the coral reefs of Florida, undertaken at the suggestion of Alexander Dallas Bache, superintendent of the Coast Survey. In the spring of 1851 Agassiz spent ten weeks on a Coast Survey schooner, studying the growth of coral reefs, their mode of existence and their fauna. Extracts from his observations were published in the *Annual Report* of the Coast Survey of 1852, and contain excellent plates, paid for by the United States Government. The complete report was published only in 1882 and forms an important addition to the classical studies of Darwin and Dana based on Pacific atolls.

He also gave his help to the publication of some books. One of them was a Boston edition of a book with the curious title of *The Footprints of the Creator, or the Asterolepis of Stromness,* by the British geologist Hugh Miller, which appeared in 1852 with a long introduction signed by Agassiz. The book was a pre-Darwinian contribution to the struggle against the doctrine of evolution and defended strongly Cuvier's point of view. Miller had found a fossil fish, named Asterolepis, which had polemical importance. It was a fish appearing in early geological strata, but nevertheless strongly developed, being about thirty feet in length. This was supposed to show that fishes had not developed gradually from reptiles but had suddenly appeared in full size, in accordance with Cuvier's theory of catastrophes. We shall return to this form of polemic when we shall deal with the Darwin controversy.

Agassiz's life before and during the Civil War was a continuation of his work at Harvard—teaching, collecting and writing. Under the direction of Francis C. Gray, a Boston merchant, he undertook in 1855 the editorship of a colossal work, a natural history of the whole of North America. In accordance with the practice of those days, a subscription list was started, in the hope that at least five hundred persons would insure the success of the publication, which would consist of ten volumes, each costing twelve dollars. The popularity of Agassiz was such that the subscription list quickly reached the number of twenty-five hundred. Two volumes of the *Contributions to the natural history of the United States* appeared in 1858, two more volumes appeared in 1860 and 1862, and then, during the Civil War, the work was interrupted, and never resumed.

The work is a curiously unbalanced combination of popular appeal and extreme specialization. The preface of the first volume defends the democratic approach, the proud proclamation of a people born to science. Europeans, it said, should understand the American ways:

> There is not a class of learned men here, distinct from the other cultivated members of the community. On the contrary, so general is the desire for knowledge, that I expect to see my book read by operatives, by fishermen, by farmers, quite as extensively as by students in our colleges, or by the learned professions; and it is but proper that I should endeavour to make myself understood by all.

The book itself is, with the exception of a few passages, and perhaps the opening "Essay on Classification," an extremely specialized text, dealing mainly with turtles, jellyfishes and radiatae, such as starfishes, and is of interest only to the extreme specialist. The learned author, in dealing with his pet topics, forgot about the fisherman and the farmer. The *Essay on Classification* is still quite readable as a mature work of Cuvier's greatest pupil. It has an interesting historical review of the schemes of classification in the best tradition of Cuvier, together with his anti-evolutionary views. In addition it contains a theory which became later known as the so-called biogenetical law, when Ernst Haeckel rediscovered and popularized it. It was not even entirely a new discovery of Agassiz—it appears, for instance, in his book on Fossil Fishes—but he made it an intrinsic part of his system. In his words:

> The changes which animals undergo during their embryonic growth coincide with the order of succession of the fossils of the same type in past geological ages.

It is not only an important principle of classification, but its dialectics seems to lead closely to the ideas underlying evolution, and it seems strange that a man like Agassiz could enunciate such a principle and at the same time oppose so vehemently the Darwinian thesis. Agassiz's "Essay," however, after a short period of popularity in which it was reprinted in England and translated into French, was soon put in the shadow by Darwin's *Origin of Species,* which appeared two years later. Darwin placed the whole problem of classification in an entirely new light. For Agassiz it was a tragical fact, and it may explain somewhat the deep personal bitterness of his opposition to evolution.

The years before the Civil War saw the beginning of the realization of another ambitious scheme of Agassiz, the foundation of a great museum to house all his collections. The problem of housing the enormous number of specimens of all kinds which were coming in year after year had become quite acute. During his early days in America, he had filled his house and an old bathhouse near the Charles River with his treasures. This was followed later by a wooden structure on the Harvard College grounds, for which the university voted an annual sum of $400. Afterward friends raised $12,000 for a permanent museum, but the plans did not take shape until 1858, when Agassiz's old benefactor Francis C. Gray left a bequest of $50,000 for the establishment and maintenance of a museum of comparative zoology. Agassiz now set out in earnest to collect the money for his museum. State aid was necessary, but the Massachusetts General Court was not famous for its financial benevolence to scientific institutions. Agassiz personally approached all the higher dignitaries of the Commonwealth, beginning with the governor and the Board of Education, and was actually able to obtain a grant of $100,000 from the legislature. An almost equal sum was raised from among the citizens of Boston, and in 1860 the building was completed and duly inaugurated in the presence of the governor and a host of dignitaries, including the ever-vigilant Jacob Bigelow. Much of Agassiz's time was subsequently spent in organizing the collections and the laboratories. Later another wing was added through a gift of the banker, George Peabody. The enormous building is still standing on Oxford Street near the site of Agassiz's home, and retains in its name, Museum for Comparative Zoology, a memorial to Agassiz's master, Cuvier, the originator of comparative anatomy.

9

Agassiz remained active until his death in 1873. Among the achievements of his later days was a journey to Brazil, the country of his first scientific love, undertaken in 1862. He was accompanied by several assistants and volunteer workers, including Mrs. Agassiz, who collaborated with her husband in an interesting account of this trip, published in 1869. In the last year of his life he conducted an experiment, a school for natural history held at the seashore during the summer months. From fifty to sixty persons assembled for this purpose on the little island of Penikese, off the southern shore of Massachusetts. The enterprise was discontinued after Agassiz's death, but the example it offered led to the establishment of other more permanent schools, of which the Marine Biological Laboratory at Woods Hole, on the coast of Massachusetts not far from Penikese, is one of the most famous in the country.

Such a dynamic and striking figure as Agassiz was the source of a large number of anecdotes and stories. Every one of his biographies contains a few, and more are to be found in the books and letters written by those who came in contact with him. Well known is the answer he gave to a lyceum in the West, which offered him a large sum for a lecture, at a time when Agassiz was very busy with some scientific work: "I cannot afford to waste my time in making money." This story is also told of other great men, but it is supposed to have taught money-grabbing Americans that there are other values than gold. The following story by Agassiz's friend Whipple illustrates some of Agassiz's scientific philosophy:

> Some thirty-five years ago, at a meeting of a literary and scientific club of which I happened to be a member, a discussion sprang up concerning Dr. Hitchcock's book on "bird-tracks," and plates were exhibited representing his geological discoveries. After much time had been consumed in describing the bird-tracks as isolated phenomena, and in lavishing compliments on Dr. Hitchcock, a man suddenly rose who in five minutes dominated the whole assembly. He was, he said, much interested in the specimens before them, and he would add that he thought highly of Dr. Hitchcock's book, as far as it accurately described the curious and interesting facts he had unearthed; but, he added, the defect in Dr. Hitchcock's volume is this, that "it is dees-creep-*teeve,* and not compar-a-*teeve.*" It was evident throughout that the native language of the critic was French, and that he found some difficulty in forcing his thoughts into English words; but I never can forget the intense emphasis he put on the words "descriptive" and "comparative," and by this emphasis flashing into the minds of the whole company the difference between an enumeration of strange, unexplained facts and the same facts as interpreted and put into relation with other facts more generally known. . . . The critic was, of course, Agassiz.

Many stories deal with Agassiz's wonderful gift as a teacher. Here is an abstract of some lectures at Penikese, which Agassiz gave in the year of his death. It contains some very good advice:

> Never try to teach what you do not know yourself and

know well. If your school board insist on your teaching anything and everything, decline firmly to do it.

Lay aside all conceit. Learn to read the book of nature for yourself. Those who have succeeded best have followed for years some slim thread which has once in a while broadened out and disclosed some treasure worth a life-long search.

You cannot do without some specialty. You must have some base line to measure the work and attainment of others. For a general view of the subject, study the history of the sciences.

Select such subjects that your pupils cannot walk without seeing them. Train your pupils to be observers, and have them provided with the specimens about which you speak.

In 1847 I gave an address at Newton, Mass., before a Teachers' Institute conducted by Horace Mann. My subject was grasshoppers. I passed around a large jar of these insects, and made every teacher take one and hold it while I was speaking. If any one dropped the insect, I stopped until he picked it up. This was at that time a great innovation, and excited much laughter and derision. There can be no true progress in the teaching of natural science until such methods become general.

You should never trifle with Nature.

10

In Asa Gray, Agassiz's colleague at Harvard, we meet an interesting counterpart to the dynamic Swiss prophet of European science in America. Gray was an oldtime Yankee, a self-made man with only a backwoods education as a start in life, a quiet and undramatic student of plants. Agassiz and Gray were both hard and inspiring workers; they belonged essentially to the classifying school of taxonomists, which reached its highest development in America through their work, yet both were able to progress from mere taxonomy into the deeper realms of science. Agassiz was superlative at arousing the enthusiasm of many people; Gray, at the most, mediocre. But when the depth of scientific insight, the future development of natural science, the very dialectics of science were tested in the Darwinian controversy, Gray showed himself the man of greater insight.

Asa Gray was born in 1810 in Sauquoit, South of Utica, New York, of parents who had come from the backwoods of New England to settle in the wilderness of upper New York State. His father was a hard-working farmer and tanner, who allowed his son to absorb whatever education rural eastern New York could offer in those days. This included five years of study at the small medical college in Fairfield, from which Gray graduated in 1831, perhaps the most distinguished graduate of these ephemerous provincial medical schools. However, the medical work of the young physician served chiefly to provide a background for the study of botany, which soon became his passion. Gray later told how during the winter of 1827–1828 he read an article on botany in the *Edinburgh Encyclopedia* and in the following spring was able to determine, with the aid of Amos Eaton's *Manual,* the exact name of an early claytonia, the spring beauty. This little triumph determined Gray's career—the physician turned botanist.

One hundred years ago the task of a botanist in America was still mainly that of the describer and the classifier. The situation throughout the country was typified by conditions in New England, where Bigelow's *Florula* and Eaton's *Manual* —the works of an amateur and of a semiprofessional—comprised the total literature for this field. In the southern and middle Atlantic states this study was somewhat more advanced, but with the opening of the western territories new material was constantly brought to light and confused the unsatisfactory condition of botanical knowledge even more. No authoritative floras for the whole of explored America existed; the only two volumes boasting this title written around the turn of the century by Michaux and by Pursh. Both were foreigners, and both floras were highly insufficient; Pursh was even accused of extreme unreliability. The best authority in the field was John Torrey of New York, professor in the College of Physicians and Surgeons since 1827. He was also professor at Princeton University; at the time of Gray's graduation he was thirty-five years of age. In 1824 and 1826 Torrey had published some parts of a *Flora* of the Northern United States, and was collecting new material when Gray approached him.

In his Fairfield days Gray had begun to collect plants, which he sent to Torrey for classification. John Torrey became interested in the young physician, and made him his assistant in 1833. Gray later obtained a position as curator of the New York Lyceum of Natural History, which Torrey had helped to found. The result of this collaboration was that Gray became thoroughly familiar with Torrey's work on the classification of the plants of North America, and also with the difficulties connected with this ambitious enterprise. He decided to make it his lifework.

This period, during which the classification of North American plants was beginning to be taken seriously, was also the period during which the ancient and artificial Linnaean system of classification was being replaced by the so-called natural systems. The Linnaean system was built simply on the number of stamens and styles, as offering a convenient means of grouping plants. Its efficiency in establishing the relationship of a plant as a living entity to other plants can be compared to the efficiency of listing all people named Smith on the same page of a directory in order to establish the relationship of each Smith to the community in which he works. The natural system of classification was based on the morphology of the plants and sought to establish their relationships as based on the kinship of various kinds of parts and not only the sex characteristics. The leading scientists espousing the natural system of classification were Antoine De Jussieu, Lamarck and A. P. De Candolle; their work was supported by specialists who described and classified special families or special genera. De Jussieu's *Genera plantarum secundum ordines naturales disposita* dated back to 1789. A modified natural system was proposed by Augustin de Candolle in his *Théorie élémentaire de la botanique,* which was published in 1813. This was followed by the immense *Prodromus,* written by De Candolle and his son, the first volume of which appeared in 1824. By this time the advantages of the natural system had begun to be apprehended even in America. Torrey's *Flora* was still based on the Linnaean system, but under the influence of De Jussieu's methods Torrey abandoned the work before the second volume was ready for the press. Torrey decided to let his material accumulate and it was at this period that his collaboration with Gray began.

In 1836 Gray published the first of his influential textbooks, the *Elements of Botany*. It was based on the writings of De Candolle, and opposed the artificial system on which Eaton's *Manual* was based. At that time he was already working with Torrey on his material. In 1838 the first parts of the *Flora of North America* appeared, in which volumes Torrey and Gray began to present the results of their collaboration: a comprehensive picture of the plants existing in the explored parts of the continent. It at last brought order into a chaotic situation, which had become more pronounced every year, due to the exploration of new territories and the intensified study of regions already known. It was indeed a worthy companion to De Candolle's *Prodromus*. Parts of the *Flora* were published at irregular intervals by Torrey and Gray between 1838 and 1843.

Gray, in the meantime, had accepted a professorship at the newly founded University of Michigan, in behalf of which he sailed to Europe in 1838, where he remained for a year. He visited the botanical celebrities and their collections in many countries; in this way not only collecting valuable information on the classification of species, but at last achieving that liberal scientific education which had come so easily to Agassiz. After his return to America, he obtained a continuation of his leave of absence from the University of Michigan; before it expired, he accepted an appointment at Harvard which gave stability to his further career.

11

Gray accepted the professorship at Harvard in 1842. Instruction in botany had not been fully satisfactory in Cambridge under Peck and Nuttall, but there was nevertheless the precious Botanical Garden, and the beginning of a fine collection. A full professorship in natural history had at last been made possible by a special grant, and Gray became the first Fisher professor, a position he held for forty-five years.

At Harvard Gray continued the ambitious work he had

undertaken as assistant and collaborator of Torrey, the creation of a systematic American Flora. The scope of such an undertaking was truly overwhelming. A long series of great transcontinental voyages began in the thirties, and each expedition returned with important collections of plants. Torrey, in close touch with Gray, continued to work on the results of special expeditions. Gray, while devoting his talents to the classification of one group of plants after another, always kept his main goal, the composition of what he called the Synoptical Flora of North America, in view. He contributed an unbroken and widely read series of monographs, combined with reviews and bibliographical notes, to the *American Journal of Science* and the *Proceedings of the American Academy*. His knowledge of plants was encyclopedic; as a pupil wrote:

> to see Gray run through a bundle of newly arrived plants was a revelation to the cautious plodder. Every character he had ever met seemed vivid in his memory and ready to be applied instantly; and the bundle was "sorted" with a speed that defied imitation. It seemed like intuition, but it was vast experience backed by a wonderful memory; perhaps it could be called genius.

With Asa Gray, North American taxonomy reached a status comparable to that of Europe. On some subjects, for instance the complicated family of the Compositae, his knowledge probably surpassed that of any contemporary botanist.

Gray's textbooks have been widely praised for their accuracy and clarity. Among them are *How Plants Grow* (1858) and *Lessons in Botany;* and his great popular reputation came rather from these books than from his monographs. Many a youngster received his botanic inspiration from "How plants grow—Gray," as the saying was. But perhaps he was best known for his *Manual of Botany,* a textbook which listed every known plant in the northern United States, with keys directing the inquirer to the exact name. The first edition of Gray's *Manual* appeared in 1848, Gray's latest revision dated from 1867, and it has been published and republished ever since, with modifications which for a long time barely affected the fundamental character of the work. For half a century or more the *Manual* was a standard household book in every family with an interest and a love of wild flowers. Even now, a century after its publication and despite all excellent floras of more modern date, it still stands on many shelves, a popular monument to solid scholarship.

It would be unfair to recognize Gray only as a taxonomist. Classification is but the first step to the understanding of relationships, and a man of Gray's stature could not rest content as a classifier alone. The natural system of De Candolle, which underlay his basic method, and which he modified and enriched, itself suggested an infinity of relationships, morphological as well as physiological or geographical. Gray's contributions to the geographical distribution of plants began on a large scale when he edited the botanical results of the Wilkes expedition. Part of the work appeared in 1854 in a beautiful volume, with many plates, indicative of the new type of scientific government publications for which Agassiz had pleaded. Gray originally intended to go with the Wilkes expedition, but changed his mind when it was delayed. We may well speculate what would have happened if Gray had joined the expedition in view of the experiences of his contemporary, Darwin, on a similar expedition.

Gray's *Statistics of the Flora of the Northern United States,* which appeared in 1856, was an extensive study in geographical botany. He came to the conclusion that the interchange of alpine species between America and Europe must have taken place in the direction of Newfoundland, Labrador and Greenland rather than through the polar regions. He also pointed out that the special resemblance of the North American flora to that of Europe is due not to any large proportion of peculiar genera, but must be found in the similarity of species. Later he extended this work to compare the North American flora with that of Japan, and came to the conclusion that a large proportion of extra-European types of North America are shared with eastern Asia; and that no small part of them are unknown in western America. This led him to speculations on a former continuity of territory between America and Asia; and in this way showed how classifying botany leads to geographical botany, and from here to geology and even paleontology.

Gray's understanding of the structural and geographical relationship of plants led him to an early association with Charles Darwin in England. When Darwin published his *Origin of Species* in 1859, he found in Asa Gray a skillful supporter of his views. This highly interesting chapter of Gray's work will be discussed in a special chapter.

12

The last of the three leading American naturalists of this period was Silliman's son-in-law, James Dwight Dana, professor of geology in New Haven. Like Asa Gray, who was three years older, he was born the son of pioneers from Massachusetts, who had settled in western New York State. Dana's youth reminds us of Gray's, but Dana's father was more prosperous and the son was able to attend Yale. It is noteworthy that Dana's love of natural sicence was stimulated by his teacher at Utica Academy, who had been a pupil at Rensselaer Institute and had learned from Amos Eaton to teach with the aid of laboratory methods. Gray, who taught for a while at Utica Academy, came there after Dana had left, but both men knew of each other from the early New York days. While at Yale, Dana enjoyed the classical curriculum only moderately, but he profited richly from the lessons of Silliman. After his graduation in 1833, he served a year and a half as teacher of mathematics to midshipmen on board the ship *Delaware* of the United States Navy. When he returned to America, he joined Silliman as his assistant and continued his studies, which at that time centered around mineralogy.

The result of these studies, the System of Mineralogy, had already appeared in 1837. In this book the 24-year-old scientist had collected such an amount of sound research, presented in such excellent form, that it became a standard work almost overnight. America had as yet little choice in mineralogy texts. After Parker Cleaveland had published his book in 1816, Charles Upham Shepard of Yale had written a text, which had appeared in 1832–1835, but both books were easily superseded by Dana's work. Successive editions of the

System of Mineralogy appeared with the regularity of clockwork; in 1941, more than a century after its first publication, it was republished in a revised edition. Few scientific texts in the world can boast of such sustained appreciation.

It is typical of this period in American natural science that Dana changed his methods of classification in successive editions of the *System of Mineralogy*. He originally used the system of the German mineralogist Mohs, which, though better than the classification of Hauy which Cleaveland had adopted, still was based on external physical characteristics, such as hardness, luster and specific gravity. It included a dual Latin nomenclature analogous to the Linnaean system in zoology and botany, which Dana abandoned in 1850. However, in the third edition of 1850, he changed his system to a modern one based on chemical principles, and it remained the foundation of Dana's classification in all subsequent editions. "Not to change with the advance of science is worse than fickleness," he wrote, "it is persistence in error."

It is of interest that as early as 1835 Dana had already made glass models of crystals, which were probably the first models of this kind. It was during this period that the pedagogical reformers, notably Holbrook, insisted upon the use of visual models for educational purposes.

From 1838 to 1842 Dana took part in the Wilkes expedition, largely through the influence of Asa Gray, who could not go himself. For four years the young geologist collected geological and zoological material in the southern Atlantic and the Pacific. Like Darwin a few years earlier he studied the formation of coral reefs and atolls; like Nuttall, he explored the natural history of California, then a part of Mexico. In the next ten years he devoted himself primarily to the preparation of reports on the scientific work of the expedition, a task which involved an enormous amount of work.

His first report was on Zoophytes, the "animal plants," which have flower-like structures—the builders of the coral reefs are Zoophytes. This report appeared in 1846; it was followed by that on Geology (1845) and on Crustacea (1853–1854), all issued in magnificent quarto volumes accompanied by folio atlases of plates. The care bestowed on the publication was, as we have seen, a concession by Congress to the pressure exercised by scientists and their friends for the dignified presentation of government scientific publications, but the number of copies authorized by Congress was so small that from the beginning they were almost bibliographical rarities. Not even Dana himself received a copy. The work on these reports, combined with his other duties, finally broke his health and though he lived and was very active until 1895, he never fully recovered.

In 1850 Dana was appointed professor of natural history in Yale College, a title which was subsequently changed to professor of geology and mineralogy. From this time until his death his activity was identified with the Yale Department of Geology, which under his leadership acquired the luster of not only national, but international leadership.

13

Dana's name is primarily associated with his remarkable textbooks, which were internationally appreciated and still are reprinted in revised editions. The *System of Mineralogy* of 1837 was followed in 1862 by the *Manual of Geology* and by the *Textbook of Geology* of 1864. In order to keep abreast of the newest developments Dana had the custom of revising his texts regularly. Revising and improving his own opinions came easy to Dana, who was willing to learn, possessed a deep sense of objectivity and was liberal when opinions were at variance. "When a man is too old to learn, he is ready to die; or at least he is not fit to live," he once said to a friend. He possessed that objectivity in the search for truth which characterizes all sound men of science. Though by background and education Dana was inclined to oppose the theory of evolution, he frankly changed his position when already advanced in years because of the incontrovertible evidence.

Since its first publication the *Manual of Geology* has been an indispensable book of reference for any American geologist, and has also been widely used abroad. Here America could read for the first time a consistent history of its geological origin. Dana's treatment was eminently a dynamical and historical one, as any modern treatment should be; he embodied in his book the different contributions made by

Lyell, the Rogers brothers and other geologists. His *Manual* therefore passed far beyond the stage of mere classification; geology appeared as a modern science in the sense of a comprehensive system of interrelated facts. Here was a theory of the origin of continents, of the formation of oceans, of volcanoes and of coral reefs. Dana maintained the theory of the permanence of continents and oceans, which held that our present continents and oceans were outlined as areas of relative elevation and depression in the crust of the globe from the very beginning of the earth's history. He also described the formation of mountains as the result of internal contraction causing wrinkles in the crust involving upward and downward folds. Dana compared it to the smooth skin of an apple which becomes wrinkled when the apple dries and shrivels. It was an impressive book; and if less revolutionary and pioneering than Lyell's *Geology,* in Dana's opinion it was "all the sciences combined in one," the proud statement of a man living in an age and a country where geology had temporarily become the science most influencing the progress of other sciences.

14

These three leading naturalists—Agassiz, Gray and Dana—lived and worked in New England, two at Harvard, one at Yale. Indeed, New England was no longer second to Philadelphia in scientific standing, as it had been in former years. Industrialism had stimulated popular interest; there was money to finance research. Science was no longer an academic recluse, but had its devotees in a much wider circle. This does not mean that science had reached the masses—such a situation is impossible in a class society. It means that there were many industrialists, merchants, mechanics, professional men, farmers and industrial workers who had respect for science or at any rate for some fields of science. Obscurantism and indifference were not dead—nor are they even today—but they were unable to stem the tide of progress in the appreciation of science. Nor was America alone listening to what New England had to say in science. Europe too began to listen. Agassiz always felt that American scientists often put too high a value on scientific recognition from foreign sources, to the neglect of native appreciation. In opposition to this subservient attitude he argued that, at the risk of being scientific provincials, American scientists should judge their work in the light of the results achieved and the value of their current activities. In addition to the New Englanders, Gray, Agassiz, Dana, Wyman, Jackson, Peirce, several other American scientists such as Bache, Henry, Maury and the Rogers brothers were appreciated in America and in Europe alike. Franklin's work on electricity was the first outstanding scientific contribution of America to be recognized abroad. Now the principal contributions of Americans to science lay in the field of natural history, with Joseph Henry following Franklin in making startling discoveries in electricity.

This leadership in research could never have been accomplished without the preliminary spadework of many other professional and amateur scientists, many of whom were authorities in a special field of geology, mineralogy, zoology or botany. Louis de Pourtalès, Jules Marcou and Arnold Guyot were some of the men who came to America with Agassiz. There were several other naturalists doing their share in exploring the resources of the continent. Charles Upham Shepard, who as lecturer at Yale wrote a text on geology and from 1844 to 1877 taught chemistry and natural history at Amherst, was an excellent mineralogist. Edward Tuckerman, who taught botany at Amherst, was a specialist on lichens. Tuckerman Ravine on Mount Washington is named after him. Many another specialist was active in New England colleges.

Essex County, Manasseh Cutler's ancient field of exploration, remained a favorite place for botanists. Salem, Beverly and Marblehead, important towns in pre-Civil War days, were in this county, where the cultural tradition was very much alive. The physician Andrew Nichols lectured on botany at Salem as early as 1816. From Salem too came Charles Pickering, who sailed with the Wilkes expedition. When in his youth he studied the plants of his native soil, he was accompanied in these explorations by William Oakes of Danvers, later of Ipswich, who became the first really critical botanist of that region.

Oakes, who studied at Harvard under Peck, started as a lawyer, but turned to botany, spending his life in his beloved Essex County. He prepared lists of native plants and collected an excellent herbarium. He loved to explore the White Mountains, of which he prepared a sketch; it appeared after his death in 1848 and was his most elaborate publication. Large selections of Oakes's extensive collections have been sent to all parts of the world. Charles Lyell, on his second visit to the United States, climbed Mount Washington with this "amiable and accomplished naturalist." Without the work of men of Oakes's type the achievements of Gray and Agassiz would have been impossible.

Chapter 11

Science and Religion

> Until Darwin, what was stressed by his present adherents was precisely the harmonious co-operative working of organic nature, how the plant kingdom supplies animals with nourishment and oxygen, and animals supply plants with manure, ammonia, and carbonic acid. Hardly was Darwin recognized before these same people saw everywhere nothing but *struggle.* Both views are justified within narrow limits, but both are equally one-sided and prejudiced. The interaction of dead natural bodies includes both harmony and collisions, that of living bodies conscious and unconscious struggle. Hence, even in regard to nature, it is not permissible one-sidedly to inscribe only "struggle" on one's banners.
>
> —FREDERICK ENGELS, c. 1880

1

THE DARWINIAN CONTROVERSY which after 1859 raged in New England as well as in other parts of the world was the continuation of a long-standing struggle between a constantly developing science and the orthodox interpretation of the Bible. In the eighteenth century the struggle had centered upon Newton's interpretation of the laws of mechanics and astronomy. The issue was primarily whether the universe was governed by "immutable" laws or subject to free interference by the Deity. In the early nineteenth century the battleground shifted to geology, and the main issue became the interpretation of the Mosaic account of Creation, with the question of the historicity of the Deluge as a secondary point of controversy. Questions of organic evolution became gradually more prominent in the world at large, though they did not reach New England until deep in the nineteenth century. After the publication of the *Origin of Species* there was another change, and evolution became the cardinal point of dispute. Previous controversial questions were never entirely forgotten, and became part and parcel of the entire struggle between modern science and orthodox religion; even nowadays they form topics of discussion and dogmatic assertion. However, the center of gravity changed in the course of time, and the school and pulpit of one generation accepted much which an older generation had rejected as materialistic and heretical.

Of course, the struggle was not fought in a social and political vacuum, and in its own, somewhat esoteric way, it reflected the struggle between existing economic interests. The defense of science in colonial New England was undertaken primarily by representatives of the merchant aristocracy, who had always inclined toward religious tolerance, and their allies in the colleges, while the orthodox position was held in large majority by the farmers and mechanics. Newton was defended by Ezra Stiles and John Winthrop; Harvard College was a center of Newtonianism. The development, however, was never entirely rectilinear, and many secondary factors influenced the position of scientists, ministers and their followers. The strong anticlericalism of the Continent, fostered by the French Revolution, made many a representative of the New England merchant aristocracy suspicious of anticlerical interpretations of science. Early nineteenth-century Unitarianism was theologically liberal, expressing the mode of thinking of the New England mercantile class, but because of this very connection it had at the same time a conservative social cast. Where Unitarians might agree on a more liberal interpretation of the Bible in the sense of the Newtonian theory, as a rule they were not too favorably inclined toward the anticlerical implications of geology and biology. Similar influences made themselves felt in other groups of Protestants, the countryside harboring a traditional distrust of the liberalism of aristocratic Harvard.

In New England there was never a deep social-economic foundation for anticlericalism. With the exception of a few Tory ministers, most of the clergy had supported the Revolution. The building of the new republic was accomplished in full collaboration with the churches. The old theocratic rule of the Puritan elders had disappeared gradually and without bloodshed; the separation of Church and State in the beginning of the nineteenth century was accomplished in a peaceful way. The relations between the people and the Church had remained as a whole harmonious in the United States. No battlecry such as Voltaire's *"Ecrasez l'infame"* was ever raised as an issue in the political struggles of the country. People felt sympathetic or at the most indifferent to their churches and no basis existed for a mass movement which would take up science as a means to combat the clergy.

In early republican America Jeffersonian democracy breathed an atmosphere of freethinking, both in the field of religion and in the field of social relationships. It had, however, only a few spokesmen among the leading men in New England, and they were kept isolated. The overtone of intellectual life in Boston, New Haven and other cities along the north Atlantic Coast was conservative. It remained conservative when the new industrialism began to replace the older system of shipping and commerce. Although the new factory system and changing modes of communication stimulated the growth of new ideas, the New England ruling class remained as a whole very much on its guard against what was considered as freethinking.

Jacksonian democracy touched New England much more deeply than Jeffersonian democracy ever did. A struggle for the readjustment of old conceptions into the framework of a new age of industrialism set in, and affected all classes in its own way, not only the older mercantile class, the mechanic and the farmer, but also the new classes, the industrial bourgeoise and the working class. There were many men and women willing to draw consequences from the new scientific ideas. There were also men and women willing to fight for the *status quo,* whether in the field of social relationships or in the field of ideas. Religion was often a means of sanctifying this *status quo.* The new social ideals of the working class were opposed not only by the employers, but by a large section of the clergy. The propaganda of the abolitionists also met with strong resistance from the New England Cotton

Whigs, and these industrialists, who were interested in Southern cotton production, were again supported by a considerable section of the clergy. Ten-hour-day and antislavery propaganda tended to strengthen a conservative ideology among the ruling classes, though at the same time these very classes revolutionized the whole economy of the country by means of factories and railroads.

This situation was reflected in the public attitude towards Harvard College. The fact that it was tainted in many eyes with liberal opinions on religious issues did not mean that it was also socially liberal. Harvard stayed conservative socially and actually maintained a cautious position in theological controversies. This social conservatism made it, more than Brown and Yale, the object of distrust for many farmers, workers and small middle-class elements as a stronghold of the ruling class. This occasionally even reflected on liberal religion, which was for the same reason also distrusted.

The reception by Harvard of the ideas of Ralph Waldo Emerson shows how cautious the college authorities were in matters of liberal religion. Emerson, in July 1838, some years after he had abandoned the Unitarian Church, delivered his address to the senior class of the Harvard Divinity School. "That masterpiece of revolutionary rhetoric," writes Professor Morison, quoting the words of the address, "seemed to float in from the New England countryside sweet with the breath of the pine, the balm of Gilead, and the new hay." Soon the auditors realized that they were hearing a devastating criticism of all organized religion; the Church was contrasted with the Soul, at the end they heard the Messianic prophecy of a "new Teacher" who would "see the world to be the mirror of the soul, shall see the identity of the law of gravitation with purity of the heart." Harvard authorities were upset, and Emerson was severely criticized; Unitarianism was not willing to let such unorthodox philosophy go unchallenged, even if Emerson's mysticism did not threaten to disrupt the foundations of a society based on wage labor and chattel slavery. Emerson lost any chance of receiving a professorship from Harvard, and it was not until thirty years later, when the issues had changed, that the college atoned for its previous lack of liberalism by giving the aging philosopher an honorary degree and electing him an Overseer.

2

In the early nineteenth century only a few daring radicals expressed doubt concerning the Mosaic account of creation. In the beginning, it said, God created Heaven and Earth, on the sixth day He created Man, and on the seventh day He rested. The date of this Event was known and was computed by counting the number of generations from Adam; it was about 4000 years before the beginning of the Christian Era—4004 B.C., to be exact.

The liberal point of view among the orthodox Protestants was expressed by the widely read texts of the English divine William Paley, whose numerous books were reprinted in many editions. He established the authority of the church in its utility; the motive for virtuous action was the expectation of future reward. Particularly popular was his *Natural Theology,* published in 1802, in which he demonstrated the existence and perfections of God from the evidences of design in the adaptation of creatures and objects to nature. His book is a small scientific encyclopedia with a large number of pictures, all chosen to demonstrate divine power, wisdom and benevolence. Even the inclination of the earth's axis with respect to the ecliptic was evidence of this benevolence, for where would we be without the seasons? This book by Paley had several American editions, including one in Boston in 1826, and we find enough references in the works of pre-Civil War scientists of New England to show that Paley's natural theology enjoyed a great authority.

The *Evidence of Christianity* (1794) was another of Paley's books from which critical minds could draw rational answers, when plagued by doubt concerning the truth of Revelation and the literal interpretation of the Bible. Paley's answers, however, became more and more inadequate with the progress of geology. The Huttonian and Wernerian theories, which in those early years of the American Republic competed with each other in Europe, created considerable disturbance among those scientists who were theologically inclined. Both theories required enormous lengths of time to account for the building and subsequent convulsions of the beds of rock. Hutton was already preaching the doctrine of "uniformitarianism," later identified with Lyell's position. It was obvious that by only admitting those geological processes which can still be observed today, the Mosaic six days would prove to be an insufficient period. In New England, where Federalism, allied to a socially conservative religion, constituted the background of college thought, official geology proceeded to tackle the touchy subject of religion and geology with great hesitation.

Benjamin Silliman, the first professional geologist in New England, received the greater part of his geological training during the winter of 1805–1806 at Edinburgh, where Hutton had worked until his death in 1797. Silliman, however, was inclined at first to favor Werner's "Neptunist" doctrines, which did not seem to clash so conspicuously with the Mosaic report as the "Vulcanic" theories of Hutton. Both Werner's theory and Genesis began with the same kind of watery chaos, from which land eventually emerged. Hutton's theory, however, began to improve its academic standing with the accumulation of geological data. Maclure, in 1822, could write that Werner's views were "fast going out of fashion," and Maclure's influence on American geologists was considerable. By that time public interest in the new sciences had definitely begun, and all kinds of questions were asked. Silliman was forced to take a more definite stand on the theological sides of the issues emerging from geology. For a professor at strictly Congregationalist Yale, it was a difficult choice to make:

> Professor Silliman was embarrassed in this conflict by his sincere respect for the teachers of religion, and his reluctance to lower the estimation in which they were held. Hence, whatever he published on this theme, is characterized by the utmost forbearance and courtesy. For his own part, he felt that the Bible was a revelation from God. . . . He was impressed with the observation of Cuvier, that the cosmogony in Genesis "considered in a purely scientific view, is extremely remarkable, inasmuch

as the order which it assigns to the different epochs of creation is precisely the same as that which has been deduced from geological consideration." At the same time, Professor Silliman judged that it was no part of the object of the sacred writer "to enter further into details than to state that the world was the word of God; and thus he was naturally led to mention the principal divisions of natural things, as they were successively created."

Silliman's rationalization was as follows. The Bible, as he saw it, was a code of moral instructions rather than a scientific textbook. Therefore, it should not be surprising to find natural phenomena described as they appear to the layman rather than in the more analytical terms of science. Seen in this light, however, the Scriptures appear to be in remarkable agreement with the results of geology. Both science and the Holy Script declare that the world had a beginning, and thus prove the existence of a Creator. Both give an identical account of the order of events by which sea and land, living creatures and finally human beings, appeared. It is only necessary to conceive a day in the Mosaic account as a geological period of indefinite length to arrive at a full correspondence between the two records. And though Silliman disagreed with the common view that the different deposits on the earth's surface were laid down during the deluge at the time of Noah, he claimed that there was enough geological evidence to believe that the deluge actually occurred.

Silliman was applauded by Hitchcock, and from the conservative side attacked by Moses Stuart of Calvinist Andover Seminary. The opposition from the more progressive side could not originate in the New England of the thirties. It came from that old freethinker, Thomas Cooper, the English-born friend of Priestley and Jefferson, and president of South Carolina College. "Old Coot" had to use Silliman's edition of Bakewell's *Geology* in his classes, which annoyed him greatly. He expressed this annoyance in a pamphlet of 1833, entitled *On the Connection between Geology and the Pentateuch,* in which he held "that the Old as well as the New Testaments were by no means meant as infallible guides in mineralogy and geology." He concluded that it was well for Professor

Silliman "that his useful services to science have placed his reputation on a more stable foundation than his absolute unconditional surrender of his common sense to clerical orthodoxy." His personal correspondence with Silliman ended with a letter in which, according to Silliman, "he reviled the Scriptures, especially of the Old Testament, pronouncing them in all respects an unsupported, and in some respects, a most detestable book." The controversy brought calumny and abuse upon the venerable head of Cooper, and despite an eloquent defense before the South Carolina House of Representatives he was forced to resign as president of the college, at the age of seventy-four. However, he continued to teach chemistry.

3

Silliman's piece of exegesis, which essentially reproduced ideas expressed before by such men as Cuvier, was quite sympathetically received in New England, at any rate in the academic and the more liberally inclined religious circles. When Silliman lectured in Boston in 1835 and 1836 the Unitarian clergy were friendly towards his views and Silliman wrote that he was thanked "warmly for the manner in which they say delicate points were treated."

Silliman's pupil, Edward Hitchcock, for a time president of orthodox Amherst College, was equally concerned with the dangers geology harbored for traditional religion. Hitchcock had been a minister, and Amherst was even more than Yale under the influence of orthodox Calvinism. The college authorities aproved of revival meetings, many of which were held at Hitchcock's home. For him the teachings of science were not so much preparation for a better material world as an illustration of the principles of "natural theology." Hitchcock decided to publish his views and around 1835 began to write extensively on the subject of Genesis, religion and geology. Like Silliman's articles of some years earlier these were attacked by the literal-minded Moses Stuart of Andover Seminary. A series of articles followed, in which Hitchcock and Silliman were joined by Professor James L. Kingsley of Yale, a classicist with scientific leanings, who had assisted Silliman, years before, in studying the Weston meteor.

By this time Lyell's *Elements of Geology* had appeared, introducing his theory of "uniformity." Lyell himself was a religious man, and took care to end his book with the statement that "in whatever direction we pursue our researches, whether in time or space, we discover everywhere the clear proofs of a Creative Intelligence, and of His foresight, wisdom and power." This theory, however, made Hitchcock and other men of his type extremely unhappy. Silliman was more reserved in his opinion, and when in 1837 Hitchcock placed Cooper and Lyell in the same class, he remarked that "the evidence ought to be much stronger to justify placing Mr. Lyell in the same company." Lyell's position in the United States was enormously strengthened by his two visits in 1841–1842 and 1845–1846, which convinced many a good Christian scientist in America that the theory of "uniformity" could be reconciled with religion. The problem kept Hitchcock constantly on the alert, and eventually led him to the publication of a volume entitled *The Religion of Geology.* It appeared first in 1851, was reprinted with an addition in 1859. This book was the culminating document of the pre-Darwinian controversy between natural science and theology in America. It brought Paley up to date. Here Hitchcock, at that time president of Amherst College, expounded the favorite theory of all "natural" theologists that by the study of scientific laws we can learn something about the nature of God, and that geology is consistent with a belief in the full inspiration of the Bible. He analyzed the Hebrew word for "day" and concluded that it could be used to denote any period. He had at the same time to modify Silliman's belief that the "days" of Genesis really represented indefinite periods, since the order of the fossils in the rocks does not correspond exactly with the order indicated in the Bible. There might have been a long period between the creation "in the beginning" and the six demiurgic days. The Mosaic account, wrote Hitchcock, pays scant attention to all the different creations preceding the creation of man, because they are irrelevant to the main purpose of the Bible, which is a moral one.

The whole scheme of creation, in Hitchcock's eye, revealed one "great system of benevolence and wisdom." God, in all these eons of time, had been improving the universe for the benefit of man, the end for which all was established. The soil was weathered and disintegrated so that man might grow plants, the strata of the earth were broken up so that he might find minerals, water was placed where man could reach it:

> We should expect . . . that this element . . . must be very unequally distributed, and fail entirely in many places; and yet we find it in almost every spot where man erects his habitation.

The subject kept many pens busy, and books and pamphlets continued to appear regularly. In 1856 James Dwight Dana at Yale took issue with those who attacked natural theology by attacking science. Dana, in an article reprinted as a pamphlet, *Science and the Bible,* represented a modified, and more modern, point of view. Dana was interested first of all in vindicating science, rather than—Paley fashion—presenting science as an adjunct of theology, useful primarily for the added light it throws on revelation. This mode of reconciling science and religion received the approval of several outstanding scientists, including Agassiz and Peirce. Dana's standpoint constituted not only a defense of the conclusions of science, but, by attacking the literalists, also a defense of religion. Such literalists, claimed Dana, not only did a disservice to the Bible, they actually led people to believe that science and religion were hostile; in such a state of mind, they might be inclined to abandon the Bible altogether.

4

When the *Origin of Species* appeared in 1859 the battle front shifted to the fields of evolution and biology. This does not mean that the problems of biology had not previously been drawn into the controversy between religion and science. On the continent of Europe the question of evolution had been vividly dramatized by the debate of 1830 in the Paris Academy between Geoffroy Saint-Hilaire and Cuvier. Agassiz, who felt it his main goal in life to be a second Cuvier, defended Cuvier's anti-evolutionary principles wherever the opportunity offered and in his edition of the *Asterolepis of Stromness* he spared no efforts to assert his belief in special creation and in Design. In the days before Darwin the semi-popular side of this controversy was exemplified by the debate on the merits of a book called *The Vestiges of the Natural History of Creation,* published anonymously in 1844 in England, and soon followed by a *Sequel.* The author was a Scotch writer and publisher, Robert Chambers, whose publishing house was widely known through such publications as *Chambers's Cyclopaedia of English Literature.* In this book Chambers marshaled an able defense of the "principles of progressive development," which, at the time of its publication, was sufficiently unpopular in Great Britain to impress upon the author the necessity of anonymity—an anonymity he maintained during his whole life.[1] Darwin, in the introduction to the *Origin of Species,* criticized the accuracy of the information and the want of scientific caution in the earlier editions, but found that it had "done excellent service in this country in calling attention to the subject, in removing prejudice, and thus preparing the ground for the reception of analogous views." That Chambers in his enthusiasm occasionally went too far is shown by his argument that spontaneous generation shows that life can be created from lifeless substances.

Chambers's book found an immediate and wide circulation, and was reprinted repeatedly. The first American edition—it was the period of the cheap American reprints without any copyright legislation—appeared in 1845, issued by the respectable firm of Wiley and Putnam in New York. It was promptly attacked by scientists as well as church leaders; the scientific attack was led by Hitchcock and later by Agassiz. The revolt was against Chambers's thesis "that any of the lower animals were concerned in the origin of man." The frequency with which the *Vestiges* was mentioned in their

[1] The anonymity was far from strict, and it was widely rumored that Chambers was the author. See, e.g., art. Chambers, W. and R., in *New American Cyclopedia* IV, p. 684, published in 1858. The name of the author was made public in 1881.

work shows how deep the influence of the book must have been. In those days Asa Gray took the anti-evolutionary side, and in a discussion in Boston's *North American Review* called the *Vestiges* an ingenious scientific romance and a hypothetical history of creation. His criticism was qualified, however, and in a review of the *Sequel to the Vestiges of Creation* he conceded that the *Vestiges,* notwithstanding some errors, had been, and would continue to be, of incalculable value to science and general knowledge. But he remained critical in a crudely mechanistic way: "How can a baboon originate anything above a baboon, with only a baboon's mechanism?" This was in 1846. Gray gradually changed his position to a more dialectical one, and in 1854 he wrote to the British botanist Hooker that "scientific Systematic Botany" rests upon species "created with almost infinitely various degrees of resemblance among each other," and raised the question of variation. Gray's work on the geographic distribution of plants may also have influenced the modification of his ideas. Hooker showed Gray's letter to Darwin, who was at that time quietly collecting material in support of his general thesis concerning the origin of species. Then, in 1855, Darwin wrote his first letter to Gray, in which he guardedly expressed some of his ideas; when Gray showed sympathetic interest, the correspondence continued. In 1856 Darwin confided to Gray that he had "come to the heterodox conclusion that there are no such things as independently created species —that species are only strongly defined varieties." Both Darwin and Gray had reason to pursue very quietly the direction in which their thoughts were heading; Darwin because he was afraid that the author of the *Vestiges* would popularize his ideas before they had matured—"and this would greatly injure any chance of my views being received by those alone whose opinions I value"; Gray because of the narrow spirit prevailing in his New England circle and above all because of the strong anti-evolutionist feelings of his dynamic friend Agassiz. On September 5, 1857, Darwin wrote to Gray the letter, now famous in the history of the Darwinian theory, in which he outlined his theory for the first time. It is still an interesting document, which gives a clear idea of the main character of Darwin's thinking. Darwin called it "a *most* imperfect sketch," and told Gray that his "imagination must fill up very wide blanks":

> I am convinced that intentional and occasional selection has been the main agent in the production of our domestic races; but however this may be, its great power of modification has been indisputably shown in later times. Selection acts only by the accumulation of slight or greater variations, caused by external conditions, or by the mere fact that in generation the child is not absolutely similar to its parent. Man, by this power of accumulating variations, adapts living beings to its wants—may be said to make the wool of one sheep good for carpets, of another for cloth, etc. . . .
>
> I think it can be shown that there is such an unerring power at work in *Natural Selection* (the title of my book), which selects exclusively for the good of each organic being. The elder De Candolle, W. Herbert, and Lyell have written excellently on the struggle for life; but even they have not written strongly enough. Reflect that every being (even the elephant) breeds at such a rate, that in a few years, or at most a few centuries, the surface of the earth would not hold the progress of one pair. I have found it hard constantly to bear in mind that the increase of every single species is checked during some part of its life, or during some shortly recurrent generation. Only a few of those annually born can live to propagate their kind. What a trifling difference must often determine which shall survive, and which perish!

It was this letter, read at the Linnaean Society in England on July 1, 1858, on the occasion of the presentation of the papers by Darwin and Wallace, which established the priority of Darwin. When, in 1859, the *Origin of Species* appeared, Gray was the first American scientist to meet the objections against the iconoclastic theory.

5

The American public was prepared for the new theory by a paper by Joseph Hooker on the Flora of Tasmania, which appeared in the *American Journal of Science* of January 1860. In this article the British botanist placed his considerable professional authority behind "the ingenious and original reasonings and theories of Mr. Darwin and Mr. Wallace." Gray added a note: "To those who have intelligently observed the course of scientific investigations, and the tendency of speculation, it has for some time been manifest that a re-statement of the Lamarckian hypothesis is at hand." Hooker, as well as Gray, had found in his studies on the geographical distribution of species a bridge to the theory of evolution. Then, in the next number of the *American Journal,* in March 1860, Gray began to review Darwin's book. In thirty-one pages of cautiously worded approval, Gray made a plea for scientific tolerance. He compared Agassiz's point of view with the ideas of Darwin, carefully explaining both doctrines:

> [The theory] of Agassiz differs fundamentally from the ordinary view only in this, that it discards the idea of common descent as the real bond of union among the individuals of a species, and also the idea of a local origin,—supposing, instead, that each species originated simultaneously, generally speaking over the whole geographical area it now occupies or has occupied, and in perhaps as many individuals as it numbers at any subsequent period.
>
> Mr. Darwin, on the other hand, holds the orthodox view of the descent of all the individuals of a species not only from a local birth-place, but from a single ancestor or pair; and that each species has extended and established itself, through natural agencies, wherever it could; so that the actual geographical distribution of any species is by no means a primordial arrangement, but a natural result. He goes farther, and this volume is a protracted argument intended to prove that the species we recognize have not been independently created, as such, but have descended, like varieties, from other species. Varieties, in his view, are incipient or possible species: species are varieties of a larger growth and a wider and earlier divergence from the parent stock: the difference is one of degree, not of kind.

Gray summed up: "The theory of Agassiz regards the

origin of species and their present general distribution over the world as equally primordial, equally supernatural; that of Darwin, as equally derivative, equally natural."

Gray was not too well pleased with his own review, and wrote to Darwin that it did not "exhibit anything like the full force of the impression the book has made upon me. Under the circumstances I suppose I do your theory more good here, by bespeaking for it a fair and favorable condition."

It was Gray's intention to rid Darwin's theory of its apparent materialism by showing that evolution by natural selection did not exclude Design from nature. At the time he wrote to Darwin that he was determined to baptize the *Origin of Species nolens volens,* which would be its salvation: "But if you won't have it done, it will be damned."

Gray's attitude was that Darwin's theory had no special bearing on religion, that natural selection and evolution were not necessarily atheistic and that the mystery of creation remained as much of a mystery as before. He asked for study and acceptance of Darwin's work on its scientific merits. He who wanted to believe in Design could continue to believe in it even after accepting evolution. This remained Gray's position throughout the whole controversy.

Gray's many subsequent articles on the same subject were later published under the name *Darwiniana,* and also propounded in his *Natural Science and Religion.* Gray himself always remained a faithful member of the Congregational Church. Darwin, on the other hand, did not like to enter into religious controversy, and in a letter to Gray expressed doubts as to his tactics:

> I grieve to say that I cannot honestly go as far as you do about design. I cannot think that the world, as we see it, is the result of chance, and yet I cannot look at each separate thing as the result of design. . . . With respect to the theological view of the question, this is always painful to me. I am bewildered.

Darwin remained steadfast in this detached point of view, which led his pupil Thomas Huxley to agnosticism; he also discouraged attempts to connect his name with modern materialism, which Karl Marx was developing in the same period. He was an outstanding example of the "pure" scientist as developed by the nineteenth century, who pursues his science without asking for its bearing on religious or social forces—"truth for truth's sake." In the same period New England's more oustanding scientists clung more firmly to a religious point of view. This gave to their philosophy a somewhat archaic flavor, but it had also a tendency to bring it closer to the people. There was as yet no room for modern or even "shamefaced" materialism in a country with a still undeveloped labor movement. It was the period where the warriors of the North in their fight against slavery were inspired by that beautiful religious anthem, the "Battle Hymn of the Republic," composed by a New England author.

6

Agassiz's bitter and sometimes acrimonious opposition to Darwin's theory came as no surprise to those who had followed the trend of his thinking since the days when he had decided to follow in the footsteps of his master Cuvier. Already in 1844, in his work on fossil fishes, he had written that the successive development of creatures, especially of fishes, "shows conclusively the impossibility of referring the first inhabitants of the earth to a small number of branches, differentiated from one parent stock by the influence of the modifications of exterior conditions of existence." He reiterated this point of view in later works, so that Gray knew well what a formidable antagonist he was creating by his defense of Darwin's book. Agassiz's position on evolution did not mean that he was religious in the ordinary sense of being a church member—he was never affiliated with any organized religious body, though he probably sympathized most with the Unitarianism of his Harvard friends—nor does it mean that he was afraid to take a position against which church opposition could be expected. The "biogenetic law," which he formulated in 1857, and which also can be traced to his earlier work, contains an element of revolutionary dynamics against which conservative minds might well take issue. At least once he had met with serious attacks from the pulpit. This occurred around 1852 when he declared that man had sprung not from a common stock, but from various centers, and that the original circumscription of these primordial groups of the human family corresponded in a large and general way with the distribution of animals and their combination into faunae. That did away with Adam and Eve, and a section of the clergy resented it. But Agassiz believed in Design, and never tired of showing his belief in every field in which he worked and lectured. A personal element may have entered when Darwin's book appeared just two years after Agassiz's masterful "Essay on Classification" and almost totally eclipsed it. The mainspring of his opposition to Darwin, however, lay in Agassiz's faithful adherence to the teachings of Cuvier, whose enormous work in classification and comparison—above all, comparison—he felt it to be his life's task to continue. This adherence to Cuvier harmonized with his natural religiosity, but there was more involved than religion. The essence of the struggle lay, for Agassiz, in the defense of that attitude in science which was guided by the search for Design.

Agassiz took issue with Darwin's theory in the third volume of his *Contributions to the Natural History of the United States,* the great work in which he emulated Cuvier. In the *American Journal* of November 1860 his remarks were reprinted, ending with the temperamental conclusion:

> I shall therefore consider the transmutation theory as a scientific mistake, untrue in its facts, unscientific in its method, and mischievous in its tendency.

Agassiz used his endless knowledge of zoology in his attempts to refute the hated theory. He marshaled example after example to show that there was no reason to believe in gradual evolution, that there were unaccountable jumps and gaps, and that similarity in form does not imply a common descent. This position can be summarized in the following words, taken from a book which Agassiz published in 1863, the result of a lecture course:

> It is my belief that naturalists are chasing a phantom, in their search after some material gradation among created beings, by which the whole Animal Kingdom may have

been derived by successive development from a single germ, or from a few germs. It would seem, from the frequency with which this notion is revived,—ever returning upon us with hydra-headed tenacity of life, and presenting itself under a new form as soon as the preceding one has exploded and set aside,—that it has a certain fascination for the human mind. . . . But I nevertheless insist, that this theory is opposed to the processes of Nature, as far as we have been able to apprehend them; that it is contradicted by the facts of Embryology and Paleontology, the former showing us norms of development as distinct and persistent for each group as are the fossil types of each period revealed to us by the latter; and that the experiments upon domesticated animals and cultivated plants, on which its adherents base their views, are entirely foreign to the matter in hand, since the varieties thus brought about by the fostering care of man are of an entirely different character from those observed among wild species. And while their positive evidence is inapplicable, their negative evidence is equally unsatisfactory, since, however long and frequent the breaks in the geological series may be in which they would fain bury their transition types, there are many points in the succession where the connection is perfectly distinct and unbroken, and it is just at these points that new organic groups are introduced without any intermediate forms to link them with the preceding ones.

Agassiz's arguments, from time to time, offer interesting topics of discussion in formal and dialectical logic: "If species do not exist at all—how can they vary and if individuals alone exist, how can the differences which may be observed prove the variability of species?" he asked in 1860. It was an argument essentially as old as Zeno and is a favorite with those who deny the reality of change. How far Agassiz was led into metaphysical idealism can be seen from a passage in his lectures of 1863:

When the first Fish was called into existence, the Vertebrate type existed as a whole in the creative thought, and

> the first expression of it embraced potentially all the organic elements of that type, up to Man himself.
>
> To me the fact that the embryonic form of the highest Vertebrate recalls in its earlier stage the first representation of its type in geological times and its lowest representatives at the present day, speaks only of an ideal relation, existing, not in the things themselves, but in the mind that made them.

We see that Agassiz, in his opposition to Darwin, finally joined hands with the most archaic Platonism. But for Agassiz Darwinism was just a fad. It reminded him, he wrote in 1867, of what he "experienced as a young man in Germany, when the physio-philosophy of Oken had invaded every center of scientific activity; and yet, what is there left of it? I trust to outlive this mania also."

Agassiz continued the struggle until his death in 1873. His last message appeared in the *Atlantic Monthly* of January 1874, and although by this time he was willing to grant Darwin more scientific merit than in the first hot flush of opposition, he remained adamant in rejecting the main elements of the theory of evolution. But with his death opposition to Darwin lost its main scientific proponent, not only in New England, but throughout the whole world. Darwinism gradually became established in the universities, and eventually even Alexander Agassiz, Louis Agassiz's son and fellow zoologist, adopted the fundamental principle for which Darwin had labored.

7

Louis Agassiz's opposition impeded the acceptance of Darwin's views among New England scientists. Most naturalists kept very quiet, partly because of the traditional influence of natural theology, and partly because they were impressed by the forceful personality of the Cantabrigian Saint George fighting the dragon of evolution. We have seen how even Asa Gray could come to the defense of Darwinism only after having it baptized. A few men of science shared Gray's position, but their support was most restrained; among these were

Joseph Henry, William Barton Rogers and Jeffries Wyman. Most American naturalists locked themselves up in their studies and continued their specialized work without giving public support to either side. In a publication of 1863 Wyman took a neutral position: "We arrive at no reasonable theory which takes a position intermediate between the two extremes," but his feelings were expressed in a private letter of 1871:

> It is a curious fact that the opponents of evolution have as yet started no theory except the preposterous one of immediate creation of each species. They simply deny. After many trials I have never been able to get Agassiz to commit himself to even the most general statement of a conception. He was just the man who ought to have taken up the evolution theory and worked it into a good shape, which his knowledge of embryology and palaeontology would have enabled him to do. He has lost a golden opportunity, but there is no use in talking of that.[2]

Jeffries Wyman was professor of anatomy in the Lawrence Scientific School, a part of Harvard College. He was an accomplished anatomist with an enormous stock of information in many fields of medicine, physiology and natural history. Born in Chelmsford in 1814, and a Harvard graduate with a M.D. of 1837, he had spent a year of study in Paris and London. He contributed in his own way to the discussion on evolution by his work on anthropoid apes. In his paper of 1847 he gave one of the first scientific descriptions of the gorilla. In those days little was known about the structure of anthropoid apes, though gorilla, chimpanzee and orangutan had been the object of some scientific study, going back to the seventeenth century. The skeleton of a gorilla, however, seemed new to science; it was brought to the attention of Wyman by his friend Thomas S. Savage after a visit to Africa.

Wyman's studies of anthropoid apes led him to compare them with one another and also with man. He regarded the

[2] Wyman once, in a talk with Asa Gray, recalled a conversation with Agassiz, in which the latter said that Humboldt had told him that Cuvier missed a great opportunity in taking sides against Geoffroy Saint-Hilaire.

chimpanzee as holding "the highest place in the brute creation," giving the gorilla a lower position. He determined the differences and the similarities between the skeleton structures of the anthropoid apes and man. Then he came to the controversial conclusion:

> It cannot be denied that the Negro and the Orang[3] do afford the points where man and the brute, when the totality of their organization is considered, most nearly approach each other.

This was in 1850. The implication of this statement, which gave comfort to the slave masters, was only mildly softened by the author's declaration that:

> the difference between the cranium, the pelvis, and the conformation of the upper extremities in the Negro and the Caucasian sinks into comparative insignificance when compared with the vast difference which exists between the conformation of the same parts in the Negro and the Orang.

Wyman's careful investigations of the anatomy of Negro races have great merit, though his way of stating the question of the relations between ape and man was unfortunate, to say the least, and reminiscent of the kind of "science" which the slave masters of all ages have produced as a foundation for their practices.[4] We recognize, however, that Wyman at the same time placed before the public certain facts pertaining to a common descent of man and animal. A later statement shows that Wyman's thoughts actually led in this direction; shortly before his death in 1874 he observed that man "must have gone through a period when he was passing out of the

[3] "Orang" means here any anthropoid ape.

[4] Agassiz occasionally also indulged in similar observations about the Negro. In a letter to S. G. Howe of 1863 he called the Negroes "indolent, playful, sensual, imitative, subservient, good natured, versatile, unsteady in their purpose, devoted and affectionate," and this from the time of the Pharaohs until the present. It must be added that Agassiz was at the same time in favor of absolute equality of Negro and white before the law.

animal into the human state, when he was not yet provided with tools of any sort, and when he lived the life of a brute." The material collected by Wyman was used by Darwin in his *Descent of Man.*

Agassiz was joined by Edward Hitchcock, who was still fighting the battle for natural theology, by John Torrey and by his friend Arnold Guyot. In Cambridge he was joined by James Russell Lowell, the poet and author, who was at first an abolitionist, but gradually turned into a conservative. "Such a mush seems to me a poor substitute for the Rock of Ages," was his remark concerning evolution. He hated science. "I hate it as a savage hates writing," he once said, "because I fear it will hurt me somehow." Agassiz was also supported by Francis Bowen, professor in Harvard College.

Francis Bowen has the unusual distinction of being a man not admitted—though only temporarily—to the faculty of Harvard College because of his reactionary political views. He was an editor of the *North American Review* and a Lowell Institute lecturer on "metaphysical and ethical science to the evidences of religion." In 1850 he defended Daniel Webster's plea for a Compromise on the slave question, and in articles and lectures attacked Kossuth and the struggle of the Hungarians against domination by the Czar. At that time he was chosen professor of ancient and modern history at Harvard. Public opinion in Boston, which had rallied in favor of the Polish rebels in 1830, and of European liberalism in the Revolution of 1848, became so incensed against Bowen that the Board of Overseers, which at that time included representatives of the state government, refused to confirm the nomination. However, when public resentment had quieted down in 1853, Bowen was quietly reappointed, now to the chair of "natural religion, moral philosophy and civil policy," where he continued to teach for many years.

In 1861 Bowen came to the aid of the anti-evolutionists. "There can be no mistake," he declared, "as to the character of such a scheme of cosmogony as this [Darwin's]. Creation denied, or pushed back to an infinite distance, and a blind and fatalistic principle watching over a chaos of unmeaning and purposeless things, and slowly eliciting from them, during an eternity, all the order and fitness which can characterize the organized world." He attacked Darwin because of his denial of Final Causes—"that is, the doctrine that purpose, intention, or design, is nowhere discoverable in organic nature." Such a position helps explain the negative attitude which natural scientists in the later nineteenth century began to assume toward philosophy, to the disadvantage of both science and philosophy.

8

James Dwight Dana took no part in the controversy between Agassiz and Gray. He did not read the *Origin of Species* until 1863 and then was not too favorably inclined. In his *Manual of Geology* and in other papers he had indorsed the current scientific faith in immutability. He believed that a species was "a specific amount or condition of concentered force, defined in the act or law of creation." Gray, who reviewed the *Manual* in 1863, concluded that one who held this view could hardly be expected to welcome such a theory as Darwinism.

Like Gray and Agassiz, Dana was no longer a young man when Darwin dropped his bombshell upon the scientific world. He was trained in that school of Christian apologetics which found the best evidences of Divine power in frequent interruptions of the cosmic scheme. As late as the 1870 edition of his *Manual* he still believed that the attempt to establish evolution was vain. "There are no lineal series through creation," he wrote, "corresponding to such methods of development." During the next years, he began to alter his position somewhat—Agassiz died during this period—but it was not until the fourth edition of the *Manual,* in 1895, that he expressed his conversion to Darwin's principles. He added, however, that "the intervention of a Power above Nature was at the basis of Man's development." He was vague about including man in the evolutionary chain. It was not until Dana's death in the same year that a letter was published in which he showed that he too was willing to recede from his position.

It would be wrong, however, to see Dana's position with respect to evolution exclusively as a passive one, first of nonacceptance, later of gradual acceptance. Dana had his own theory, which did not attack the doctrine of the invariability of species, but had, like Agassiz's biogenetical principle, an evolutionary flavor. The study of the Crustacea, which Dana undertook during and after the Wilkes expedition, suggested to him a broad general principle, which he called cephalization. He enunciated it first in his report on the Crustacea, and later discussed it more fully in a number of papers, mostly in the *American Journal of Science* between 1863 and 1866.

In the Crustacea each segment of the body bears a pair of jointed appendages, appropriated to different functions. The more developed Crustacea, such as the crabs and lobsters (Decapoda), have more segments with appendages cephalic in function—sensory and oral—than the lower ones, such as the sand fleas. Moreover, in the crabs, which form the highest division of the Decapoda, the posterior part of the body is greatly reduced in size, and most of its segments have no appendages at all. It is as if the whole body were almost absorbed into the head.

This fact and many others of a related nature suggested to Dana the broad principle that the grade of different animals in comparison with each other is shown by the "degree of structural subordination to the head and of concentration headward in body structure." There is indeed a tremendous advance in cephalization as we pass from the lower, and generally earlier forms of animal life, to the higher types. A protozoan has not even a mouth—the earliest cephalic structure to be developed—while the higher mammals have well-developed brains and sensory organs. The principle has been criticized, especially since Dana's distinction between "high and low" organisms was not always clear; moreover, he occasionally went too far in giving undue weight to mere analogies. However, it is undeniable that there is an evolutionary element in the principle of cephalization. More precisely, it asserts that the development of animal life is proceeding in a definite direction. In the course of geological times there occurs, therefore, an irregular process of growth and perfection of the central nervous system. The brain, once it has reached a certain level in the process of animal development, is not subject to retrogression, but can only progress further.

We must understand that it is doing an injustice to these nineteenth-century naturalists as well as to biology to see the question of evolution too much from one side, that of Darwin. Dogmatic Darwinism also had its shortcomings. Dana, and even Agassiz, despite their prejudices against the particular tenets involved in the Darwinian theory of evolution, both contributed valuable ideas to the general picture of evolution, as we see it now—perhaps we should say to the general dynamical development of living matter. Dana contributed his principle of cephalization, Agassiz his biogenetical law and his emphasis on comparison rather than classification, and Gray not only his defense of Darwin's position, but also his theories on the geographical distribution of plants. And if we see the guiding idea in this development that evolution of living matter proceeds in a definite direction, then, as the Russian geologist Vernadsky has pointed out, Dana's contribution deserves to be compared with that of Darwin.

With Dana's willingness to accept Darwin's conclusions, evolution became an accepted mode of thinking in New England scientific circles. Organized religion was never fundamentally attacked; "natural" theology had time and opportunity to gradually adjust itself to scientific evidence. This adjustment, however, was not performed without considerable loss of prestige. Post-Civil War scientists rarely bothered to study the theological implications of their results. The era of the specialist had arrived; and the increasingly secularized Protestant churches for many years no longer took a strong polemical interest in the relations between religion and whatever branch of science was currently popular in the public eye. The scientist, on the other hand, often took little interest in the more general social and philosophical implications of his work. Science, losing contact with theology, also lost contact with philosophy as well as with human relationships, to the detriment of all. It has become the task of the twentieth century to reverse this last trend.

Chapter 12

Sky and Ocean

Lo! The unbounded sea!
On its breast a Ship, spreading all her sails—an ample
Ship, carrying even her moonsails;
The pennant is flying aloft, as she speeds, she speeds so
stately—below, emulous waves press forward,
They surround the Ship, with shining curving motions
and foam.

—Walt Whitman, 1865

1

This was the age of the packet boat and the clipper ship—representing the final evolution of the search for speed and efficiency of the sailing vessel. From the days of the privateers American shipbuilders had been engaged in building faster and faster vessels. Now, with the danger of pirates and British men-of-war a thing of the past, quick performance no longer had to be combined with smallness of craft. The tonnage of the ships could be increased almost indefinitely, from the 100-ton schooners of the early years of the century, to the 700-ton Cunarder *Unicorn* of 1840, and the 4555-ton *Great Republic* of 1853. The turbulent expansion of the American economic system set its mark upon ship construction, stimulating ship-owners as well as captains to surpass each other's records in design, in tonnage and in speed.

The New England clipper period began when in 1843 Enoch Train, a Boston merchant, ordered the Newburyport shipbuilder Donald McKay to build a ship for him "on the New York model." Train wanted to own a packet line from Boston to Liverpool. Donald McKay was a native of Nova Scotia, who had been trained in a New York shipyard; in 1840 he came to Massachusetts to become a partner of a Newburyport shipbuilding firm. McKay built the *Joshua Bates* for Train. It was a vessel of 620 tons, which was perhaps somewhat of a comedown for its builder, who had already achieved a reputation as a constructor of larger ships. However, when he saw the graceful performance of his new liner, Train was so pleased that he invited McKay to come to Boston. "We need you," he wrote, "and if you want any financial assistance in establishing a shipyard, let me know the amount and you shall have it." McKay accepted, and established his own shipyard in East Boston.

McKay built a series of ships in rapid succession for Train's White Diamond Line, all of which became famous as fast-sailing Yankee clippers. Their names reveal the pride both owner and builder took in these new marvels of the sea: *The Flying Cloud, Ocean Monarch, Lightning, Cathedral, Chariot of Fame.* The record for speed set by *The Flying Cloud* has rarely, if ever, been surpassed by any sailing vessel.

McKay built several more majestic ships of over 2000 tons, among them *The Great Republic,* which caught fire; but by 1855 clipper ship building had reached its apogee. The rather sudden end has been ascribed to different causes—the financial panic of 1857, the speculative character of the whole enterprise and the gradual perfection of the steamboat. Whatever the cause or causes of the eclipse, the prominence of the sailing vessel as a major means of ocean transportation came to an end.

The clipper ships, with their magnificent performance and wonderful beauty, have been the subject of a considerable amount of lyric prose. Every line of it is deserved, but we must also be willing to acknowledge the reverse side of the picture. They were speculative as a financial enterprise, and technically an anachronism, so that they have been called "the most overadvertised type in maritime history." Labor conditions on board ship in those days were poor, often outrageous. Discipline had to be maintained among an international crew by cruel and inhuman methods, and Shattuck, in his *Sanitary Survey* of 1850, found it necessary to quote what old Sam Johnson had to say about conditions of sailors in his day—1778:

> As to the sailor, when you look down from the quarter-deck to the space below, you see the utmost extent of

> human misery; such crowding, such filth, such stench! A ship is a prison, with a chance of being drowned—it is worse, worse in every respect—worse air, worse food, worse company.

It is difficult to state what the influence of clipper ship construction was on naval engineering and on the development of technique and technical education in general. The best we can say is that its influence was indirect: the creation of big shipyards with the consequent division of labor, the consistent construction of ships from models rather than by more intuitive methods, the accumulation of information on the behavior of hulks of different shapes under different velocities. The clippers helped create the public attitude necessary to the introduction of an age of speed. But the clippers left few direct imprints on the history of science and invention, and modern naval engineering owes more to the primitive experiments of John Fitch and Robert Fulton than to the highly specialized skill of Donald McKay.

2

The disappearance of the clipper ship signified the final triumph of the steamboat, and with it the triumph of iron as construction material. Steamboats, we have seen, had their early start in America on the large inland waterways in the West. Then, in 1818, the *Savannah,* a New York built ship, with side wheels and combining sails with steam, crossed the Atlantic from New York to Liverpool in twenty-six days, and returned safely. This established the possibility of crossing the Atlantic by steam, if supplemented by sails, but since the steam engines on the *Savannah* were small, the public was not very much impressed with the practicability of ocean steam navigation. Such mode of transportation was not resumed for twenty years.

The first regular passages by steam were made in 1838 by the *Sirius* and the *Great Western,* the former making the trip from London to New York in seventeen days, the latter from Bristol to New York in fifteen days. When, in 1839, Samuel Cunard founded his North American Royal Mail Steam Packet Company, he chose Boston as his terminus in the United States. Bostonians, who in 1835 had seen the opening of three main railroad arteries, now saw their city with triumph and pride as the end of a transoceanic steamboat line. The chances to compete successfully with New York increased. When Boston Harbor froze during the bitter winter of 1844, the local merchants had a channel cut from the wharf to the sea to allow the passage of the Cunarder *Britannia*—a glorious feat commemorated by a picture which still may be seen in sitting rooms of old New England.

These early Cunard liners took a little less than fifteen days to sail from Liverpool to Boston, including a stop at Halifax. They were thus faster than the packet ships, especially for westward passages in the fifties; even a great Train packet averaged forty days on such trips. Train himself, in the last period of the clippers, was one of the incorporators of a company to "navigate the ocean by steam," but he did not intend to compete with his own clipper ship business. Train expected that freight would continue to be carried by sailing vessels. This persistent belief in sailing ships was typical of New England, where steam engines met with stronger opposition than in other sections, especially in the field of water transportation. There was no life on the New England rivers comparable to that described in Mark Twain's *Life on the Mississippi.* New England did not fully participate in the age of steam until Civil War days, despite the fact that one of the greatest single improvements in steam engines since Wyatt was made by Corliss in Providence.

3

The seaboard population of the thirties and later differed in many respects from that of the early republic. The small towns, once so prosperous, were either declining or being transformed into factory towns, and Boston was gradually assuming a position as center of all commerce in New England. This decline of the smaller towns began with the Embargo and the War of 1812, when many merchants began to move into Boston. Between 1810 and 1839, for instance, Newburyport declined in population from 7634 to 6375. The seaport of Salem struggled valiantly for many years against increasing odds, but Nathaniel Hawthorne, who became surveyor of its port in 1846, found plenty of time for daydreaming. The Middlesex Canal, followed later by the railroads, brought inland traffic more and more into the commercial orbit of Boston, and only towns such as Providence, New Haven or Portland, with their own outlying tracts of hinterland, could survive a deadly competition with the growing New England metropolis. Salem survived as a factory town after the Naumkeag Steam Cotton Company was opened in 1848, while Derby Wharf decayed, and even the great whaling center of New Bedford escaped oblivion only through its transformation into a textile town. Many coastal towns, however, remained important shipbuilding centers.

Boston, in the meantime, grew from a city of 43,000 inhabitants in 1820 to over 100,000 in 1842, and the annual arrivals from foreign ports into its harbor increased from an average of 789 between 1800 and 1810, to 1473 between 1835 and 1841. Its whole geography changed—its hills were reduced, its waterline pushed deep into the ocean and the Back Bay; its suburbs expanded and became parts of the metropolitan center. New Englanders became more than ever convinced that Boston was the hub of the universe; New England culture, in the arts, the sciences and engineering, became more and more concentrated in Boston and its environs. The decline of New England agriculture and the depopulation of the countryside made this tendency even more pronounced. There was a large influx of foreign population; the immigrants either stayed in Boston or moved to the neighboring factory towns; but since these factory towns were often ruled and owned by Bostonians, the immigation only increased the hegemony of Boston over New England.

4

Astronomy, navigation and meteorology, the ancient scientific favorites of the New England coastal towns, remained a subject of constant interest to many townspeople. But the time when laymen could accomplish distinguished work in these fields was to a certain extent past, and professionals from the populous centers, especially from Boston and Cambridge, began to assume leadership in research. A large field remained open to amateurs and nonprofessionals, and still exists—but the center of gravity for work in astronomy, meteorology and theoretical navigation had necessarily to shift to the growing scientific centers. Bowditch moved from Salem to Boston, the older Bond from Portland to Boston. Comparatively little scientific vitality remained in the smaller coastal towns, with the exception of New Haven, where the enormous development of Connecticut industries provided a base for the continuation of scientific life at Yale College.

To a Boston captain of these days belongs the distinction of one of the most famous discoveries in the field of navigation, the discovery of the so-called Sumner line. Thomas Hubbard Sumner was the son of a Bostonian architect and Congressman, and graduated from Harvard College in 1826. The sea attracted him, and he shipped as a sailor on the Canton trade, and eventually became a captain—rather an unusual career for a man with a Harvard sheepskin. Sumner's fame is based on a discovery he made when sailing near the southeast coast of Ireland in December 1836. Weather was bad, he was running ENE under short sail, and he had only been able to determine his position by dead reckoning. We have seen how the determination of the exact position at sea kept the navigators of many centuries busy; we have also seen that Bowditch devoted much of his time to the perfection of the method of lunars. The gradual introduction of chronometers on board ships facilitated longitude computations, but we have seen how slowly Yankee skippers and Yankee shipowners became convinced of the need for chronometers. Sumner, in 1836, had a chronometer, and this enabled him to simplify considerably the determination of his position. In his own words:

> At about 10 A.M. an altitude of the sun was observed, and the chronometer time noted, but, having run so far without any observation, it was plain the latitude by dead reckoning was liable to error, and could not be entirely relied on.
>
> Using, however, this latitude, in finding the Longitude by Chronometer, it was found to put the ship 15′ of Longitude, E, from her position by dead reckoning; . . . but feeling doubtful of the Latitude, the observation was tried

with a Latitude 10′ further N, finding this placed the ship ENE 27 *nautical* miles of the former position, it was tried again with a Latitude 20′ N of the dead reckoning; this also placed the ship still further ENE and still 27 *nautical* miles further; these three positions were then seen to lie in the direction of *Small's light.* It then at once appeared that the observed altitude must have happened at all three points, and at *Small's light,* and at the ship, at the *same instant of time;* and it followed, that Small's light must bear ENE, if the Chronometer was right. Having been convinced of this truth, the ship was kept on her course, ENE, the wind being still SE, and in less than an hour, Small's light was made bearing ENE ½ E., and close aboard.

The Latitude by dead reckoning, was erroneous 8 miles, and if the Longitude by Chronometer had been found by this Latitude, the ship's position would have been erroneous 31½ *minutes* of Longitude too far W, and 8 *miles* too far S. The ship had, from current, tide, or error of log, overrun her reckoning, 1 mile in 20.

Thus it is seen, that an observation taken at *any* hour of the day, and at any angle between the meridian and E and W points, is rendered practically useful, inasmuch as the Chronometer can be depended on.

This was the discovery of the process by which a single altitude taken at any time is made available for determining a line in the chart—now generally called the Sumner line—on some part of which the ship is situated. Or, in more detail, and in Sumner's own words: "One altitude of the sun, with the true Greenwich time, determines 1) the True bearing of the land, 2) the errors of longitude by chronometer, consequent to any error in the latitude, and 3) the sun's true azimuth; when two altitudes are observed, and the elapsed time noted, the true latitude is projected; and if the times be noted by chronometer, the true longitude is also projected at the same operation."

Sumner's concept of the line of position is very close to that held by the modern navigator, by which the observer is placed on a circle of equal altitude with the geographical position of the observed body as the center. The method found almost immediate acceptance, and some British navigators even claimed that they had known it long before under the name of "Cross Bearing of the Sun." Whatever truth there might have been to this claim, Sumner made the method widely known. On his return to Boston he found the Harvard mathematician Benjamin Peirce willing to endorse its correctness; scientifically inclined sea captains tried his method out and found it satisfactory; one of them even placed Sumner in the same category as Bowditch and the Harrisons, the inventors of the chronometer. All this was duly recorded by Sumner in a booklet, published in 1843 by a Boston firm, which was republished and revised several times. From these booklets Sumner's method passed into the textbooks of navigation, where it can easily be found. Some modifications have been introduced, notably by the French, but the method in present use follows that of Sumner in considerable detail.

Sumner himself did not live long to enjoy his well-deserved fame. He fell mentally sick and after a short time ended his days in the McLean Asylum in Boston.

5

Captain Sumner's discovery of his method to find a ship's position at sea was only the most famous of a whole series of improvements and discoveries which pressed modern science into the service of American navigation. In the thirties and forties the conception that scientific teamwork under government supervision was required to aid navigation became at last officially accepted. One of the results of this change of attitude was the Wilkes expedition, which, as we have seen, served as a training school for many American naturalists. Another result was the systematic support now afforded to the Coast Survey.

The United States Coast Survey, reorganized in 1832, developed under Ferdinand Hassler and Alexander Dallas Bache into a magnificent scientific organization. In his twenty-four years of service, beginning in 1843, Bache, with the financial support of Congress, organized the continuation of the trigonometric survey of the Atlantic Coast and the Gulf of Mexico, the study of the Gulf Stream and of the large maze of secondary streams, the investigation of the tides and weather conditions, and began a systematic study of the magnetic force of the earth. He also instituted a succession of studies on the depth of the oceans and the forms of animal life to be found in deep sea. Another of Bache's innovations was the wide use of scientific instruments. The electric telegraph was pressed into service for the determination of longitudes.

Many of the ablest officers of the navy and of the army were brought into the service of the Coast Survey. A corps of surveyors, geographers and hydrographers was established, to whom the services of the best talent obtainable in the country were added. Agassiz and Peirce, Bond and Mitchell, worked for the Coast Survey.

The astronomical and magnetic work was considerably facilitated by the establishment of the Harvard Astronomical Observatory. It seems at first an astonishing fact that New England, traditionally interested in astronomy, waited until 1839 to found an observatory. But the whole idea of appropriating relatively large amounts of money for scientific purposes made slow progress. The geological surveys, which really prepared the public for such appropriations, were only a few years old. The colleges were dominated by men who placed the classics above all other knowledge and thought of science solely in terms of Euclid and Newton, a subject for which a blackboard, a piece of paper, and a pen were deemed sufficient, if supplemented by an occasional amateur's telescope. Many people had not yet any understanding of the importance of natural science.[1]

The support of science had been repeatedly brought to

[1] This lack of understanding existed in other countries as well. The amusement with which many people looked at the collective exertions of scientists is reflected in the description by young Dickens of the meetings of the Mudfog Association for the Advancement of Everything (in the *Sketches by Boz*). The sketches were written in 1836, at the time when the British Association for the Advancement of Science was only a few years old. Dickens carried some of this satire over into the *Pickwick Papers*. An American counterpart is the funny Dr. Obed Battius in Fenimore Cooper's *Prairie* (1827), said to be a take-off on Thomas Nuttall.

the attention of Congress, especially by John Quincy Adams —who, after the death of Jefferson and John Adams, was practically the only national political figure who thoroughly understood the importance of science to modern society— but without success. When, as late as 1832, the act to continue the Coast Survey was passed by Congress, the proviso was appended that "nothing in the act should be construed to authorize the construction or maintenance of a permanent astronomical observatory." Money for scientific purposes from private sources was slow in forthcoming. When in the forties and fifties Yale and Harvard began to receive relatively large sums of money for scientific purposes, the old shipping interests contributed some money to astronomy; for the larger purpose of applied science and engineering, the money came from more modern sources: Scheffield, Storrow and Lawrence, the big sponsors, were financially interested in railroads and textiles.

New England slowly acquired during the thirties some observatorial facilities. A five-inch reflecting telescope was donated to Yale College and installed in 1830. It was placed in the steeple of one of the college buildings, where low windows effectively concealed every object as soon as it attained an altitude of thirty degrees above the horizon. Denison Olmsted and the young tutor Elias Loomis worked with it as well as circumstances allowed, and in 1835 were able to announce the return of the famous comet of Halley, weeks before the news arrived from Europe—an event which dramatized for the public the use of telescopes and of observatories in general. In 1836 Albert Hopkins at Williams College erected a small observatory with a reflecting telescope and some other instruments, which he paid for himself. When Loomis became professor at Western Reserve University in Ohio, he was able to establish an observatory there. One of the best equipped places in all America, however, was the Dorchester home of William Cranch Bond, the Boston instrument maker.

At Harvard as well as at Yale the plan for an astronomical observatory had been for many years a subject of discussion. John Quincy Adams had tried to convince not only Congress, but also the Harvard Corporation of the need for such an observatory, pointing to the many magnificent institutions of this kind in Europe. It had also been a favorite scheme of Nathaniel Bowditch. In the thirties Harvard purchased some land and a house which might serve as an observatory. The decision to establish the Observatory was taken in 1839, and this by the simple procedure of inviting William Cranch Bond to move to the Harvard Yard with his own instruments. Harvard College offered him the house—known as the Dana House on Quincy Street—but no salary; the American Academy of Arts and Sciences appropriated $1000 for the purchase of some more instruments for magnetic observations. In the words of the biographer: "It was the day of small things, of pennies not dollars, in the College treasury." The college began to grasp only slowly the requirements of a modern age.

Bond accepted. At that time he was engaged, at his Dorchester home, in geodesic and magnetic work for the Federal Government. The Wilkes expedition was under way, and much of Bond's observations were related to this expedition; one of his tasks was the correct determination of a zero longitude for final reference to Greenwich. Bond continued this work in the Dana house, and also began to contribute to a series of magnetic observations made in collaboration with the Royal Society in London. This early research at the Cambridge Observatory was therefore simply the continuation on a somewhat larger scale of Bond's former work, with added academic dignity. Bond was assisted by his sons and, in more theoretical questions, by the professors Peirce and Lovering. The academic position of Bond, however, was not always clear, since he was a "simple" instrument maker without a classical education and held only the title of observer. It was "Professor Lovering and Mr. Bond," as President Quincy expressed it. The situation was somewhat clarified in the course of years, when Harvard received a large gift which made it possible to offer Bond a salary. He was given the title of Director of the Observatory.

6

This clarification was largely due to growing public interest in astronomy. The return of Halley's comet in 1835 had made a considerable impression. Interest was dramatically heightened by the appearance of a comet in the last days of February, 1843. It could originally be seen even in broad daylight, and continued to dominate the sky during the whole month of March, presenting a magnificent spectacle every clear evening. It had a brilliant tail up to fifty degrees in length, a result of its small perihelion distance from the sun. Such comets had inspired previous generations with fear and awe. Many people, who believed that the ancient accounts of enormously brilliant comets had been greatly exaggerated, now found out that there was truth in old stories. The astronomers enjoyed a heyday of popularity, and their explanations were widely read. The public discovered that its brilliant luminosity on February 28 was caused by the fact that the comet had almost grazed the sun's disk, and that its tail had been turned almost directly toward the earth. Its appearance remained a subject of popular discussion for a long time.

Most New Englanders of 1843 had progressed beyond the stage of superstition or apprehension at the appearance of the comet. There were some exceptions, notably the Messianic sect of the Millerites, who had been told by their prophet that 1843 would see the end of the world and now saw their expectations confirmed in this fiery apparition. Nothing happened, however, to discourage rationalistic opinion concerning the nature of comets. There was, on the contrary, a growing desire to know more about the structure of the heavenly bodies. Thus the appearance of the "Great Comet" led to an important scientific result: a $25,000 endowment for the Cambridge Observatory, brought together by citizens of Boston, Salem, New Bedford and Nantucket, supported by the American Academy as well as by insurance societies and other organizations. The money was to be spent on a telescope for the observatory. Harvard College purchased a favorable site for the erection of an efficient observatory; it was on a hill north of the college, which at that time was free of trees, houses, smoke and other disturbing elements. The present observatory still stands on this site.

The money offered for the purchase of a telescope was a large sum for its day, and considerable care was taken to obtain the best instrument available. A refracting telescope was chosen, to be made in Germany. This fifteen-inch "Great Refractor," then one of the best instruments in existence, was brought to America and placed in position in 1847. Other instruments had already been put in place, some from the old Dana House, some newly purchased. In 1847 the Observatory was ready to assume its place as an equal with the best observatories in Europe.

Around the Great Refractor cluster most of the traditions and the sentiments of the early days of the Harvard Observatory. For many years it was the most famous astronomical eye of America. It was in regular use for observations on planets, comets and nebulae, and much important work of a routine nature was accomplished. More spectacular discoveries were also made: In September 1848 Bond discovered the eighth satellite of Saturn, called Hyperion, and in November 1850 the third or so-called Dusky Ring of Saturn. They belong among the first astronomical discoveries made in the United States. New instruments were gradually added to the collection, and the budget increased. After Daguerre's invention, photography was added to the tools of astronomical research. William Bond died in 1859, and was succeeded by his son, George Phillips Bond. When the second Bond died in 1865, the Observatory already had an established international reputation. Jules Verne, in 1866, made it a part of international folklore in his tale of the trip to the moon in twenty-eight days:

> There scholars of the greatest distinction gathered; there the powerful telescope was operated which made it possible for Bond to determine the Andromeda nebula and Clarke to discover the satellite of Sirius[2].

[2] "There [in Cambridge] are savants of the highest distinction united; there functions the powerful looking glass which permits Bond to resolve the Andromeda nebula and Clarke to discover the satellite of Sirius."

This "Clarke" was Alvan Clark, who started as an engraver and portrait painter, but then, aided by his sons George and Alvan, built up the famous astronomical instrument works of Alvan Clark and Sons in Cambridge. Here, in 1861, the Clarks discovered the companion of Sirius while testing an 18½ inch refractor ordered

7

The appearance of the Great Comet of 1843 had such an influence on American astronomy that it has been said that in America astronomy only began to exist in 1843. This, of course, is highly exaggerated, but there is no doubt that 1843 constitutes a milestone in intensity of interest and in the standing of astronomy in the community. The era of the astronomer-gentleman of colonial and Federalist days had come to an end; the professional was taking the place of the amateur. This transition is also characterized by another enterprise which was stimulated by the appearance of the comet, the *American Nautical Almanac*. Up to this period American navigation had to rely on foreign publications, one in English, one in French and one in German; the best known being the *British Nautical Almanac* first published in 1767 by Maskelyne. The publication of the *American Nautical Almanac* was mainly the work of Charles Henry Davis, the son of a Solicitor General for Massachusetts, who had entered the navy in 1823. From 1844 to 1849 he was assistant in the Coast Survey, where he did distinguished work in surveying the waters around Nantucket and discovering several unknown shoals. The *American Ephemeris and Nautical Almanac* was authorized by Act of Congress in 1849 as a government enterprise, and Davis became its first superintendent. He remained until 1856 the head of this enterprise, when he was ordered to naval service in the Pacific. In 1863 he became a rear admiral.

Davis was a good man of science with a specialized knowledge of ocean currents. His range of interest was at the same time wide enough to successfully organize such a vast work as the computation of the many tables of an *American Nautical Almanac*. He met, as usual, with a good deal of diffidence, and initial skepticism, but was able to overcome the difficulties in due time—mainly by selecting the right men to work on the *Almanac*, men such as Peirce and the Cam-

---

by the University of Mississippi, but because of the war sold to the Chicago Astronomical Society. Until a few years ago the ruins of the works, marked by a desolate mounted telescope tube, could still be seen near Brookline Street in Cambridge.

bridge astronomers. So as to be near these men Cambridge was selected as the site of the office of the *Nautical Almanac;* responsibility for all mathematical work was intrusted to Peirce. The first volume, for 1855, was published in 1853. It was good computational work; the publication has been called "a true Minerva birth, taking at once a stand equal, and in some respects superior to that of its three rivals." Davis was succeeded as superintendent by Joseph Winlock, who later succeeded George Bond as director of the observatory. In 1866 the headquarters were removed to Washington, where the *American Ephemeris and Nautical Almanac* is still published yearly.

A number of distinguished persons—or persons who would later distinguish themselves—helped Davis and Peirce with the computational work. There was Sears Walker, a Harvard graduate, who worked at the Washington Observatory, John D. Runkle, who later became president of the Massachusetts Institute of Technology, and Maria Mitchell, the woman astronomer of Nantucket. To the Cambridge office also came Simon Newcomb, then a young man, but later one of America's leading astronomers. Newcomb came to Cambridge when Joseph Winlock was superintendent, and always dated his "birth into the world of sweetness and light" from that frosty morning in January 1857 when he "took his seat between two well-known mathematicians before a blazing fire in the office of the *Nautical Almanac* at Cambridge, Mass." Newcomb continued his studies with Peirce, remaining at the university until 1861, when he went to the Naval Observatory in Washington.

8

In that period the central figure in mathematics and in theoretical astronomy was Professor Benjamin Peirce of Harvard, who established his authority not only in New England, but throughout the whole of the United States. He was born at Salem in 1809, the son of the Benjamin Peirce who represented Salem for several years in the Massachusetts Legislature and later became the librarian and historian of Harvard University. As a boy Benjamin, Jr., made the acquaintance of Nathaniel Bowditch, who stimulated his interest in astronomy and in mathematics. The young man entered Harvard in 1825, and since Bowditch was now residing in Boston it was possible for young Peirce to assist him in preparing his Laplace translation for the press. In this way he became familiar with a type of mathematics quite advanced for America and considerably beyond that treated in the new textbooks of Farrar and Davies. The influence of Laplace's astronomy dominated Peirce's work throughout most of his life.

After two years of teaching at George Bancroft's short-lived school in Northampton, Peirce returned to Harvard in 1831 as a tutor in mathematics. Two years later he became professor of mathematics, a position he held until his death in 1880. He was for many years the sole representative of mathematics at the college, and one of the few men in the United States who understood the meaning of modern mathematics. This isolated Peirce from his colleagues, who respected him and admired him, but also considered him in possession of some kind of esoteric knowledge. In due time he developed into a curious professional figure equally famous for his great scholarship and his inability to make himself understood in class.

This aloofness certainly in no way affected Peirce's willingness to write textbooks for school use. There was still a wide field for such work in the United States, since Farrar, Davies and Day had merely scratched the surface. In 1835 Peirce initiated his series of schooltexts by publishing a *Treatise on Plane Trigonometry,* followed next year by a *Treatise on Spherical Trigonometry*. In the same year a *Treatise on Sound* appeared. These books were followed by texts on algebra, geometry, calculus, and even the theory of complex functions, still a novelty in the United States. All these books, even the elementary ones, showed some originality of approach, also a novelty in American mathematics. Their originality, however, was not always an improvement, as in his axiom: "Infinitely small quantities may be neglected." Peirce used his own notation, which was somewhat more concise than the traditional one. In 1855 his *Physical and Celestial Mechanics* appeared; it was the first American textbook in this field, and one of the most original of Peirce's books.

Together with his colleague Lovering, Peirce published in 1842 the *Cambridge Miscellany,* a periodical exclusively devoted to mathematics, again a novelty in the United States. Indeed, it was so novel that it lacked a sufficient market to keep it going, and only five numbers appeared. The fare of this journal was similar to that of several other scientific journals published in the United States before the Civil War, only a few of which managed to survive. This illustrates the organizing ability of the Sillimans, father and son, who were able to keep the *American Journal of Science* going throughout the years.

Peirce's scientific fame is based mainly on two series of original investigations, one in planetary mechanics, the other in algebra. The later investigations, culminating in Peirce's *Linear Associative Algebras* of 1872, lies outside the time limit set for this book. It will be sufficient for our purpose to characterize it as the first major original contribution to mathematics produced in the United States. Present-day algebra, especially the theory of hypercomplex numbers, considers Peirce's work as one of its classics.

The other series of investigations was intimately connected with the astronomical work which engaged Peirce's attention when, as a young man, he corrected Bowditch's proof sheets. Celestial mechanics was familiar territory to the Salemite mathematicians. Peirce was always interested in the determination of the orbits of comets and planets; in 1842 he worked on Encke's returning comet, in 1843 on the Great Comet. He also assisted Bond in determining the latitude of the new observatory. Then, in 1846, an event happened which, even more than the Great Comet, excited the imagination of scientists and the public alike—the discovery of the new planet Neptune.

The story of this discovery has often been told and is a classic of the history of science. The perturbations of the planet Uranus were not fully accounted for by the action of the other bodies in the solar system. To explain these irregularities, the influence of a new planet was assumed and by a remarkable extrapolation its position determined. This was done independently by Adams in Cambridge, England, and by Le Verrier in Paris. Le Verrier, in 1846, indicated the position where the supposed planet was to be found, adding that it was large enough to be recognizable by its disk. He wrote to the atronomer Galle in Berlin, and Galle, by comparing a star map with the heavens, found the planet on the very first night of his search.

The announcement of the finding of the new planet, named Neptune, a calculation based on belief in the universality of the law of gravitation, and the immediate discovery of the planet as soon as the theoretical position was pointed out, caused a profound sensation throughout the world. It therefore came as a surprise and almost as *lèse-majesté,* when shortly afterwards Professor Peirce proclaimed in the American Academy that Galle's discovery of Neptune in the exact place predicted by Le Verrier was in reality a stroke of good luck—a happy accident. Peirce pointed out that Le Verrier had based his calculations on two fundamental propositions, one limiting the mean distance of the supposed planet, the other limiting its mean longitude, and that the two were incompatible. "Neptune," he concluded, "cannot then be the planet of M. Le Verrier's theory, and cannot account for the observed perturbations of Uranus under the form of the inequalities involved in the analysis."

> It is related that Mr. Edward Everett, being present at the meeting, actually addressed the academy upon the subject, and begged that so utterly improbable a declaration might not go out to the world with the academy's sanction. "It may be utterly improbable," retorted Peirce, "but one thing is more improbable still, that the law of gravitation and the truth of mathematical formulas should fail!"

In the ensuing discussion Peirce stated that the problem of the perturbations of Uranus actually had two solutions, the one found by Le Verrier, the other corresponding to the actual case. We now know that Le Verrier and Adams both made several incorrect assumptions, and that Peirce's criticism of Le Verrier's inconsistent propositions was so pertinent that both Le Verrier' propositions were not realized in the actual planet. There actually was a stroke of good luck in the discovery of 1846, although it must be conceded that even though the data and methods employed were not sufficient to determine accurately the planet's orbit, they were sufficient to determine the direction of the unknown body, which was the one thing needed to insure its discovery.

During the next years Peirce, supported by Walker, devoted considerable energy to an investigation of the motion of the new planet. One of the results was Peirce's computation of the perturbation of Neptune and Uranus produced by the action of other planets.

From 1849 to 1867 Peirce was consulting astronomer for the *Nautical Almanac*, and from 1852 to 1867 director of the longitude determinations of the United States Coast Survey. After Bache's death he succeeded him as superintendent of the Coast Survey, a position he held from 1867 to 1874. It is therefore incorrect to think of Benjamin Peirce as a "pure" mathematician in the European sense, a conception which might be held because of his best-known work of a later date, his linear associative algebra. It is true that with Peirce original mathematical research became a recognized academic activity in the United States, but in Peirce's work itself this research was originally still deeply associated with astronomy.

Peirce's name appears also in textbooks on the theory of probability and the theory of errors in connection with his so-called criterion for the rejection of doubtful observations. Here he proposed to determine the limit of error in a series of observations, beyond which all observations involving so great an error may be rejected, provided there are a definite number of such observations. It showed Peirce's mastery of the classical theory of errors and has been widely used after Benjamin A. Gould, his pupil and friend, computed the necessary tables. Peirce's theory has also been subjected to considerable criticism, in which Winlock, of the Harvard Observatory, refuted Airy, the Astronomer Royal. More recent criticism has pointed to some more serious weaknesses in Peirce's criterion; as a matter of fact, this field by its very nature is full of open questions.

Although there is little resemblance between their personalities, Peirce, like Agassiz, was the source of a rich fund of anecdotes. Some deal with his deplorable ways of teaching, which strengthened the popular belief in the intrinsic difficulty of mathematics. It must be said that by this unwillingness to talk clearly he rendered a real disservice to science. After all, the American public of the '60s had only few men who could tell what modern mathematics was. Other anecdotes are kindlier, such as the reminiscences of one of his pupils, the Harvard mathematician W. E. Byerly:

> The College was small, the Faculty was small, but distinguished and picturesque. "There were giants in those days," bearded giants mainly, though Agassiz and Childs were beardless, Sophocles, Longfellow, Lowell, Asa Gray, Benjamin Peirce. There are giants in the faculty now, but they are more or less lost in the crowd. Then, poets, discoverers, philosophers, and seers, in soft hats and long cloaks, looked their parts, and we newly-fledged freshmen gazed at them with admiration and awe.
>
> The appearance of Professor Benjamin Peirce, whose long gray hair, straggling grizzled beard and unusually bright eyes sparkling under a soft felt hat, as he walked briskly but rather ungracefully across the college yard, fitted very well with the opinion current among us that we were looking upon a real live genius, who had a touch of the prophet in his make-up.
>
> I can see him now at the blackboard, chalk in one hand, and rubber in the other, writing rapidly and erasing recklessly, pausing every few minutes to face the class and comment earnestly, perhaps on the results of an elaborate calculation; perhaps on the greatness of the Creator, perhaps on the beauty and grandeur of Mathematics, always with a capital M. To him mathematics was not the handmaid of philosophy. It was not a humanly devised instrument of investigation, it was Philosophy itself, the divine revealer of TRUTH.
>
> I remember his turning to us in the middle of a lecture

on celestial mechanics and saying very impressively: "Gentlemen, as we study the universe we see everywhere the most tremendous manifestations of force. In our own experience we know of but one source of force, namely will. How then can we help regarding the forces we see in nature as due to the will of some omnipresent, omnipotent being? Gentelmen, there must be a GOD."

We see that Peirce's philosophic thinking was not far removed from that of Agassiz, or of Gray. His best-known contribution to philosophy is, however, his definition of mathematics: "Mathematics is the science which draws necessary conclusions." In this expression the divorce of mathematics from astronomy and from the other physical sciences found an abstract expression. We see here connections between Benjamin Peirce's mode of thinking and that of his talented son Charles Saunders Peirce and his contribution to mathematical logic. This son was also one of the founders of pragmatism, New England's contribution to philosophy.

9

The growth of the Harvard Observatory, the establishment of the *American Nautical Almanac,* the development of magnetic, geodesic and meteorological observations, was made possible through the work of a considerable number of very able scientists of different backgrounds. Peirce and Walker had an academic education, Davis was trained in the navy, Bond was without any academic education at all. Joined with these men were other scientists, some of whom deserve special attention here. We think of the Nantucket Mitchells—already discussed in Chapter 3—of Joseph Lovering and of Benjamin Apthorp Gould.

Joseph Lovering, born in Charlestown, was a Harvard graduate of the class of 1833. He was for a while Peirce's assistant, and in 1838 succeeded Farrar in the Hollis chair at Harvard. He occupied the chair of natural philosophy for fully fifty years until his retirement in 1888. He was, unlike his colleague Peirce, an executive officer and a popularizer of science, but he lacked Peirce's understanding and originality. Most of his scientific work consisted of the collection and discussion of masses of observations, many in connection with the investigations in terrestrial magnetism undertaken by the Cambridge Observatory as part of a world-wide campaign, many others in a study of the periodicity of the aurora borealis. He was interested in the history of science and contributed to the Memorial History of Boston, published in 1881, a paper of science in and around Boston during the preceding centuries, a sketchy collection of data, but the only attempt ever made.

Benjamin Apthorp Gould—not to be confused with Augustus Addison Gould, the physician and naturalist—was the son of the principal of Boston Public Latin School, well-known teacher and school reformer, who in later life became a prosperous merchant and the owner of a fleet active in the Calcutta trade. The son was a pupil of Peirce at Harvard; he graduated in 1844 and then went to Berlin, Bonn and Göttingen, to study for four years under the great astronomers of the day, Gauss, Bessel, Encke and Argelander. When he returned to the United States in 1848, he was a doctor of the University of Göttingen with a dissertation written under Gauss, and the first European trained American astronomer.

Gould settled in Cambridge, tutoring in languages and mathematics, and publishing a new periodical, the *Astronomical Journal,* which existed until the Civil War. From 1852 to 1867 he joined with Peirce in managing the longitude department of the Coast Survey. It was here that he adapted the telegraphic method and obtained the longitude difference between Washington and Greenwich by using the first transatlantic cable. Gould, who married a daughter of President Quincy of Harvard, had his own private observatory near Cambridge, where he began what was probably his most important contribution to astronomy, his work on the Standard Catalogue of Stars. This was a venture suggested by the work of his German teachers Bessel and Argelander, who had begun such a catalogue for the Northern Hemisphere. After the Civil War Gould organized the observatory at Córdoba in Argentina, and continued Bessel's and Argelander's star catalogue with observations on the Southern Hemisphere. He spent his later days in Cambridge; in 1886 he reestablished the *Astronomical Journal,* which still exists as one of the leading periodicals in its field.

Chapter 13

The Scientific Schools

> Only one who has spent weeks poring over the old textbooks, government reports, biographies, and records of museums can begin to appreciate the comprehensive, varied, and fruitful labors of scientists in that age, so often belittled by its successors.
>
> —C. A. AND M. D. BEARD, 1927

1

IN 1845 COLLEGE education was still mainly centered upon the study of the classics, and based on an established curriculum from which no deviation was permitted. In the twenties of the century, George Ticknor, the widely traveled professor of French and Spanish literature in Harvard College, had proposed a program of departmental subdivision and electives for the college, perhaps under the influence of Jefferson's program for the University of Virginia. He had not been able to change the traditional system, which, according to Professor Morison, "would have put Harvard College a generation ahead of her American rivals," and "made her equal to the smaller European universities." Reform had to begin where it usually begins—in the lower ranks of the system, in the grade and secondary schools, which reflected more directly the changes in the economic structure of the country. Harvard, in 1825, and even in 1845, still reflected the complacent attitude towards science prevalent among the Federalist aristocracy. Natural science, in this system, was considered of relatively small importance and of little educational value in comparison with the classics and its adjunct in the sciences, Euclidean mathematics. The colleges believed in the philosophy of "training the mind," for which purpose Caesar, Plutarch and Euclid were supposedly unsurpassed. In favor of this system it can be said that it stressed the need for a university to teach fundamentals, a principle as true at present as it was in the past; the trouble was that the fundamentals taught were those of 1645 and not of 1845.

The student body showed its appreciation of this system by various kinds of mockery and even by occasional outbreaks of rowdyism. The students at Yale, for instance, had a pageant at the end of the sophomore year, called "The Burial of Euclid," with a dirge that ran:

> No more we gaze upon that board
> Where oft our knowledge failed
> As we its mystic lines ignored,
> On cruel points impaled . . .
> We're free! Hurrah! We've got him fast,
> Old Euk is nicely caged at last. . . .

The knowledge that most of his students were at best indifferent to his subject must have discouraged many a teacher of mathematics. Peirce's lack of intelligibility in teaching may have been subconsciously caused by this knowledge. We know that Farrar was a good teacher, but he was a reformer, breaking with "Old Euk's" stifling tradition. It must be added that the attitude towards mathematics prevalent in the American colleges of 1845 is still present in much of our secondary education, in spite of all the intervening educational reform.

At Yale the first two years of study were limited to Latin, Greek and mathematics. In the third year a one-year course was offered in "natural philosophy," a catchall for scientific odds and ends, with a one-term course—a third part of the college year—in astronomy. In the senior year no natural science was offered, although there was a one-term option in astronomy. The junior class, however, was expected to attend a course of experimental lectures on natural philosophy, and the senior class was expected to hear "courses of lectures on chemistry, mineralogy, geology, and selected subjects of natural philosophy and astronomy." The professors teaching science in 1845 were Silliman and Olmsted, assisted by three junior teachers (Charles U. Shepard, A. Eldridge and Benjamin Silliman, Jr.). There was also some science at the Medical School.

Harvard offered essentially the same subjects, or rather

the same dearth of subjects, although some chemistry and natural history was given in the first two years. During the second half of the senior year the Rumford professor was supposed to lecture on the "application of science to the arts"—it was for this purpose that Count Rumford had endowed the chair, knowing how much the American colleges of his day needed modern science. Jacob Bigelow, who coined the word "technology," had lectured as the Rumford professor until 1827, after which Daniel Treadwell had occupied the chair. Neither Bigelow nor Treadwell was a full-time professor. Treadwell resigned in 1845 and the lectures ceased temporarily. In 1845 the science professors at Harvard were John W. Webster, Benjamin Peirce, Asa Gray and Joseph Lovering, two of whom—Peirce and Gray—were embarking upon a career which was to lead to international distinction. The situation at Harvard was therefore perhaps a little better than at Yale, but this was potential rather than actual superiority. The Sillimans, in fact, had made Yale, despite its poor facilities, the Mecca of any young American interested in the new sciences of geology and chemistry.

Williams College had always been relatively strong in natural science, although is curriculum differed little from that of Harvard or Yale. Only a few scientific lectures were given in the first two years in addition to the regular classical curriculum. The science teachers were Ebenezer Emmons, Albert Hopkins and Edward Lasell. We saw that through the personal efforts of Albert Hopkins, Williams College already boasted an astronomical and magnetic observatory. Dartmouth and Bowdoin had also a strictly classical curriculum with fewer scientifically informed teachers than the other colleges, although the 65-year-old Parker Cleaveland was still teaching at Bowdoin.

2

No laboratory instruction existed in any of these colleges, despite the early example given by Amos Eaton at Rensselaer. College authorities were not easily swayed even by examples from their venerated Europe, and were blind to the benefits of a method used by somebody whom they might well consider a self-made backwoods teacher and ex-convict in Troy,

New York. At most students were able to witness experiments performed by the professor himself in the lecture room, as was the case in Silliman's class at Yale. Silliman had professional assistants, to whom he was wonderfully helpful, but he did not encourage students to work in his laboratory:

> Many times I have said to those who as novices have offered to aid me, that they might come and see what we were doing, and I should much prefer that they should do nothing; for then they would not hinder me and my trained assistants, nor derange or break the apparatus.

Such an attitude was due principally to the lack of laboratory facilities. Before 1817 Silliman had only one basement room, which later, after the Medical School was established, was somewhat enlarged. John W. Webster at the Harvard Medical School had a chemical laboratory and fairly sizable lecture room in the Mason Street building, the ground plan of which he proudly showed on the front page of his *Manual of Chemistry* (1826). Here again there existed little or no opportunity for laboratory practice. Students in need of such practice had to go to private laboratories, such as that of Charles T. Jackson in Boston, which enjoyed a wide reputation. In 1842 Silliman, Jr., who was his father's professional assistant, secured a room at Yale in which he could instruct a few students in practical work. However, this place was his private workshop and had no official connection with the college.

3

The development of industry, agriculture and transportation made this whole system of instruction year by year a more absurd anachronism. Schools and factories, farms and ships, government agencies and communication systems, required men acquainted with modern science and engineering. Private agencies were unable to supply this need, and college trustees began to feel more and more acutely the lag between higher instruction and the demands of the times. In the forties this discrepancy between college and society reached a new high. America was beginning to compete with Europe in technological advance, and even in science it was emerging from its lethargic provincialism. America, New England especially, was no longer a country of craftsmen and teachers; it had begun to be a country of engineers and scientists.

In 1845 American railroads were well established, and American railroad engineering had reached such maturity that Major Whistler was building the St. Petersburg–Moscow Railroad. Textile industry vied with that of Lancashire and inventions were registered every year. In hydraulics—a science basic for all New England industry as long as steam was still only hesitantly used—progress was made which in 1844 led to the Boyden turbine. America's bridge trusses were carefully studied by European engineers. Elias Howe was tinkering with his sewing machine, on which he took out a patent in 1846; Charles Goodyear was vulcanizing rubber and publishing his invention by means of his French patent of 1844. In Albany, Joseph Henry was keeping pace with Faraday in electrical discovery; the first major practical result was Morse's telegraph, inaugurated in 1844 between Washington and Baltimore. Corliss was busy in Providence revolutionizing the steam engine; in Hartford Colt was striving to perfect his revoler. The American genius for invention was beginning to assert itself.

The weak spot in New England's economic structure was agriculture, which languished under three blights—soil exhaustion, competition with the West, and the drain on labor power by the factory system. Agricultural societies did their best to combat the evils, but the odds against which they were fighting were large. Many people felt that improved scientific agriculture might help the farmer to establish himself. Agricultural technique was improving, and the books of Liebig and S. L. Dana were widely read. A section, at any rate, of the farming population was willing to look favorably upon agricultural education.

In this period America began for the first time to have an extensive group of professional scientists, several of them trained by Silliman, by Silliman's pupils, or abroad. In the fifteen or less years of their existence the geological, botanical, zoological, agricultural and trigonometric surveys had accomplished miracles for the scientific developoment of the country. Charles Lyell liked America, where he could learn as well as teach. The opinions of the Rogers brothers, of C. T. Jackson, of Joseph Henry, of Silliman and of several others were valued by the scientists of Europe—a recognition by which the standard of merit was measured for many years. Some scientists, such as the Rogers brothers and Loomis, had already gone beyond classification to theoretical analysis. The Harvard Observatory was busy purchasing its Great Refractor which would place it in the very first rank of the world's astronomical centers; it had already participated in the magnetic survey of the world organized by the Royal Society. In meteorology Matthew F. Maury at what was to become the U. S. Naval observatory in Washington and the older James P. Espy in Philadelphia were organizing their chains of observations. Asa Gray's botanical work had already become so well known internationally that Darwin began to take him into his confidence. The Wilkes expedition and the Coast Survey had shown that the Federal Government could be made to participate in large-scale scientific investigations. Curiously enough, one of the weakest fields of science was mathematics, which had not yet been emancipated from the shackles of Euclid and Newton. The very prominence of this antiquated type of mathematics in the classical college curriculum was frustrating the free development of the science. The papers on mathematics in the *American Journal of Science* were puerile, and contrast sharply with the thoughtful papers of Rogers, Gray or Henry on natural history. However, at Harvard, Benjamin Peirce was giving promise of a more vital future.

Science and technology had not yet found each other, and would find it difficult to see eye to eye for a long time to come. The principal reason was that they were historically linked to different classes in American society. Some links between the two domains did exist, mainly through men of an older and less specialized generation. Many of them had been able to study different professions in different parts of their life—men such as Jacob Bigelow, Daniel Treadwell or William Redfield. They looked upon the divorce of science from technology with disfavor, and joined with younger men who were interested in agricultural and engineering applications of the new sciences. Among them were men of money, others had political influence, some had both, and in the ensuing years they were able to reach several important achievements.

4

This was a generation which could appreciate what was going on in Europe, in the professional chambers, laboratories, observatories, medical schools, hospitals and research institutes of Germany, France and England, but also of Italy, Switzerland, the Netherlands and Sweden. Many American scientists, physicians and technical experts had gone across the ocean and later tried to practice here what they were taught in Europe. They all missed in America the research facilities with which they had become familiar in Europe; not only the laboratories, but also the specialized scientific journals. What was the matter with a republic built on the will of a sovereign people, that it lagged behind a constitutional monarchy like England, or banker's kingdom like France, or a set of autocracies like Germany? "Your nobody in England is our everybody in America," Lyell was told in Pennsylvania. This was a monstrous exaggeration in a country with slaves and disfranchised whites, but there was also a profound element of truth in it. A decisive section of the American public had to be educated to the need of science, and the economical and technical development of the thirties and forties began to provide the basis for such an education.

Few among the research laboratories of old Europe exercised such an attraction for young America as that of Justus Liebig. Liebig, a German chemist, became in 1824 professor at the University of Giessen. Here he soon established a research laboratory for organic chemistry, the first of its kind, to which he freely admitted students from any and all parts of the world. Liebig was a great investigator, who stressed theoretical as well as practical aspects of chemistry. Many American students came to study in Liebig's laboratory, where they enjoyed facilities unobtainable at home. There they worked under a great teacher of a sharply polemical mind, who fascinated and stimulated them to original research. Among Liebig's American students were Horsford, Samuel Johnson, Porter and Wolcott Gibbs, who all became leading chemists in New England. Johnson, who became a professor at the Sheffield school, has left us a description of Liebig's methods:

> [All students] were set to testing the truth of some idea or the correctness of some fact, or else to make new observations and discover new facts to lead to new ideas. It was not the novelty or the glory of discovery, but the genuineness of discovery that was regarded as of first importance. He listened patiently to their accounts of each day's progress, considered their plan of investigation, saw the apparatus or arrangements they devised, witnessed the observations they were led to, and heard the theories they imagined. He encouraged, but he criticized. He asked questions, suggested doubts, raised objections. His students were required not only to collect facts, or supposed facts, and to connect and complement them by comparison, analogies and theories, but they were made to attack their theories in every weak point and to verify or disprove the supposed facts by scrutiny from every side.

Liebig's investigations on the application of chemistry to agriculture were an especial attraction for the students from the United States, coming as they did from a country where the turbulent changes in agriculture had created very serious problems which might be solved in part by the application of scientific methods. They were also of interest to the people at home, who had at last become aware of the possibilities existing in scientific farming. The English translation of Liebig's *Organic Chemistry in its applications to agriculture and physiology* appeared in 1840. It was immediately popular in America. Liebig's other books were also easily absorbed by the American market, especially his *Familiar letters on chemistry*. They were printed and reprinted in thousands of cheap copies, edited under the convenient absence of an international copyright agreement. Here was a scientist of international reputation, one of the founders of organic chemistry, who was not afraid to open a laboratory to students, or to relate his academic work to the practical arts. Liebig has been called the earliest of extension teachers.

Liebig's influence brought practical men closer to the colleges and gave the colleges a deeper understanding of the importance of natural science. This happened at a period when the two conditions necessary for the foundation of advanced schools of science began to be realized—the existence of a public opinion sufficiently mature to recognize the need for systematic training of scientists and engineers, and the presence of scientists sufficiently schooled and eager to work at this task.

Other European influences were at work. At Paris, for instance, there was the Faculty of Science at the Sorbonne, with twenty-three professors and instructors in pure science, supplemented by the nine top leading scientists at the Collège de France. There was the École Polytechnique, with the School of Engineering, the School of Mines, the Central School of Arts and Manufactures. The Jardin des Plantes had seventeen professors in the different branches of natural history. There were scientists and architects at the School for Architecture and Design, and there were several other schools, less advanced in character, for training in special trades. Even in technical training a fair amount of stress was laid on fundamentals, on geometry and drawing, on mechanics, physics, chemistry and natural history. There were polytechnic institutes of high standing in Dresden and Freiburg. England was less advanced, but the Royal Institution in London, the Greenwich Observatory and the laboratory of the Agricultural Chemical Association at Edinburgh were model institutions of their kind. Much scientific inspiration came also from hospitals and medical schools, where many leading American chemists and physicians had received an essential part of their training.

5

Initiative for the foundation of scientific schools in New England was not taken by state governments. As in the case of astronomical observatories, the legislatures were slow or unwilling to act. The extremely confused political situation of the Jacksonian era and its aftermath, lasting until the Civil War, made specific legislative action of any sort on a subject relating to science a difficult thing. Too many different groups of the population had opinions or lacked opinions on the value of science. The need for scientific schools and observatories was felt most strongly by some professional men, who eventually were able to interest the proper college authorities. They were eventually able to enlist the help of capitalists with an eye open to the future of industry.

At Yale College Benjamin Silliman had long felt the need of enlarged facilities for scientific education and research. He was at last able to convince the college authorities that steps should be taken toward that end, when John Pitkin Norton, one of his most promising students, had returned from his study in Edinburgh, an expert on European agricultural chemistry. Norton was born in 1822 and lived for a while on his father's farm in Farmington, Connecticut, where from the beginning he showed interest in scientific management of farms. He obtained his father's permission to become a farmer, but under the condition that he should be thoroughly educated as an agriculturist. This brought him to Yale to study under Silliman, and to Boston for some special instruction. From 1844 to 1846 he studied under the well-known agricultural chemist James Johnston at Edinburgh, and later in 1846 under Johannes Mulder, the agricultural chemist at Utrecht in the Netherlands.

It is interesting that the foundation of a scientific school at Yale was more closely related to the needs of agriculture than of industry, whereas the foundation of a similar school at Harvard was primarily the work of men schooled in industry. It may be that Congregationalist Yale in small New Haven was always closer to the agricultural section of the population than Unitarian Harvard near Boston, where industrial interests had more influence. At any rate, in 1846 the Yale authorities approved the appointment of Norton to a newly created chair of agricultural chemistry and animal and vegetable physiology—"it being understood," wrote the Corporation, "that the support of this professor is in no case to be chargeable to the existing funds or revenues of the College." Yale, as did Harvard in 1839 when Bond was appointed, was graciously yielding to the necessity of increasing its interest in science, but was not willing to pay for it.

When Norton came back from Utrecht—curiously enough, he did not study under Liebig—he was probably the best trained agricultural chemist in the United States. He had already published some papers, one a careful study of the oat, with hundreds of analyses. He possessed the zeal of the emancipator:

> He felt that he was now entering upon an extensive field of usefulness—that he was laboring in the service of his country—and that on his return he would possess a power for good which would perhaps belong to no other one of his countrymen.

The School of Applied Chemistry started in 1847. The teachers were Norton and Benjamin Silliman, Jr., the son of the veteran geologist. Young Silliman had received a careful education, had graduated from Yale in 1837 and was for some years his father's assistant. He taught chemistry, mineralogy and geology and had a private laboratory, where he trained students who wanted special experimental practice in analytical chemistry. He accepted the appointment at the new Scientific School, although there was hardly any financial future to it. The school had few funds and relied mainly on Norton and the Sillimans for support. They paid the rent of a house, formerly occupied by Yale presidents, and equipped it with laboratory facilities. Silliman, Jr., soon left for a teaching position in Louisville, Kentucky, where he stayed until 1854, when he succeeded his father as professor of chemistry[3], in a distinguished career that only ended with his

[3] Silliman Jr.'s analysis of some oil samples taken from Oil Creek, a tributary of the Alleghany River, published in 1855, is of historic importance because of its connection with the opening of the Titusville, Pennsylvania, oil field in 1857 by Colonel Edwin L. Drake, president of the Seneca Oil Company. This was the first commercial exploitation of petroleum in the United States.

death in 1885. In Silliman's absence Norton had to organize the school practically singlehanded. Students registered and received laboratory instructions; among them was Samuel W. Johnson, who later succeeded to Norton's position at Yale. In 1852 the first six men graduated.

In 1852 Norton died at the age of thirty years. Because of his youth, his endless endeavor, self-sacrifice and iron determination to make Yale's Scientific School a success, he deserves a place among the pioneers of science in America. He was one of those relatively rare figures in the academic field who seriously tried to bridge the gap between town and gown, as well as between farm and city. His prize essay, written for the New York Agricultural Society, was published in 1850 under the title of "Elements of scientific agriculture, or the connection between science and the art of practical farming," and contains the lessons Norton had learned with Johnston and Mulder.

In 1856 a public effort was at last made to raise a permanent endowment for the Yale Scientific School. Pamphlets were published, lectures were held, the aid of newspapers and periodicals invoked. The result was that at last the angel was found willing to place the young institution under his wings. It was a financier by the name of Joseph E. Sheffield, the father-in-law of Norton's successor, John Addison Porter. Sheffield had been a promoter of the old Farmington Canal between New Haven and Northampton, and the prime mover in the construction of a railroad to take the place of the canal. He was financially interested in other railroad construction and was the chief projector of the New York-New Haven Railroad. Sheffield donated a building on Grove Street to the Yale Scientific School, he provided the money for scientific equipment, and several times donated large sums to the school. By 1861, when his gifts amounted to over $100,000, the College gave to the school the name of Sheffield Scientific School, and as such—abbreviated as "Shef"—it is known today. There were two schools, the Chemical School and the School of Engineering. Sheffield continued to help the school financially; when he died, in 1882, the total of his gifts and bequests to the school amounted to more than $1,000,000.

6

The Scientific School at Harvard started with less financial headaches than the Sheffield School. Here men such as Bowditch and Peirce had been stressing the need for better scientific instruction, and in 1847 a "Scientific School of the University at Cambridge" was organized. The President of Harvard was Edward Everett, who, ten years earlier, as governor of Massachusetts, had authorized a number of successful scientific surveys. The main task of the school was to provide instruction in mathematics, astronomy, chemistry and civil engineering. The practical emphasis was not on agriculture, as at Yale, but on industry and shipping. There were no special funds, and the venture depended on the voluntary services of the professors plus whatever the university could spare of its existing resources. At this juncture financial help was offered.

Charles S. Storrow, who was a Harvard graduate and had studied at the École Polytechnique at Paris, was engaged at that time in laying out the new textile town below Lowell on the Merrimack. One of the financial sponsors of this enterprise was Abbott Lawrence, after whose family the new town was to be named. Storrow discussed the need for a new scientific school with Lawrence, and as a result Lawrence offered the sum of $50,000 to the Harvard Overseers for the purpose of endowing a scientific school. The money was to be applied to the foundation of professorships of geology and engineering, to the erection of a new laboratory, and a suitable building for the other departments. The Harvard Overseers, "in commemoration of this act of enlightened munificence," resolved to give the projected school the name of Lawrence Scientific School.

Whereas the promotion of chemistry was the main goal set by the originators of the Sheffield School, engineering was the chief end held in view by the founder of the Lawrence School. Abbott Lawrence, in his letter to the treasurer of Harvard College, made this point sufficiently clear:

> Where can we send those who intend to devote themselves to the practical application of science? How educate our engineers, our miners, machinists and mechanics? Our

country abounds in men of action. Hard hands are ready to work upon our hard materials; and where shall sagacious heads be taught to direct those hands?

Lawrence proposed three fields of study: engineering (meaning civil engineering); mining, including metallurgy; and the invention and manufacture of machinery. It comes therefore somewhat as a surprise to see that one of the first appointments to the school was Louis Agassiz, a geologist and as such qualified for the school in Lawrence's eyes, but also as distinctive a representative of the "pure" scientists as the world at large could offer. The reason for this appointment was that Agassiz, after a sojourn of two years in this country, had just lost his salaried position in the service of the King of Prussia, and Harvard saw in the new endowment a chance to connect the brilliant Swiss investigator permanently to its staff. The appointment was a fortunate one for Harvard and for America, but it did not contribute to the fulfillment of the ideals foremost in the mind of the founders of the Scientific School.

More appropriate to this goal were the two other decisions taken—the assignment of Eben Horsford, the new Rumford professor, to teach chemistry, and that of Henry Lawrence Eustis to teach engineering. Eustis, like his colleague Norton at the Sheffield School, was a graduate of West Point, still the only school from which to obtain academically respectable engineering teachers. Eustis's appointment was accompanied by the official establishment of a School of Engineering at the Lawrence School.

Like Asa Gray and James Dwight Dana, Eben Norton Horsford was born in upper New York State of New England stock. His youth, like that of Gray and Dana, was that of a boy in a pioneering country with a taste for natural science. He studied under Amos Eaton at Rensselaer Institute and graduated in 1837. During the following years he worked on the geological survey of the State of New York, studied Daguerre's photographic process under Morse, and taught at the Albany Female Academy.

From 1844 to 1846 Horsford went to Giessen to study with

Liebig, where he extended his chemical knowledge into the field of organic chemistry, at that time one of the frontiers of science. On the strength of his work in this field, and a strong recommendation of Liebig, he was after his return to Cambridge elected to the Rumford professorship, left vacant by the resignation of Daniel Treadwell. One of his first duties was to organize the laboratory of the newly founded Lawrence Scientific School.

Horsford is principally known for several practical applications of chemistry, in particular of the so-called Rumford baking powder, and for his quixotic researches on the Norsemen in America. He erected the Norumbega Tower near the Charles River not far from Boston, to commemorate the early Vikings, who, he claimed, had built a fortress there in the time of Leif Ericson. The fortune which enabled him during the latter part of his life to indulge in this historical work resulted from his ability in industrial chemistry. He invented the phosphatic yeast powder, the object of which was to return to bread the phosphates lost in bolting the flour. In 1856 he founded the Rumford Chemical Works in Rhode Island to manufacture this yeast powder, and eventually it became a financial success. The demands of this business became so great that in 1863 he was obliged to resign the Rumford professorship. During the remaining part of his life he devoted himself mainly to the management of his company, but continued to live in Cambridge until his death in 1893. He was one of Wellesley College's benefactors.

The Lawrence Scientific School benefited from his enthusiasm and productive knowledge for sixteen years, and under him many chemists of distinction were educated. He led the laboratory work, where, like Johnson at Yale, he applied the methods he had learned when studying with Liebig.

7

Despite considerable endowments, Agassiz's abundant energy, Horsford's laboratory work and Eustis's teaching, the Lawrence Scientific School did not particularly flourish. For a long time it remained some sort of stepchild to Harvard University. During these years when the departments of chemistry, botany, astronomy and geology at the college itself were emerging as strong and important branches, very few changes or improvements were made in the Engineering School. No engineering laboratory was established until 1892.

There were two scientists who watched the development of the Lawrence Scientific School, and of the institutes of applied science in general, with the greatest interest. They were the two Rogers brothers—William Barton Rogers and Henry Darwin Rogers. One of their deepest concerns was the improvement of the teaching of applied science and engineering, of polytechnic education. Their interest dated back to their childhood in Philadelphia and in Williamsburg, Virginia.

There were four Rogers brothers, who all became excellent scientists—long the most remarkable scientific family in the United States. The father, Patrick Kerr Rogers, was a native of northern Ireland, who had joined the Irish Rebellion of 1798. This was the rebellion organized by the United Irishmen, who cherished Tom Paine and the democratic ideals of the American and French Revolutions and had Protestants as well as Catholics in their ranks. After the suppression of the rebellion many United Irishmen, including Rogers, sailed to America and settled in Pennsylvania, at that period one of the states most tolerant of democratic ideals. Here they became the target of bitter attacks by the Federalists, who saw in the new immigrants, to use the words of a Federalist senator, with few exceptions "United Irishmen, freemasons, and the most God-provoking democrats this side of hell."

In this atmosphere of bitter party struggle the United Irishmen naturally gravitated toward the ranks of Jeffersonianism. This fact may have supported Rogers in his career as a scientist, since Jefferson's influence was strong in the scientific world outside of New England. Rogers studied medicine and chemistry at the University of Pennsylvania, and settled after graduation first in Philadelphia and later in Baltimore as a physician. In 1819 he accepted the professorship of chemistry and physics at William and Mary College, Jefferson's alma mater, and taught there until his death in 1828. Rogers was a cultured man who kept alive the scientific interest aroused by Jefferson in the natural resources of Virginia. In this way he laid the groundwork for the later geologic survey of the state, in which his sons participated.

Rogers's wife also came from the ranks of the United Irish. The four sons inherited their father's interest in natural science, notably in geology and chemistry. They received a careful education. The oldest of the brothers, James Blythe Rogers, became a chemist and eventually succeeded Robert Hare at the University of Pennsylvania. William Barton Rogers, the second son, born in 1805, became a geologist and the founder of the Massachusetts Institute of Technology. The third son, Henry Darwin Rogers, also became a geologist and eventually was appointed to a professorship in Glasgow. The youngest brother, Robert Empie Rogers, was a chemist and a physician; he succeeded his brother in the chair of chemistry at the University of Pennsylvania after Jame's death in 1852. They were a good-looking and closely knit set of men, who often collaborated with each other in their scientific work; the published correspondence between William and Henry casts important light on the scientific growth of America during pre-Civil War days.

William Barton Rogers graduated in 1822 from his father's college and after teaching for a while in Baltimore succeeded his father at Williamsburg. In 1835 he became the professor of natural philosophy and geology in the University of Virginia, where he taught until he resigned in 1853. During 1835 he was also appointed official geologist of Virginia, in charge of the survey of the state which was authorized by the legislature.

For several years Rogers prosecuted the survey with great vigor, at the same time fulfilling his duties as a teacher. As a surveyor he had to be both scientist and organizer; all his brothers from time to time assisted him in field and laboratory work. Every year, until 1842 when the survey was discontinued, Rogers published a report. He was a popular teacher, able to fascinate an audience by his accounts of astronomy and geology, then so popular in the country. One of his pupils wrote reminiscently:

> [I have] seen his lecture hall crowded with students from other departments, including those of law and medi-

cine; indeed, so crowded with young men, eager to hear the presentation of the subject by the professor whom they so greatly admired, that not even standing room could be found in the hall. All the aisles would be filled, and even the windows crowded from the outside with eager listeners. In one instance I remember the crowd had assembled long before the hour named for the lecture, and so filled the hall that the professor could only gain admittance through a side entrance leading from the rear of the hall through the apparatus room.

At this period of his life he is described as tall in stature, "with a figure of the type known to us through the pictures of Henry Clay"; a man without arrogance but of singularly commanding face, and without a peer among his fellow scientists in addressing an intelligent audience. In those days his brother Henry, who had become state geologist of Pennsylvania, acting in close collaboration with William, was able to show that the time had come for American geologists to abandon their purely descriptive attitude and to venture into dynamic geology. The result was their theory of the unfolding of the great Appalachian Mountain chain, which was presented before the Boston meeting of the Association of Geologists in 1842.

Rogers was elected in 1844 to the chairmanship of the faculty of the University of Virginia, an office which, according to the democratic principles as Jefferson understood them, constituted the presidency. This position heightened Rogers's administrative and organizational interests, and afforded him the opportunity to evaluate Jefferson's educational ideals in an expanding America. These ideals stressed the full use of science and technology for the betterment of man's state. Under the influence of his father Rogers had from his childhood looked upon the pursuit of science not only as a search for truth, but also as a method of ameliorating the condition of the human race. He was critical of the poor facilities offered by the American college for the study of modern science and engineering. Both he and his brother decided to devote their efforts to the modernization of polytechnic education.

8

William Rogers, supported by his brother Henry, thought of Boston not only as the most desirable place for his future scientific activity, but also as the best site for the polytechnic institute he was planning. He had great love and respect for the University of Virginia; in his report of 1845 on the state of the university, he praised its elective system, then still a novelty in the United States. But the intellectual atmosphere at the university was deteriorating markedly, due to the increasing arrogance of the slaveowners and their followers. Rogers bitterly hated slavery; he admired the European liberals of 1848 and their representative Kossuth; he was acutely aware of the moral decline of the South. Boston, with its flowering arts and sciences, the center of machine- and tool-conscious New England, and a leader in the struggle against slavery, held an appeal which both brothers tried to follow. In 1846 Henry Rogers vainly tried to establish himself at Harvard in the position which Horsford obtained—eventually he went to Glasgow—but in 1853 William Rogers resigned from the University of Virginia and settled in Boston.

The ideas of the Rogers brothers were partially realized in 1847, when the Lawrence Scientific School was opened. Yet they were still dissatisfied. They foresaw that a school for science and engineering could not flourish as a part of Harvard, where the classical and humanistic tradition was too strong. The appointment of Agassiz at the Lawrence School had already shown that the Harvard ideas concerning the requirements of a school of engineering were vague. The appointment of Lieutenant Eustis, a good teacher but uninspired scholar, with only a moderate practical experience in military engineering, increased their disappointment in the Lawrence School:

> The more I think of our plan of a Polytechnic School, the more confident I feel of its rapid and great success. The Lawrence School never can succeed on its present plan in accomplishment of what was intended. It can only, as now organized, draw a small number of the body of students aside from the usual college routine. It should be in reality a school of *applied science,* embracing at least four

professorships, and it ought to be in a great measure independent of the other departments of Harvard. Besides, Cambridge is not the place for such a school. It should be *Boston*. (William to Henry, October 3, 1847.)

Perhaps the fact that Henry Rogers did not receive an appointment at the Lawrence School added to the disappointment. It may also be that William Rogers, an admirer of Kossuth, an enemy of the counter-revolution in Europe and of cotton slavery at home, felt ill at ease in the complacent Whig atmosphere of Harvard, where Francis Bowen was accepted and Ralph Waldo Emerson rejected. However, when in 1853 William Rogers finally settled in Boston, after his marriage to a Boston lady, he was thoroughly at home in the distinguished community of scholars and scientists clustered around Harvard and the American Academy of Science. His presence added to the international reputation which New England science was rapidly acquiring.

The scientific work of Rogers in his Boston days included a study of electrical discharges in vacuum tubes in connection with the Ruhmkorff coil. He also continued his geological and chemical studies, and identified himself closely with the educational and public projects of the Commonwealth. In 1861 he accepted the office of Inspector of Gas and Gas Meters in Massachusetts, organizing a superior system of inspection; in 1863 he took photogrammetric measurements of the illumination of the State House in Boston by means of electric (carbon) lights. His most important activity, however, remained the improvement of technical education.

In 1859 Rogers opened the public discussion of his plans for a real polytechnical school, which was to combine original research upon the largest possible scale with agencies for the diffusion of popular knowledge. Now well established in the community, he was able to interest a number of prominent citizens of Boston. The legislature of the Commonwealth was not overenthusiastic; the idea of technological instruction beyond the simplest requirements of civil engineering still met with little public approval. The Civil War also interfered with educational reform, though it emphasized the need for technically trained men. Rogers, however, was able to find the strong support of outstanding men in the community. Their agitation had effect, and already in 1861 the Legislature of Massachusetts granted the charter of the new Institute. A part of the newly made land in the Boston Back Bay was set aside for a building. Financial gifts began to come in, and in 1865 the Massachusetts Institute of Technology was opened, with Rogers as its first president. Laboratory methods were introduced, and full courses in the different fields of engineering were offered.

This was the beginning of "Boston Tech," which introduced a new and modern type of education in the United States, only partly similar to that of the existing scientific schools and the Rensselaer Institute. William Barton Rogers devoted his later years to the consolidation of the Institute, helping it through a period of considerable financial difficulties. It is during this period of his life that he is best remembered at the Institute, as a frail-looking but energetic man with a crop of gray hair, whose paternal mien may tend to obscure Rogers's considerable stature as a scientist. He stayed at the head of his educational venture until 1870, when he retired. After his retirement, his main interest remained concentrated on the growth of the Institute. For a while he even reassumed the leadership. His end was worthy of the man; he died while addressing the graduating class of 1882:

> President Walker, with words of eloquent and glowing tribute, by which Mr. Rogers was visibly moved, invited him to speak. His voice was at first weak and faltering, but, as was his wont, he gathered inspiration from his theme, and for the moment his voice rang out in its full volume and those well-remembered, most thrilling tones. Then, of a sudden, there was silence in the midst of speech; that stately figure suddenly drooped, the fire died out of that eye . . . and, before one of his attentive listeners had time to suspect the cause, he fell to the platform instantly dead.

9

The early attempts at agricultural and engineering education invariably met with one paralyzing obstacle—the lack of funds. It killed the Gardiner Lyceum at Hallowell, Maine, it crippled the Sheffield School, it handicapped the Lawrence School, for many years it prevented the realization of Rogers's ideals. When funds were forthcoming, they came from private capitalists. It was through the initiative of Morrill, a United States Senator of Vermont, that substantial Federal aid to agricultural and engineering education at last was granted.

Justin Smith Morrill was not a scientist or even a college graduate, but he was a man of good common sense who understood the problems of the ordinary dirt farmer. He was worried about the irresponsible way in which the endless public lands in the West were dissipated by donations to local, private and often predatory interests. He also understood that the very cheapness of the public lands and the facility of their purchase and transfer encouraged the traditional American system of bad farming and waste of soil. He came to the conclusion that this could only be arrested by the spread of a more scientific knowledge of agriculture. The tragedy was that the existing colleges were not much good for this purpose; in his own words:

> Most of the existing collegiate institutions and their feeders were based upon the classic plan of teaching those only destined to pursue the so-called learned professions, leaving farmers and mechanics and all those who must win their bread by labor, to the haphazard of being self-taught or not scientifically taught at all, and restricting the number of those who might be supposed to be qualified to fill places of higher consideration in private or public employments to the limited number of the graduates of the literary institutions. The thoroughly educated, being most sure to educate their sons, appeared to be perpetuating a monopoly of education inconsistent with the welfare and complete prosperity of American institutions.

Some agricultural and technical education had already been established, and many petitions for improvement of such education had been received by Congress before Morrill, in 1857, introduced a bill which would reserve a certain portion of the public lands for the purpose of maintaining agricultural colleges. From the beginning it met with opposition from the Committee on Public Lands, and many Congressmen and Senators, but it was eventually passed by both the House and the Senate. It was vetoed by President Buchanan, despite the support of agricultural societies and influential friends, among them Marshall P. Wilder, the Boston horticulturist. Under the Lincoln administration Morrill tried again. Southern opposition being eliminated, the bill passed in 1862 and was approved by President Lincoln. This was the famous Morrill Act, establishing the so-called land-grant colleges.

The grant of lands—thirty thousands acres for each Representative and Senator—was promptly accepted by the Northern states. The Southern states were included in its terms when they came back to the Union.

Though the lands in many states were indiscriminately sold on the market for whatever they could bring—often less than a dollar an acre—there was also wise foresight used and some colleges benefited enormously from the provisions of the Morrill Act. Cornell University is a well-known example. The Massachusetts Institute of Technology was one of the beneficiaries. The new Massachusetts College for Agriculture at Amherst was founded under the Morrill Act. Several other colleges were founded under the same provision, while the schools received substantial financial support. Morrill was able to come to the aid of the struggling land-grant colleges with a bill to provide additional funds, which was passed in 1890 after six unsuccessful attempts. By this time more than fifty colleges were firmly established under the Morrill Act, and the future of agricultural education guaranteed, with great benefit to other forms of scientific and engineering education.

10

By the time of the Civil War, New England had a well-developed industrial system, and an efficient corps of engineers, technical experts and scientists. But it was not New England alone which waged war upon the South. New England played its role in the mobilization of science and technology as part of the Union. The great battle accelerated a process which had already started before, in which local characteristics gradually found their place in a more general national pattern. New England science, after the Civil War, became more and more an integrated part of the scientific life of the United States as a whole.

The victory of the North (and, it should not be forgotten, of the West) over the South was the victory of the modern industrialism over a slave-owning agricultural society. This meant that the North was able to mobilize far greater forces of science and engineering than the South. This was decisive in a war like the Civil War, which was already in many respects a mechanized war. Among the innovations in technical and scientific warfare which must be associated with the Civil War are large rifled cannon, "ironclad" warships, rotating turrets for naval guns, the use of railroad, telegraph, photography and even of balloons for communication and intelligence, and the machine production of shoes and uniforms. The South, even if it had possessed scientific and technical talent to compete with the North, simply had no sufficient manufacturing facilities.

The scientific forces of the North were also mobilized. The army's medical service was entirely reorganized. The Smithsonian Institution tested materials and armaments. The United States Coast Survey gave hydrographical and topographical information. And the crowning recognition of the importance of scientific research to the cause of the Union came in March 1863, when President Lincoln organized the National Academy of Science. Philadelphia science, New York science, New England science and the science of the Middle West found at last a common center. The foundation was a token that an era in the nation's scientific growth had come to an end. New England science had now fully grown into American science, taking an honorable part in the scientific development of the world.

Postscript (1991)

It was in part through reading *The Flowering of New England, 1815–1865,* that interesting book by Van Wyck Brooks, that the thought of writing a book on early New England science occurred to me. Brooks's book was about the literary scene in the period of Prescott and Emerson; was it possible to write a parallel account of the science, and related technology, of the same period, the early years of republican New England? After all, that was the time of Slater, Bowditch and Agassiz. And so I went to the libraries, found the Vail collection at M.I.T. particularly useful, traveled to places where things had happened—such as Salem and Pawtucket, the Saugus iron works, the old Middlesex canal—and talked to persons with historic recollections. This undertaking offered a chance to get better acquainted with the spirit of my new country, especially of Yankee New England. It was not unpleasant to be, in a modest way, a kind of inverse Motley. The Yankee diplomat had stimulated the Dutch to look at their own history; could a person who was Dutch-bred have a reciprocal effect on the Yankees?

At that time, in the 1940's, there existed, so far as I could see, only one book that was a kind of precedent for my project, and that was *Men of Science in America,* by Bernard Jaffe (1896–1986), New York high-school science teacher and author of books on science, particularly chemistry and its history (for example, *Crucibles: The Story of Chemistry from Ancient Alchemy to Nuclear Fission* [1930–1976]). His *Men of Science in America,* published in 1944, covered the whole extent of "America," meaning the United States of America and the British plantations from which the states grew, and also the whole period of their existence from Thomas Harriot to Edwin Hubble. Though written in the form of a series of biographies, Jaffe's book captures much of the scientific life of the different periods. Jaffe also had an eye for the many ways in which science and society are related. His book is a nice pioneering effort and should not be forgotten, even if we now possess not only a much more varied literature on the subject, but also a more professional approach to it.

Yankee Science in the Making deals only with New England. We now have in Professor Robert V. Bruce's *The Launching of Modern American Science, 1846–1876* a detailed and careful account of this field in the whole of the mid-nineteenth-century United States. Although Bruce's account centers on only a thirty-year period, it reveals much new information on the subject of "Yankee Science" and places it in the context of American science as a whole. Bruce's book is flanked by other informative and scholarly works, such as those by Greene, Reingold, Kohlstedt and others. Yet it seems that the Jaffe book is still the only one attempting a view of the whole of science in "America" from the beginning of European settlement.

Hardly any of these publications goes beyond what became and what is now the U.S.A. Not even Canada is included in the term "America." But there is room for a history of American science and technology in the true sense of the term. There are already accounts of science in Mexico and Canada. There is much similarity to be noted among scientific experiences in the various vast territories between Patagonia and the Canadian Arctic, especially in colonial days when the inspiration usually came from London, Paris or Madrid. Enormous differences also exist, with the U.S.A. having forged ahead into a position of leadership. Historians should not forget the precolonial days of science on this continent, either. But a beginning has been made. I hope that my *Yankee Science* may still remain of some value for its attempt at a broad perspective on the subject.

The reader should keep in mind that the text of the book is that of the 1962 edition. Although this does not affect the main narrative, minor changes have occurred. The old manufacturing section of Lowell, for instance, has become a historic landmark, the Slater mill in Pawtucket is a museum, and sections of the Middlesex canal have passed into the loving care of the Middlesex Canal Association. All these events show how more and more sections of our public have become aware of the history, not only of science, but also of technology.

(A few titles of more recent writings are given in the Supplementary Bibliography on page 531.)

Notes

D.A.B. = *Dictionary of American Biography; N.A.C.* = *The New American Cyclopedia,* 16 vols., 1863; E.S.S. = *Encyclopedia of the Social Sciences,* N. Y., Macmillan, 1924.

Preface to the Revised Edition

Ample information on the European exploration of the New England coast before 1620 in *H. F. Howe* 1943, see also *H. F. Howe* 1953, to be supplemented by the cartographic material in *J. Winsor's* Narrative and critical history of America (Boston, 1884–89) and the detailed (but, I believe, little known) studies on "crucial maps" by *W. F. Ganong* in the Proc. and Trans. of the Roy. Soc. Canada 1929–37. In the article of 1931 (3d ser., 25) pp. 200–202 are references to the word Norumbega (at present attached to a place near Boston, thanks to the late Professor Horsford). The works of Champlain and John Smith are available in modern editions. On Thomas Brattle see *J. Lovering* 1881, p. 491, on Deshayes *P. G. Roy* in the (Canadian) Bull. Recherches histor. 22 (1916) pp. 129–138.

On Watson, the Vaughan brothers, Gardiner, Wilder, Fessenden, Colmar and Holmes see D.A.B., on Fessenden also *P. G. Perrin* 1925. On the Gardiner Lyceum: *E. H. Nason* 1909; *N. E. Stevens* 1921. Among its principals was Benjamin Hale, who later taught chemistry and mineralogy at Dartmouth until in 1835 the trustees abolished the office. (Hale had become an Episcopalian.) See *L. C. Newell* 1925.

Chapter 1

The quotation is from J. Bigelow's inaugural address as Rumford professor in Harvard College. *Bigelow* 1816, p. 276.

1

On Bacon any history of philosophy can be consulted, and there are several good biographies. Collected works in 12 vols. by Spedding, London 1857–1872. On the engineering achievements in England before and under James I see *Smiles* 1861, 1.

2

On the crafts cultivated by the Pilgrims see *J. L. Bishop* 1866,

I p. 301; on their opinions and learning *V. L. Parrington* I; on their early history any textbook of American history.

3

On the present character of the New England towns see the *WPA Guidebooks* dealing each with one of the six states (Massachusetts, New Hampshire, Rhode Island, Vermont, Maine, Connecticut).

4

See *J. G. Palfrey* 1858 preface, the descriptive chapter in the *WPA Guidebooks,* e.g. Massachusetts, Ch. "The Natural Setting" pp. 9–17, and *P. E. Sargent* 1917, Ch. "Old New England" pp. 24–29. Also *I. B. Crosby* 1928.

5

On agriculture: *P. C. Bidwell-J. I. Falconer* 1925; *N. S. B. Gras* 1940; *J. Schafer* 1936.

6

On early manufactures see *J. L. Bishop* 1866, I esp. pp. 122, 431, 471, 476, 506, 510, 623; *W. B. Weeden* 1890; *R. Burlingame* 1941. On mining and metallurgy see *C. R. Harte* 1944; *H. C. Keith–C. R. Harte* 1935; *J. W. Swank* 1892. See also *J. A. Eliot* 1762. Ethan Allen did not keep his forge for a long time; for pleasant details see *C. R. Harte* 1944, p. 142. One of the earliest books on surveying in this country came from the Salisbury district: *S. Moore* 1796. On the younger Winthrop see Ch. IX of *S. E. Morison* (Builders) 1930; *H. G. Lyons* 1941; *W. Haynes* 1939; on early American chemistry *W. Haynes–L. W. Bass* 1935. The history of the Saugus iron industry in *E. N. Hartley* 1957.

7

See e.g. *J. C. Miller* 1943, Ch. I; *J. W. Swank* 1892, esp. p. 479; *A. Smith* 1776, Book IV, Ch. VII ("Of Colonies") in Everyman's Library ed. II, p. 79. Interesting are also the ideas of the British economist *J. Gee* 1738.

8

On craftsmen: *W. A. Dyer* 1915. Thomas Danforth (1703–1786), a pewterer born in Taunton, Massachusetts, opened shop in

Norwich, Connecticut, in 1733. He was the ancestor of the two largest pewtering families in America, the Danforths and the Boardmans. See *J. B. Kerfoot* 1924; *P. R. Hoopes* 1913. *M. Karolik* 1942 concludes: "Judging by all these facts, the social status of the craftsman and merchant was the same as that of the statesman and soldier." Pope's orrery in *I. B. Cohen* 1950, p. 154, Plate 5.

9

For ref. see Ch. 3.

10

On Boston in colonial days see *J. Winsor* 1881, I, II; on Newport *G. C. Mason* 1884; *I. B. Richman* 1905, Chs. V, VI; *C. Bridenbaugh* 1938; on Philadelphia *C. and J. Bridenbaugh* 1942. Isaac Newton himself presented some books to the early Yale library, including a copy of his own *Principia.* Description of Newport partly based on *W.P.A. Guide to Rhode Island.* Berkeley's poem "Westward," etc., quoted in N.A.C. III p. 170.

11

British American science from 1735 to 1789 is discussed in detail by *B. Hindle* 1956. A solid survey of the literature by *W. J. Bell* 1955. A general survey of New England science in *F. G. Kilgour* 1949.

On early academics: *W. T. Davis* 1897, pp. 1826–1887; art. J. W. Dickenson; on Cotton Mather see *C. Zirkle* 1932. The chronology of this paper is reprinted in *Isis* 21 (1934) pp. 218-220, and gives the following record. In 1716 Cotton Mather reported certain observations on Zea Mays (corn): (1) wind pollination; (2) hybridization (variety cross); (3) resemblance of some of the progeny to the male parent, and also reported a cross between *Cucurbita Pepo* and *Cucurbita maxima* (?) (squash). This is the first record of spontaneous hybridization (letter to J. Petimer, F.R.S., September 24, 1716). See also *C. Zirkle* 1935. On the Mather family see *T. J. Holmes* 1940; also review by *G. Sarton, Isis* 33 (1941) pp. 254–260.

On Stiles see D.A.B., N.A.C., also *E. Stiles* 1901; on early Harvard *S. E. Morison* 1936; on early Yale *L. W. McKeehan* 1947; *L. Tucker* 1956; on Elizur Wright *P. G.* and *E. Q. Wright* 1937, also *E. Wright* 1828. The essay on comets by Oliver is discussed by *J. C. Greene* 1954.

12

See *F. E. Brasch* 1931; *M. Kraus* 1942; *R. H. Heindel* 1938; *R. S. Bates* 1945, from which book quotation on Royal Society. Compare also *C. Zirkle* 1932; *F. G. Kilgour* 1938, 1939, 1949.

13

See *F. E. Brasch* 1916, 1928, 1940; also *S. E. Morison* 1934; *F. Cajori* 1890, 1928, pp. 48-49; *D. E. Smith-J. Ginsburg* 1934; *J. C. Greene* 1954. On Winthrop and Rumford see *G. E. Ellis* 1871, pp. 32-33. On the relations between Franklin and Winthrop see *C. Van Doren* 1938. As to Dr. Spencer, Franklin in his *Autobiography* calls him Dr. Spence, but the name seems to be Spencer; see *I. B. Cohen* 1941, esp. p. 49, also *I. B. Cohen* 1943. Spencer was an itinerant experimenter, who gave performances open to the public. On colonial mathematics see *L. G. Simons* 1924, 1931, 1936, 1936.[2] On the Venus transit of 1769 see *A. E. Lownes* 1943, *H. Woolf* 1953. The present accepted value of the sun's parallax is 8.790 seconds. The instruments at early Harvard College have been described by *I. B. Cohen* 1950 (with photographs), see also *N. H. Black* 1930. Mathematics at Harvard: J. L. Coolidge 1924, 1943, also *L. G. Simons* 1924, 1931.

14

On the lethargy in British science in the eighteenth century see *A. S. Turberville* 1933, pp. 209-286; on British scientists of that period see any history of science, e.g. *W. C. Dampier* 1943 or *W. T. Sedgwick-W. H. Tyler* 1939. On Winthrop's library *D. E. Smith-J. Ginsburg* 1934, p. 54, compare *F. G. Kilgour* 1939. Winthrop's library is now in Allegheny College, Meadville, Pennsylvania. The English authors are represented by Newton, Barrow, Halley, Keill, Oughtred, Cotes and Maclaurin; the Dutch by Huygens and Gravesande; the French by Ramus, Descartes and Lalande; the Germans by Wolf. On the backwardness of American science compare *S. Newcomb* 1876. The controversy between Jefferson and Raynal (also Buffon) in Ch. 7 of *E. T. Martin* 1952.

15

On American thought of this period see *M. Curti* 1943. On Winthrop lectures see *J. Winthrop* 1759; *A. D. White* 1896. Adams's marginal notes in *More Books,* published by Boston Public Library, March 1926, December 1930. See also on the fear of Godless colleges *D. Wayman* 1942. Cotton Mather's praise of Newton quoted by *M. Kraus* 1942, p. 270. For *Prince* v. *Winthrop* on the lightning rod see *A. D. White* 1896, I, p. 366. For Winthrop and Calvinism see *E. Stiles,* II, p. 334. *P. Miller* 1939; 1953 also analyzes the impact of science on the Calvinistic mind.

16

On New England medicine *S. A. Green* 1881, 1881.[2] On medicine in Massachusetts see *H. R. Viets* 1930; on Mather and medicine *O. T. Beall-R. H. Shryock* 1954. On the inoculation controversy: *A. D. White,* 1896; *J. T. Barrett* 1942. Compare further the general histories of medicine: *F. R. Packard* 1931; *F. H. Garrison* 1917; *N. S. Davis* 1877; *C. Singer* 1928; also *E. H. Clarke* 1876. On Holyoke see D.A.B. Benjamin Gale was the pupil, son-in-law and successor of Jared Eliot in Killingworth. This paper is in *Phil. Trans.* 55 (1761) pp. 193-204.

Some examples of epidemics in colonial New England: *C. R. Hall* 1934, p. 38; *S. A. Green* 1881,[2] p. 538. Other epidemics in *L. Shattuck* 1850. *G. C. Mason* 1884, p. 237, reports cases where inoculation hospitals became popular resorts. In 1774 a smallpox hospital was opened on Fisher's Island; in 1788 one on Ram Island, and another was opened on Dutch Island along the Connecticut coast. Parties were made up to pass a few days at a hospital pleasantly situated at the seaside, especially on Ram Island.

On early anatomy classes at Harvard see *S. E. Morison* 1936, also *Harvard Med. School* 1906. There is some evidence that there was a medical society in Boston in 1735; see *C. R. Hall* 1934, p. 48; *F. H. Garrison* 1917, p. 411.

Chapter 2

The introductory quotation is from *E. Stiles* 1783, pp. 86–87; the quotation in the text from *V. Lenin* 1918.

2

D'Alembert quoted Franklin with the words "*eripuit* etc." at the occasion of Franklin's first appearance before the Academy of Sciences in Paris, 1777. The words have been attributed to Turgot. On the national academy of those days see *G. Brown Goode* 1890, esp. p. 15, quoting *S. Blodget,* Economica (printed 1806), p. 22. Blodget urged various educational projects upon Congress. General information in *Early history* 1942.

3

See *John Adams, Works,* I, p. 29, II, p. 9, IV, pp. 260–261. John Adams's scientific interests are illustrated in a letter to Benjamin Waterhouse, concerning the education of his son John Quincy Adams in Paris. In this letter Adams writes that he "attempted a sublime flight" and after the books of Euclid in Latin, plane trigonometry, algebra and conic sections, tried to give him "some idea of the differential method of calculation of the Marquis de l'Hôpital, and the method of fluxions and infinite series of Sir Isaac Newton" (April 24, 1785, *Works* IX, pp. 530–531). Adams, bitterly disagreeing with Condorcet, agreed with him when he promised to "show that liberty, arts, knowledge, have contributed to the suavity and melioration of manners." See *More Books,* Boston Public Library, March 1926, December 1930. The story of Winthrop's plans for an Academy in *D. E. Smith–J. Ginsberg* 1934, p. 47 is unsubstantiated (see B. Hindle, *Isis* 41, 1950, p. 197).

Some parts of the charter of the American Academy of Arts and Sciences are reprinted in *R. S. Bates* 1940; see also *R. S. Bates* 1945. Professor Howard M. Jones, in his presidential address of 1944 to the American Academy, has drawn again our attention to the wisdom of some of this ancestral philosophy: *H. M. Jones* 1944. The diary of Manasseh Cutler has been published: *W. P. and J. P. Cutler* 1888. The opinion of Silliman, Jr., on Feron in *B. Silliman* 1874–1875, p. 76. Silliman called a paper in *Memoirs* I by *S. Tenney* on the springs of Saratoga the earliest record we have of these mineral springs, first observed by surveyors thirteen years before. Tenney recognized their alkaline character.

4

On Cutler see D.A.B. and N.A.C., also *W. P. and J. P. Cutler* 1888. Memories of Cutler and Putnam exist at Marietta, Ohio. On the early trips to the White Mountains see e.g. *F. W. Kilbourne* 1916; *S. Morse–J. Anderson* 1930. See also *J. H. Huntington* 1876–1878. Full bibliography: *A. H. Bent* 1911. A quotation on the horrible wilderness between New England and Canada, from Edward Everett, 1851, is quoted in *J. L. Ringwalt* 1888, p. 24. On the trip of Samuel Williams see *R. Bates* 1940; *R. Bates* 1945; *I. B. Cohen* 1950, p. 51.

Father J. F. Lafitau, missionary among the Iroquois, announced his discovery of ginseng in a letter to the Duke of Orleans, published in 1718, reprinted at Montreal in 1858.

5

See *S. A. Green,* 1881, 1881[2]; *W. L. Burrage* 1923. The history of the *Harvard Medical School* was published in 1906; see also *T. F. Harrington,* 1905.

In its earlier days, most of the work for the Medical Society was done by Holyoke's relative and pupil, Nathaniel Appleton, who later became a busy Boston physician. He was the chairman of the committee which in 1790 brought out the "Medical Communications," an organ which would be issued regularly for 124 years.

On Holyoke, the Warrens, Tufts, Appleton, see D.A.B. or N.A.C. The *Historic Guide to Cambridge,* 2 ed., Cambridge, Massachusetts, ed. by the H. Winthrop Chapter of the D.A.R. 1907, states on p. 97 that the medical examinations ordered in 1775 were the first of this kind in America, and that the examinations were severe. On this period see also *H. Thursfield* 1940; *S. E. Morison* 1936, pp. 99–100; *L. C. Duncan* 1931. Descriptions of Boston medical men by D. W. Holmes in J. Winsor, see *S. A. Green* 1881.[2] On the Connecticut Academy and the Yale Medical School see *S. Baldwin* 1901, *E. C. Herrick* 1840, *E. A. Smith* 1914; on New Hampshire see *O. P. Hubbard* 1880; on Nathan Smith: *E. A. Smith* 1914.

6

On manufacture *J. L. Bishop* 1866; on invention during the Revolution *I. B. Cohen* 1945; on the manufacture of gunpowder *O. W. Stephenson* 1924–1925. On *Bushnell* see D.A.B.; *M. C. Tyler* 1941; *R. Burlingame* 1938, p. 146; *A. P. Stokes* 1914. The machine has been described by *C. Griswold* 1820, and is mentioned by the mathematician Meusnier in the *Paris Mémoires de l'Académie* 51 (lu 1784).

7

On J. Jeffries see D.A.B. and N.A.C.; art. "Aerostation," N.A.C.; *J. Milbank* 1943. Holmes on Jeffries in *S. A. Green* 1881,[2] p. 563. There was considerable interest in aerial navigation in Boston, and the articles on this subject in the *Boston Magazine* for 1784 seem to be the earliest printed material on the subject in America (*J. Milbank* 1943, p. 107).

8

On Rumford see his biography, *G. E. Ellis* 1871; supplemented by *J. A. Thompson* 1935. Mr. Thompson calls Rumford an "organizer of marked ability and an excellent showman. But he was unbelievably cold-blooded, inordinately egotistic, inherently a snob." Rumford believed in his heart in a benevolent despotism, says Professor Bragg: *W. Bragg* 1930. This paper presents the excellent ideas of Rumford on models, museums and lectures. See further *B. Jaffe* 1944, Ch. VIII; *D. S. Jordan* 1910; *C. R. Adams* 1950. Rumford's works have been published: *Rumford* 1870–1875. Professor S. Brown of M.I.T., who is writing a biography of Rumford, has reconstructed several of his instruments, now at the American Academy of Sciences in Brookline, Mass. See *S. Brown* 1940; 1953; 1955; and Rumford Bicentennial 1953.

On Thompson's devastation of Long Island during the Revolution see *L. M. H. Lowry, C. D. Lowry, J. R. Miner* 1937, p. 268. The authors claim that Thompson was exclusively motivated by military necessity. This paper is preceded by a reprint of the first recorded scientific observations of Thompson, dealing with an aurora borealis observed in 1772 at Bradford, Massachusetts. The opinion of Marx on Thompson in *Marx* 1887, p. 613. Rumford's computation of the mechanical equivalent of heat in *Rumford* 1798. The Latin quotation in "*Dulces moriens . . .*" is from Virgil, *Aeneid* X, 781: "Dying he sees in remembrance his beloved Argos." Virgil expresses the sorrow of a warrior who dies far from his country. The quotation was used in reference to Rumford by *J. Bigelow* 1816. Here Bigelow, speaking to a generation which had in part known Thompson, claimed that "he was decidedly attached to the cause of American liberty." The quotation from De Candolle in *A. Gray* 1863, p. 9; here also some observations about Rumford's second marriage.

Chapter 3

The motto of this chapter is from *J. P. Brissot* 1791, p. 127. It means: "Nothing is difficult for mortals," and "The audacious race of Jap(h)et" (mankind); the two lines are disconnected and taken from Horace, *Odes* I, 3, 27 and 37. Brissot continues: "If these verses are applicable to some people then it should be to the free Americans. No danger, no distances, no obstacle detains them. What should they fear? All peoples are their brothers; they want peace with all." Brissot, returned to France, became the leader of the Girondins, and lost his head in 1793. The book shows how the new United States inspired, by its example, the revolutionaries in Europe. It also indicates the economic roots of Girondism: Brissot admires the merchants, but detests the embattled farmers who followed Shays.

2

On the history of New England shipbuilding: *H. I. Chapelle* 1935; *A. Laing* 1944; *H. Hall* 1884; *S. E. Morison* 1921. Some interesting information in *G. C. Homans–S. E. Morison* 1930. On Hacket see *H. I. Chapelle* 1935, p. 92; on the North River shipyards see *L. V. Briggs* 1889; on Barker *H. I. Chapelle* 1935, p. 114, also D.A.B.; on Peck *H. I. Chapelle* 1935, p. 134. The *Constitution* was built at Boston under the supervision of George Claghorne and launched in 1797. The plans were drawn by Josiah Fox, an Englishman, under the direction of Humphreys of Philadelphia.

3

For the chemical paper in the *Memoirs* of the Amer. Acad. see Ch. 2, 3. On the question of French influence on early republican shipbuilding see *H. Hall* 1882; *H. I. Chapelle* 1935; *A. H. Clark* 1910, p. 8; *G. C. Cutler* 1930, p. 49; a summary in *R. Burlingame* 1938, p. 320.

4

On lift models see *C. K. Stillman* 1933; *A. Laing* 1944; compare *S. E. Morison* 1921, p. 102. On Bentley and Pickering see D.A.B. On longitude determination see *R. T. Gould* 1923; *H. C. Brearley* 1919; also *S. E. Morison* 1921. The quotation from Bowditch: *N. Bowditch* 1802.

5

Background material for this section as well as for the whole Ch. III gives *R. H. Brown* 1943. On James Bowdoin, Jacob Perkins, E. M. Blunt see D.A.B. On Perkins and the Perkins family see *G. and D. Bathe* 1943; furthermore the *British Biographical Dictionary;* also *H. P. and M. W. Vowles* 1931.

6

On D. Willard, Cutler, Prince, Bentley, Read, Holyoke, see D.A.B. and N.A.C.; also *E. A. Holyoke* 1833. On Prince (1751–1836) also *C. W. Upham* 1837. Prince was an excellent instrument maker and was much consulted by colleges on apparatus and libraries. He left a library of 3500 volumes, with engravings, curiosities, instruments, etc. On Read also *D. Read* 1870. In a letter written to Benjamin Lincoln, Read wrote: "With this boat, by means of a crank and without a fly-wheel, I rowed myself . . . with great rapidity, across an arm of the sea, which separated

Danvers from Beverly." In 1790 he applied for a patent, but the paddle wheels were already invented. On Read's contribution to the invention of the steamboat see Ch. IV.

On Salem's library see *H. L. Byrstein* 1960; *H. Wheatland* 1856–1860; on Kirwan *J. Reilly–N. O'Flynn* 1930.

On Salem's shipbuilding before 1812: *J. D. Phillips* 1942.

7

There is considerable literature on N. Bowditch; apart from the articles D.A.B. and N.A.C., see *H. Bowditch* 1937. Two books have been written about him, *R. E. Berry* 1941 (the title "star-gazer" is not a very happy one, since Bowditch was above all a theoretical mind and gazed more at formulas than at stars) and *A. Stanford* 1927. The fourth volume of his Laplace translation contains a biography. The *Practical Navigator* is listed as *N. Bowditch* 1802, see also 1799. For children: *J. L. Latham* 1955.

On early insurance companies see *C. K. Knight* 1920; *J. Winsor* 1881, IV Ch. VII.

The return of Bowditch to Salem on his last voyage was in a fog, not, as the legend says, in a snowstorm [communication by Dr. Harold Bowditch].

On Adrian see *F. Cajori* 1928.

8

See *N. Bowditch* 1829–1839; *H. Bowditch* 1937. On the Weston meteor see *N. Bowditch* 1807, *O. C. Farrington* 1915; on Lissajous figures *J. Dean* 1815, *N. Bowditch* 1815. On his books *M. Munsterberg* 1938.

Serious work on the mapping of the New England coast started only with Samuel Holland and Joseph Frédéric Vallet Des Barres, under the British Crown, resulting in that extensive cartographic work, *Atlantic Neptune,* beginning about 1777. See *D. J. Struik* 1956.

9

Description of Boston in *J. Winsor* 1881; on Bulfinch see D.A.B. and *E. S. Bulfinch* 1896. On Parris, Willard, Benjamin, see D.A.B.; also *A. Benjamin* 1843, *W. W. Wheildon* 1865; on Cotting *J. Winsor* 1881, IV, pp. 32, 156. See also *J. L. Bruce* 1940; *E. C. Hultman* 1940.

10

On Bond see *W. J. Youmans* 1896.

"Boston light," built in 1715–16, was one of the oldest lighthouses in North America. See further the report *Lighthouses* 1846.

11

On Nantucket see e.g. *W. O. Stevens* 1936, esp. pp. 107, 111. On Folger see D.A.B.; also *American Journal of Science,* 1850, p. 313; *W. O. Stevens* 1936 and *H. Wright* 1949. On the Mitchells see *W. O. Stevens* 1936; *J. Winsor* 1881, IV, p. 508 (art. "J. Lovering"), D.A.B.

On whaling museums: *G. Sarton* 1931; on Lewis Temple: *S. Kaplan* 1953.

12

On Gibbs, the Brown brothers, West, I. Thomas, see D.A.B.; on almanacs see *G. L. Kittredge* 1920. On West's connection with the Venus transit of 1769 see *A. E. Lownes* 1943. Also *Gas* 1912 (Newport).

13

On Mansfield see D.A.B.; *R. S. Kirby* 1939, p. 39; *F. Cajori* 1928, p. 116; *Dupuy* 1943, pp. 33, 116; *J. Mansfield* 1801; on Davies see D.A.B. and N.A.C. The quotation is from Dupuy.

14

On the Stonington skippers see *A. Laing* 1944; *E. Fanning* 1833; *J. R. Spears* 1922. On the discovery of Palmer land is a considerable literature; see *E. S. Balch* 1902, pp. 85–93; *W. L. G. Joerg* 1930, p. 9; *A. R. Hinks* 1939; *W. H. Hobbs* 1939. *L. Martin* 1938 describes the original log of the *Hero,* and points out some discrepancies in Fanning. *A. Laing* 1944 shows special admiration for Palmer, see e.g. p. 139. On Palmer and Fanning see also D.A.B. On the Wilkes expedition see *C. Wilkes* 1940 and this book Ch. 7, as to the Bellinghausen expedition see footnote [15] of *W. I. Vernadsky* 1945. On an arctic expedition of 1860–1861 sailing from Boston and returning to it see the letter of I. I. Hayes to Silliman, Jr., *American Journal of Science* 33 (1862), pp. 263–265; also *Bellinghausen* 1945. The Life Pictorial Atlas of the World (N.Y., 1961) has "Palmer Peninsula."

Chapter 4

The motto at the beginning of the chapter is from a letter to D. Treadwell: *M. Wyman* 1888, p. 463.

1

On general technological conditions *J. L. Bishop* 1866, *H. Thompson* 1921, *C. Fraser* 1928; on agriculture *P. W. Bidwell–J. I. Falconer* 1925. On craftsmanship see Ch. I; on engineering in the U. S. A.: *J. W. Oliver* 1956; *D. Fitzgerald* 1899; *H. Howe* 1858. On early agriculture *S. Deane* 1790; *Centennial year* 1792–1892; *P. W. Bidwell–J. I. Falconer* 1925.

2

S. Smiles 1861 remains a fascinating book on the technological aspects of early British industrialization; it has a good account of the early British turnpike and canal builders. Samuel Smiles (1812–1904) was a Scottish writer, dedicated to the doctrine that courage and perseverance lead virtuous young people to success. His books were favorite prizes in British schools.

On the early industrialization of Coalbrookdale see *R. Jenkins* 1923–1924.

3

On turnpikes see *F. J. Wood* 1919, with full information. On early travel by road see *J. B. McMaster* 1893, I, p. 45. On transportation systems *J. L. Ringwalt* 1888; *B. H. Meyer* 1917; also art. "Roads" in E.S.S.

4

See *F. J. Wood* 1919. On Orr, Wilkinson, Ames, Blake, see D.A.B.; on the Wilkinsons also *G. S. White* 1836, *I. B. Cohen* 1945. On the turnpikes see also *R. S. Kirby–P. G. Laurson* 1932, *D. Stevenson* 1838.

Concerning Telford and McAdam see the *British Biographical Dictionary*. McAdam stayed in New York before and during the Revolution, but left with other loyalists after the peace of 1783. On improvement of the roadbed see *H. P. French* 1866, *T. H. MacDonald* 1928. See also art. "Pavements and roads" N.A.C. About the difficulties which turnpike builders met when their interests clashed with those of local farmers see the story of the Lancaster turnpike in Pennsylvania in *J. B. MacMaster* II, 1885, pp. 553–556.

Another technical aspect of the turnpike era were the coaches. On coaches of pre-Civil War days see art. "Coach" in N.A.C. A center of the industry was Concord, New Hampshire, where there was an Abbot-Downey coach on exhibition in the R. R. Station now at the Historical Building on Park Street.

5

On early American and European bridges and their builders see *D. B. Steinman–S. R. Watson* 1941; also N.A.C. art. "Bridge" and *R. S. Kirby–P. G. Laurson* 1932. Information concerning the Westerly Bridge obtained through the kindness of Miss S. E. Coy of the Westerly Public Library. Some information on old bridges also in the *W.P.A. Guidebooks* on the New England States. For early English bridges and their builders see *S. Smiles* 1861. On Tom Paine as a bridge builder see *T. Paine* 1945, II; *R. S. Kirby* 1930; some information also in *D. C. Seitz* 1927, *R. C. Roper* 1944. Dwight's remarks on Hale's bridge in *T. Dwight* 1821. On the Charlestown Bridge see *J. Winsor* 1881, III, p. 554. Pope describes some early American bridges in *T. Pope* 1811.

All available personal information on Palmer seems to be in *G. B. Blodgette.* Early settlers of Rowley, Massachusetts, Rowley 1933, xiii + 472 pp., see p. 261. The tablet on the Schuylkill Bridge on Market Street, Philadelphia, has a picture of Palmer's "permanent bridge." A picture of the Piscataqua Bridge in Stackpole's History of Durham I, p. 235 (letter from Dorothy Vaughan, Portsmouth, N. H.)

For the lore of covered bridges: *H. W. Congdon* 1946.

On Town and his trusses see the Town pamphlets: *Town* 1821, 1825, 1839; further *R. H. Newton* 1942, esp. pp. 281, 45; also particulars in *R. S. Kirby* 1939 and D.A.B. *D. B. Steinman–S. R. Watson* 1941 give particulars of several of the early bridge builders, such as Powers, Pratt, Burr, Long, Howe. On Pratt and his trusses see his obituary, *T. W. Pratt* 1875. On the Howe trusses see *R. S. Kirby–P. G. Laurson* 1932, p. 146; they are extensively discussed in modern textbooks on bridges such as *G. A. Hool–W. S. Kinne* 1928.

Early engineering texts on wooden bridges: *S. S. Post* 1859; Post (1805–1872) was a railroad engineer born in Lebanon, New York.

6

On iron bridges see *D. B. Steinman–S. R. Watson* 1941 and the other literature cited sub. 5. The Essex-Merrimack iron bridge described in *A. P. Mills* 1910–1911. Pope's account of this bridge was quoted by *M. Navier* 1823. Navier found that the U. S. A. had given the first example of a suspension bridge "en grand." A picture of the bridge from *J. Drayton* 1794, in *J. T. Adams* 1943.

Whipple's book is listed as *S. Whipple* 1847; W. J. M. Rankine (1820–1872) was professor at Glasgow University; the sources of his *Applied Mechanics* are partly French.

7

The plans of Penn in *Bishop* 1866, I, pp. 124, 562. The old N. E. canals are fully described in *A. F. Harlow* 1926; the Western Massachusetts canals also in *J. G. Holland* 1855. On the Cape Cod Canal see *W. B. Parsons* 1918, also *J. W. Miller* 1914. On Machin see *Appleton's Cyclopaedia of American Biography.*

8

The best study of the Middlesex Canal in *C. Roberts* 1938, with an account of the technical details contained in the Baldwin papers, as well as the commercial and financial aspect of the canal enterprise. There exist many descriptions of the canal such as *B. Walker* 1886, *M. W. Mann* 1910; see also the histories of the towns through or by which the canal passed (as Woburn, Winchester, Lowell, e.g. *F. W. Coburn* 1920). The quotations concerning the Middlesex Canal are from *E. E. Hale* 1893, pp. 121–122; *H. Thoreau* 1896, p. 78; and from the "historian of the Canal," *C. Eddy* 1843. On Weston see *R. S. Kirby* 1935–1936; also 1930.

9

On the Erie Canal: *S. H. Adams* 1953. Several of the Erie Canal constructors described by *C. B. Stuart* 1871. The influence of the Erie Canal on New England enterprise see *A. F. Harlow* 1946, Ch. I. Descriptions of the other N. E. canals can be found in town histories; the *W.P.A. Guide* of Connecticut has a picture of one of the ruins of the Farmington Canal. On the technical aspect of this canal see *C. R. Harte* 1933.

Quotations from *S. E. Morison* 1921; N.A.C., art. "Railroads." On the Hoosac tunnel see *A. Black* 1937, *A. F. Harlow* 1946.

10

On early steam navigation see *G. H. Preble* 1883; *W. Kaempffert* 1924, esp. p. 79; *J. Flexner* 1944. On Fitch see D.A.B.; N.A.C.; *J. G. Bathe* 1938; *T. Wescott* 1878; *T. Boyd* 1935; *C. Whittlesey* 1854; on Read see his book Ch. III, 5; on Morey see D.A.B., *W.P.A. Guide* Vermont, and *F. H. Getman* 1936.

Fulton's story is described in all books on American invention, especially those on steam navigation; see also D.A.B.; N.A.C.; *J. Flexner* 1944, p. 381; *H. W. Dickinson* 1913. Resentful words on the monopolistic tendencies of the Fulton enterprise in *T. Boyd* 1935. Fulton's memoir to Pitt in *S. Berstein* 1944.

Chapter 5

1

The main thesis from *Marx,* 1887, Ch. XV, pp. 365 seq.

2

For general information on early inventions in America see books such as *W. Kaempffert* 1924 or *R. Burlingame* 1938. There are many more, e.g. *J. L. Bishop* 1866; *E. W. Byrn* 1900; *W. H. Doolittle* 1902; *P. G. Hubert* 1896; *G. Iles* 1912; *H. Thompson* 1921; *E. Hodgins–F. A. Magoun* 1932. *J. W. Oliver* 1956 is a general history of technology in the States. Professor R. S. Woodbury of M.I.T. is writing a general history of tools: *R. S. Woodbury* 1958, 1959, 1960. Fulton's letter in *S. Bernstein* 1944. *A. E. Lownes* 1943 claims existence of an early steam engine in Rhode Island. On the steam engine at the Schuyler works see *L. F. Loree* 1929. See also *G. and D. Bathe* 1925.

3

On Yankee peddlers see *R. Burlingame* 1938, Ch. XII, quotation p. 183; the quotation from Crockett taken from *R. S. Kirby* 1939, p. 26; on Alger see D.A.B. On Yankee inventors: *E. Fuller* 1955.

4

The life of Whitney is described by *C. M. Green* 1956 and J. *Mirsky–A. Nevins* 1952. Also: *R. Burlingame* 1941: D.A.B.; N.A.C.; *M. B. Hammond* 1897–1898; *R. S. Kirby* 1939; *Fuller* 1947. On the present and past cotton gin *H. B. Brown,* 1938. A picture of a pre-Whitney cotton gin also in *R. C. Adams.* Pilgrims, Indians and Patriots (Boston, Little Brown 1928, 207 pp. popular).

5

On interchangeable parts see *R. Burlingame* 1938, p. 161 seq. The question of Whitney's indebtedness to France is still awaiting a satisfactory answer, and he may well have come to his ideas independently. See *J. Mirsky–A. Nevins* 1952, p. 141.

6

On North and Orr see D.A.B.; on Colt and Blanchard see D.A.B. and N.A.C. On Blanchard's profile lathe *R. Burlingame* 1938, p. 416; on arms industry *R. E. Gardner* 1930.

7

On early cotton manufacture see *C. F. Ware* 1931; on the beginning of cotton manufacture *W. B. Weeden* 1890; *A. M. Goodale* 1891. On Orr, Cabot, Brown, Slater, see D.A.B.; on Slater also *G. S. White* 1836, *W. R. Bagnall* 1890. The quotation on Slater's engagement by Brown in *White,* p. 72, on Slater's marriage in *White,* p. 102; on the happy conditions of the workers, p. 120. Data on the value of "goods manufactured by the loom" in *T. Pitkin* 1817, esp. p. 471. One of the men to lose money in the Beverly venture was the physician Joshua Fisher (1748–1833), who was its superintendent. Fisher later endowed the chair of natural history at Harvard, of which Asa Gray was the first incumbent. (*W. L. Burrage* 1923, p. 104.)

8

On Lowell, Moody, Appleton, Jackson, see D.A.B. On the Waltham and Lowell enterprises see *C. F. Ware* 1931; *N. Appleton* 1848; *S. Batchelder* 1863. "The promoters of these enterprises brought to the cotton manufacture experiences and resources unfamiliar to the owners of earlier mills. The founder, Francis C. Lowell, a wealthy Boston merchant, had observed in England the method of power loom weaving. He knew the way in which the English handled their capital and the credit advantages which their bank connections gave them. In addition, he had seen with horror the social conditions in the factory towns of the old country. Consequently, when he projected his own undertaking his threefold aim was to devise satisfactory power looms, to collect a large capital, and to provide a scheme for protecting the working population. . . . These were the fathers of American big business as Almy and Brown had been the fathers of the American industrial revolution." (*C. T. Ware* 1931, p. 61.) The patent agreement with Pollock in *C. F. Ware,* p. 262. On Lowell as a scientific center see this book, Ch. 9.

9

On early wool manufacture see *A. H. Cole* 1926; on Whittemore, Howe, Washburn, Chickering, Willard, Terry, Thomas, Jerome, see D.A.B.; on Howe also N.A.C. On clockmaking see *H. Terry* 1872; *G. H. A. Hazlitt* 1888; *J. Terry* 1889; *A. Webster* 1891; *C. Jerome* 1860; *C. W. Moore,* 1945. On silk industry see *L. P. Brockett* 1876; on brass industry *W. G. Lathrop* 1926.

General background information *V. S. Clark* 1916.

Chapter 6

The quotation is from *R. W. Emerson* 1887, p. 99 (*The American Scholar*). On this period in the U. S. A. see *D. J. Struik* 1959.

1

On teaching of mathematics at Harvard and elsewhere in eighteenth-century America see *D. E. Smith–J. Ginsburg* 1934; *F. Cajori* 1890; *L. G. Simons* 1924, 1931, 1936, 1936^2; *J. L. Coolidge* 1924, 1943. Bronson Alcott's statement in *O. Shephard* 1938. On Webster, Morse, Pike, Belknap, see D.A.B.; on Noah Webster also *E. C. Shoemaker* 1936, *H. R. Warfel* 1936; on Morse also *J. K. Morse* 1939. On early American cartography see *E. Raisz* 1937, *E. G. R. Taylor* 1941. Early American books on mathematics listed in *L. C. Karpinski* 1940. General information in *G. C. Bush* 1891, *J. Quincy,* 1840.

2

The struggle between Jeffersonianism and Federalism is described in books on American history, e.g. *J. B. McMaster* 1885. Examples of the Federalist attitude to foreign scientists from *S. E. Morison* 1936; *G. Chinard* 1933, pp. 254–255; *B. Jaffe* 1944, Ch. II. The verses of Bryant on Jefferson have often been quoted, e.g. *B. Jaffe* 1944, p. 91; those of Percival on Jefferson in *J. G. Percival* 1859. The partisan packet sloops are mentioned in *S. E. Morison* 1921, p. 231. Parrington's judgment in *V. Parrington* 1927–1930, II, pp. 277–278. On Dartmouth see *F. B. Sanborn* 1904; on Bentley and Meigs see D.A.B. On the University of Virginia see *H. B. Adams* 1888. Also *D. E. Smith* 1932, *C. G. Bowers* 1936.

On Meigs at the University of Georgia see *C. Eaton* 1940. Meigs has been called the first American scientific meteorologist after Franklin; see *C. Brown Goode* 1890. He applied his science in an original way: "using the formula for fallen bodies and applying it to the nine days' fall of Lucifer and the perverse angels, he estimated that Hell was 1.832.308.363 miles deep." (*C. Eaton* 1940, p. 16.)

3

On Waterhouse see D.A.B. and N.A.C. On his books at Brown see *J. W. Wilson* 1942; on his vaccination see *T. F. Harrington* 1905, I, p. 317 seq.; *R. H. Halsey* 1938, *J. B. Blake* 1957. On his dismissal see *B. Waterhouse* 1921. On L. Spaulding see D.A.B. Holmes on Waterhouse: *O. W. Holmes* 1892; *S. A. Green* 1881. On early natural history in the U. S. A.: *C. A. Browne* 1886.

4

On chemistry and geology in the period of Priestley and Lavoisier see any history of science, e.g., *W. Dampier* 1943 or *W. T. Sedgwick–H. W. Tyler* 1939. A good account in *C. R. Hall* 1934, *E. Child* 1940. For the history of geology in the U.S. we have the detailed account in *G. P. Merrill* 1924, a slightly modified version of *G. P. Merrill* 1906, 1920. Also *W. N. Rice* 1907.

5

On T. Dwight see D.A.B. and *C. E. Cuningham* 1942; *on Silliman G. P. Fisher* 1866; also D.A.B., N.A.C., *W. J. Youmans* 1896, and *D S. Jordan* 1910. A new book on Silliman is *J. F. Fulton–E. H. Thompson* 1947. On Gibbs see D.A.B. and *Am. Jour. Sc.* 25 (1834) pp. 214–215. The history of chemistry in the U.S. related by *E. F. Smith* 1914; see also *L. C. Newell–T. L. Davis* 1928, *B. Silliman* 1874–1875, *W. Haynes* 1939; a chronology in *W. Haynes–L. W. Bass* 1935; on Accum see *C. A. Browne* 1925. See also the centennial issue of the *Am. Jour. Sc.* in 1918. In the preface to the first number of the *Am. Jour. Sc.* "Bruce's Journal" is called "our earliest scientific journal supported by original American Communications." The preface continues by referring to the recent travels of Lewis and Clark and the many foreign naturalists visiting America. Mineralogy is "a treasury just opened." Gibbs's cabinet is called outstanding even for Europe. (Gibbs's collection still exists at Yale, but has now been mixed up with other collections.) On Silliman's religious position see Ch. 12, 2, 3. See also *F. B. Dexter* 1887; *J. L. Kingsley* 1835–1836.

6

On Dexter, the Warrens, J. Jackson, W. D. Peck, Nuttall, Gorham, Webster, see D.A.B. On the Harvard Medical School see *T. F. Harrington* 1905, also *Harvard Medical School* 1906. Biographies of Dexter, the Warrens and Jackson also in *W. L. Burrage* 1923. The plan for the *Pharmacopoeia* of 1808 was accepted in 1805. In the preface it is called the first work of its kind in the U.S.; its basis was the *Edinburgh Pharmacopoeia*, but this book was in Latin. The *Massachusetts Pharmacopoeia* was in English, and the preparations were given in Latin and English.

On Peck and the Harvard Botanical Garden see *W. M. Smallwood* 1941; also *W. J. Hooker* 1825, *E. Ingersoll* 1886; see also *No. Am. Rev.* 6 (1818) p. 417. On Nuttall see Ch. X, 2. On Webster and his trial there is an extensive literature, see e.g., *R. M. Devens* 1881. On Webster as a lecturer or a personality see *E. E. Hale* 1893, p. 154. Hale also speaks of a course by Farrar on the steam engine.

7

Jefferson's letter to Willard is from Paris, March 24, 1789; see Jefferson's *Works III*, p. 16 (Washington ed., 1853); on A. Wilson see D.A.B. and *S. P. Fowler* 1856–1860; on Maclure see *G. P. Merrill* 1924; *C. Keyes* 1925 and *G. P. Fisher* 1916. Parts of Michaux's accounts have been published by *C. S. Sargent* 1889. See further *F. A. Michaux* 1819. On Pursh see D.A.B. and *W. J. Hooker* 1825. John Torrey, the New York botanist, stated that Pursh "is such a notorious liar and plagiarist that we can put no confidence in his work" (*A.D. Rodgers* 1942, pp. 26, 27; Rodgers considers Pursh "an able botanist notwithstanding").

On the Linnean Society see *Milestones* 1930; *T. T. Bouvé* 1876, 1880. On Bigelow see D.A.B.; N.A.C.; *G. E. Ellis* 1883; *J. Bigelow* 1816, 1817–1820, 1824, 1831, 1840. On *F. Boott* see the *British Biographical Dictionary*. On Bigelow's *Florula, J. Bigelow* 1814, see review by C. Cushing, *No. Am. Rev.* 20 (1825) pp. 221–224. On the trip to the White Mountains *J. H. Huntington* 1876–1878.

8

On the Dana brothers see D.A.B. and N.A.C. On S. L. Dana also Ch. 7. On their book see *G. P. Merrill* 1920. On Parker Cleaveland, who was a distant relation of Moses Cleaveland, the founder of Cleveland, see D.A.B. and N.A.C.; personal details in *L. Thompson* 1938. On his book listed as *P. Cleaveland* 1816, see *G. P. Merrill* 1920.

On Tudor and Farrar see D.A.B. and N.A.C. On the Anthology Club and the *No. Am. Review* see *Van Wyck Brooks* 1936, pp. 16, 17. Farrar's books listed in *L. Karpinski* 1940. On Zerah Colburn see his autobiography, *Z. Colburn* 1833; also D.A.B. and N.A.C. and *S. Greene* 1953. *S. Breck* 1877 describes a performance of Colburn in 1811. One of his most remarkable performances, the splitting of $2^{32} + 1$ into factors, was first achieved by Euler (1731–1732).

On Farrar's work in modernizing American mathematics see *L. G. Simons* 1931; also *J. Farrar* 1818.

There also appeared a series of mathematical texts at Yale,

written by Jeremiah Day, who became the venerable president of the college. The titles can be found in *L. Karpinski* 1940. They were written in a spirit more traditional than Farrar's, and were very popular. Of Day's *Introduction to algebra* (1814) there were 67 editions before 1850.

9

On the history of life insurance see *C. K. Knight* 1920. Life insurance in America dates back to Franklin, and the first attempts in Boston to William Gordon (1792). Around 1790–1800 tontines were popular, at a time when many public and semipublic works were financed by lotteries. The Massachusetts Hospital Life Insurance Company was formed in 1818; the New England Life Insurance Company of Boston in 1835; the Aetna Insurance Company of Hartford was founded in 1819 for fire underwriting. On mortally statistics see this book Ch. 8. On mutual fire insurance see *H. F. Freeman and N. G. Abbott* 1935.

The quotation from Gauss in *H. Mack* 1927: letter from Joseph Gauss to C. F. Gauss, July 18, 1836. Schumacher was a Danish astronomer. Newcomb's opinion in *S. Newcomb* 1876; see also *B. Peirce* 1839. Bowditch wrote an account of continental astronomy in a series of book reviews, *No. Am. Rev.* 20 (1825) pp. 309–366, which is still very readable. Bowditch's translation listed as *N. Bowditch* 1829–1839.

On Barlow, Humphreys, Dewey, see D.A.B. and N.A.C.; on Hopkins and Emmons see D.A.B. On the Hartford wits compare *V. Parrington* 1927–1930, I, pp. 357 seq. On Parkman see *M. Wade* 1942, p. 190. Williams College and its teachers in *J. H. Denison* 1935, esp. p. 91. The proprietary medical schools described in *F. C. Waite* 1935.

Chapter 7

The quotation is from *H. Thoreau, Writings,* X, p. 160.

1

For the general political and economic situation during the Jacksonian period see *F. J. Turner* 1935, which has a special section on New England; also *C. R. Fish* 1929; *A. B. Darling* 1925; for the geological surveys see *G. P. Merrill* 1924. Also *M. Meisel* 1924–1929.

A. Hunter Dupree 1957 has studied systematically the relations between scientific work and the Federal Government.

2

See *C. A. Browne* 1943 on Thomas Jefferson; *G. Brown Goode* 1890 on J. Q. Adams. The geological surveys of the twenties by Dewey and Hitchcock were published in the *Am. Jour. Sc.* On Olmsted and Hitchcock see D.A.B., N.A.C., and *W. J. Youmans* 1896. On Hitchcock see also *C. M. Fuess* 1935, who has some anecdotes about Hitchcock, including his opposition to "that chef d'oeuvre of licensed gluttony, a New England Thanksgiving" (p. 120). Details on the organization of the Massachusetts survey in *E. Hitchcock* 1833. Jefferson and J. Q. Adams were authors of reports on weights and measures. See *C. Davies* 1871; *W. Hallock–H. T. Wade* 1906; *J. Q. Adams* 1821. On J. Q. Adams see *H. T. Good* 1934.

3

On Hitchcock see above sub. 2. His report: *E. Hitchcock* 1833. Collaborators were *T. W. Harris* (mammalia, insects), *J. M. Earle* (land and fresh water shells), *N. M. Hentz* (spiders), *A. A. Gould* (radiata), *F. A. Greene* (plants). See e.g. *T. W. Harris* 1841, 1862. Evaluation in *G. P. Merrill* 1924. Quotations from Hitchcock's report taken from pp. 166, 248. On the fossil tracks in the Connecticut Valley *G. P. Merrill* 1924; *J. Deane* 1863.

4

Final report *E. Hitchcock* 1841. See also E. Hitchcock 1838. On the Linnaean Society see Ch. 6, 6. It concentrated, of all things, on the sea serpent, one of these fabulous creatures having supposedly appeared off the New England coast. It took affidavits from eyewitnesses of the event, and sponsored a publication upon it, which appeared in 1817. Despite refutations of these stories by men like New York's Dr. Mitchill, who in 1828 exposed a number of them in his *History of Sea Serpentism,* belief in the saga remained; it is taken seriously in the N.A.C. of Civil War days, and even nowadays there exists, it seems, division of opinion. See *C. R. Hall* 1934, p. 90; *Am. Jour. Sc.* 15, pp. 351–356; *C. Lyell* 1850, I, Ch. VIII; *A. C. Oudemans, The great sea-serpent* (Leiden, 1892).

On the educational movement see this book, Ch. 8. Thoreau's opinion in *H. Thoreau* 1893, art. "Excursions"; on G. B. Emerson, A. A. Gould, T. W. Harris, D. H. Storer, see D.A.B. Also *G. B. Emerson* 1846, 1854; *R. S. Waterston* 1882–1883.

5

On Amos Binney see D.A.B. and N.A.C.; *A. Binney* 1851.

Conchology 1862. On C. T. Jackson see D.A.B., N.A.C., also the literature on the discovery of anesthesia, see this book Ch. 8. On Emerson and chemistry see *C. A. Browne* 1938. On early mining in N.E. *V. C. Perkins* 1941.

6

On Percival see D.A.B., N.A.C., *J. H. Ward* 1866; there are also a biography in *J. G. Percival* 1859 and a vivid description of Percival in *M. Rukeyser* 1942; for an estimate of his geological work by *J. D. Dana,* see D.A.B., *G. P. Merrill* 1924. Devastating criticism of his literary work by *J. R. Lowell* 1891; it seems somewhat exaggerated, especially when compared to the literary work of Lowell himself. Also *A. B. Benson* 1929.

7

On the Borden survey see art. "Borden" in D.A.B., N.A.C., *S. Borden* 1846. See also *F. Cajori* 1929. Hassler made nineteen criticisms of the work of Borden, to which Borden replied; see reports of meeting Am. Phil. Soc. of September 17, 1841. See also *No. Am. Rev.* 61 (1845) p. 455; 1842, pp. 446–457. A full description of the Borden survey by *E. G. Chamberlain* 1884–1886, with references to later surveys, such as Guyot's work. Borden on railroads *S. Borden* 1851. On the agricultural survey *H. Colman* 1837, 1840, 1841, 1857.

8

On the Coast Survey see this book Ch. 12. On the Wilkes expedition see *C. Wilkes* D.A.B. and N.A.C., also *C. Wilkes* 1940 and scattered reports in the biographies of participants, e.g., *D. C. Gilman,* 1899, Ch. V. On J. D. Dana and Gray see this book Ch. 10; on Pickering see D.A.B. and N.A.C.; *W. I. Vernadsky* 1945, footnote [15], points to the influence of Russian exploring expeditions on the launching of the Wilkes expedition. We have already mentioned the Bellinghausen expedition. Vernadsky adds th name of Krusenstern (1770–1846). It was the report on the Russian travels, published in 1827 in German, which spurred the lawyer J. Reynolds to his activity in behalf of our American expedition.

9

See *C. T. Jackson* 1836; *Milestones* 1930; Tuckerman's complaint in *A. D. Rodgers* 1942; Tocqueville quoted by *C. A. and M. B. Beard* 1927, p. 741.

On the establishment of the Am. Ass. Adv. Sc. see *M. Benjamin* 1899; *R. Bates* 1945; *H. L. Fairchild* 1924; *W. J. McGee* 1898.

10

The Owen survey in *G. P. Merrill* 1924. For the general picture of these days see B. *Jaffe* 1944, Ch. VII–XI.

Chapter 8

The quotation is from *J. G. Percival* 1859, II, from a poem "Prometheus."

1

On Eaton see D.A.B. and *E. M. McAlister* 1941, esp. pp. 180–184; his letter to John Torrey *A. B. Rodgers* 1942; also *H. G. Good* 1941. Stephen Van Rensselaer was a kind of enlightened despot (see D.A.B. and N.A.C.) like his fellow patroon, Chancellor Livingston (see Ch. 5); he prompted science and engineering. On the social side of the patroon regime consult the literature on the New York rent wars of the forties, notably *H. Christman, Tin Horns and Calico* (Holt, N. Y., 1945). On the Rensselaer Institute and Eaton: *Van Klooster* 1949, 1949.

2

General information on this period in *F. J. Turner* 1935, *A. M. Schlesinger* 1945; *A. B. Hart* 1927–1930, esp. III and IV; *V. W. Brooks* 1936.

3

On the Workingmen's Parties see *A. Whitman, Labor Parties 1827–1834* (N.Y., Int. Publ. 1943, 62 pp).

4

On the Lyceum movement see *C. B. Hayes* 1932, *S. I. Jackson* 1941; on Holbrook see D.A.B. and E.S.C. as well as *Am. Jour. Educ.* 8 (1860) pp. 229–247; on his Derby Academy see *Bailey, Cyclopedia of Agriculture* IV, pp. 367–368; on his publications and models *Holbrook* 1829.

5

On Blake see D.A.B. and N.A.C., his astronomy: *J. L. Blake* 1831 is an example of his many textbooks. On Colburn see

D.A.B. and *T. Edson* 1856; *W. Colburn* 1821, 1891; the last book has a biography. A book in the same vein was *J. Grund* 1830, with recommendations from Farrar, G. B. Emerson, and others, and with thanks to Boston educators (including Elizabeth Peabody) for suggestions for simplified technique. It follows a question and answer technique, has some history and problems selected from the Prussian text by Meier Hirsch. On President Wayland see D.A.B. and N.A.C. The school reform of 1837 is described in many books: see *A. B. Hart* 1927–1930, or *J. Winsor* 1881, III, esp. p. 659; on Mann see D.A.B., N.A.C. or *Mrs. H. Mann* 1865. See also *M. Curti* 1937.

6

For the history of the Lowell Institute see *H. K. Smith* 1898; on J. Lowell see D.A.B. Lyell's observations in *C. Lyell* 1845, I, p. 86. Lyell's rather complacent attitude towards slavery contrasts sharply with that of *C. Dickens* 1842. On Henry Rogers at Portsmouth see *E. S. Rogers* 1896, I, p. 238.

7

On Emerson see *C. A. Browne* 1938, with a list of Emerson's references to chemistry. The quotation is from Emerson's *Works,* XII, pp. 475–476. Emerson's address on the American Scholar in *R. W. Emerson* 1887, pp. 81–116. Channing's address "The Present Age" in his *Works* (Am. Unit. Ass., Boston, 1894) pp. 159–172.

8

General information on early American socialism in *M. Hillquit.* History of socialism in the United States (N.Y., London, Funk & Wagnalls, 1903, 371 pp.) On Brook Farm see a.o. *More Books* (Boston Public Library), February, March, 1937 (also separately printed). On New Harmony see *C. A. Browne* 1936. On Thoreau see *H. S. Canby,* Thoreau (Boston, 1939); on Thoreau's contribution to limnology *E. S. Deevey* 1942.

9

General background information in *A. F. Tyler* 1944; on Graham *M. V. Naylor* 1942; on Holmes D.A.B., N.A.C. and *J. T. Morse* 1896; his collected works, esp. *O. W. Holmes* 1892 and *S. A. Green* 1881; on general history of medicine see notes to Ch. 1, 15; on H. I. Bowditch see *V. Y. Bowditch* 1902, *A. A. Walkling* 1933; on Holmes and Semmelweiss see *B. Stern* 1941.

10

On the Harvard Medical School see *Harvard Medical School* 1906, *R. Fitz* 1939; *H. R. Viets* 1930, on the Massachusetts General Hospital *N. I. Bowditch* 1851. Early American dentistry in *J. A. Taylor* 1922; *Dentistry* 1941; art. "Dentistry" N.A.C.; on Greenwood see *B. W. Weinberger* 1943; on Wells *H. Wells* 1838, 1847; also *W. H. Archer* 1944, as well as D.A.B., N.A.C. In 1949 the American Dental Association sponsored a pamphlet commemorating Wells.

11

On Morton see D.A.B., N.A.C., *N. P. Rice* 1859, where (pp. 91–93) the Gould quotation can be found; Holmes quotation on p. 105, and Morton's "sucking bottles" on pp. 117–118. On the ether controversy, *C. Q. Colton* 1886, and above all *R. Fülop-Miller* 1930. Also *G. B. Roth* 1932; *C. E. Heaton* 1946.

On Stanley Howe and Dorothea Dix see D.A.B., N.A.C., as well as *H. E. Marshall* 1937; *D. Dix* 1843. See also, in connection with the care of the deaf, mute, blind, and insane, the art. "Gallaudet, S. B. Woodward, E. Todd, J. D. Fisher, L. Bridgman" in D.A.B.

12

On Jarvis see D.A.B., *E. Jarvis* 1847, 1848; on Shattuck D.A.B. On public sanitation in Massachusetts *G. C. Whipple* 1917.

13

On social statistics in America *L. Shattuck* 1850; furthermore *T. Dwight* 1811; *T. Pitkin* 1816; *L. Shattuck* 1841; *J. Chickering* 1846.

14

See *L. Shattuck* 1850; *E. J. Macdonald* 1936. *G. C. Whipple* 1917, Vol. 1, pp. 241–367, reprints large abstracts from Shattuck's report. *J. B. Blake* 1955 deals with Shattuck's role in the Boston water supply.

Chapter 9

The initial quotation is from *G. T. Poussin* 1836; preface. Translation: "When we like to cite an example of practical application of the great discoveries useful to the progress of industry, and therefore, to the people's prosperity, then we are obliged to turn our eyes to the United States of America. Providence seems to have placed there the great practical school of the world in order to let all modern discoveries bear fruit. The reason is that for this were necessary a virginal land, active, enterprising and free men; North America alone presented the necessary union of all these conditions." Poussin calls himself *"ex-major du corps du génie américain,"* his motto is "Time is money."

1

On Treadwell's press see M. Wyman 1888; also art. "Treadwell" D.A.B. and N.A.C.; *W. Kaempffert* 1924, p. 247; or *J. Winsor* 1881, IV, p. 515. It is claimed that the first sheet printed in America by any other than hand power was printed on Treadwell's press. On early steam engines see art. "N. Roosevelt" in D.A.B., also *L. F. Loree* 1929, *G. and D. Bathe* 1935; here also on Oliver Evans. Evans' work also in *G. Bathe* 1938.

2

See as an example of early English engineering books *O. Gregory* 1815, *T. Tredgold* 1838, *J. Robinson* 1822, or the extensive *J. Farey, Treatise on the steam engine,* 1827; its influence on American texts can be traced in such books as *J. Bigelow* 1831. On Baldwin see this book Ch. 4, 11, on Perkins this book Ch. 3, 75. On Allen see D.A.B. and *A. Perry* 1883, Allen's book is listed as *Z. Allen* 1829; he later wrote a treatise *Z. Allen* 1852, with a mechanistic philosophy, in which he faced the problem that "mechanical and chemical sciences are at issue in their respective doctrines, and the one or the other must be necessarily erroneous." He reduced all phenomena to matter in motion, brought about by an Intelligent First Cause. Allen's importance for fire insurance in *H. F. Freeman and N. G. Abbott* 1935.

3

On railroads of New England see *A. F. Harlow* 1946, with bibliography. A veritable encyclopedia on early railroad lore was republished in Philadelphia in 1832: *N. Wood* 1825. The directors of the Boston and Worcester Railroad, ordering an engine in 1833, wanted "the performance to be equal to that specified in p. 365 of Wood's treatise." (*E. C. Kirkland* 1945, p. 153.) On the railroad to level Boston's hill see *J. L. Bruce* 1940; *J. L. Ringwalt* 1888, p. 69. On the Quincy Railroad see art. on Bryant in *C. B. Stuart* 1871, on its relation with the Bunker Hill Monument *W. W. Wheildon* 1865; also *J. Winsor* 1881, IV, p. 117 seq. The quotation concerning Massachusette's palsied dotage is taken from the opening lines of *A. F. Harlow* 1946. On N. Hale see D.A.B. and N.A.C. as well as *E. E. Hale* 1893. The Rainhill experiment is described in books on the invention of railroads and in the biographies of Stephenson; see e.g. *S. Smiles* 1861. The quotations from the *Boston Transcript* taken from *J. E. Chamberlin* 1930, esp. pp. 84–86. On the Lowell Railroad see *F. B. C. Bradlee* 1918; also *A. B. Hart* 1930, IV, p. 426. The economic side of the railroad picture in *E. C. Kirkland* 1945; see also *G. S. Calender* 1902–1903. The story of the Hoosac tunnel in *A. Black* 1937, pp. 36–41; *A. F. Harlow* 1946, Ch. XI.

4

The cockney coachman quoted in *A. B. Hart* 1930, IV, p. 426; on the Lowell Railroad and "John Bull" see *F. B. C. Bradlee* 1918; *A. T. Harlow* 1946, p. 85 seq.; the quotation on the leveling of social life is from the *Boston Transcript,* see *J. E. Chamberlin* 1930. On West Point see *R. E. Dupuy* 1943; *S. Forman* 1950; on Thayer D.A.B. and N.A.C. The importance of the Paris École Polytechnique has been explained by *F. Klein* 1926, Ch. II; see also *C. G. J. Jacobi,* Werke 7, p. 355 (address of 1835). See also this book p. 228. On Crozet: *W. Couper* 1936.

5

On Whistler see D.A.B., N.A.C., *G. L. Vose* 1887, and biographies of his son, the painter; on McNeill and Swift see D.A.B. See also *A. F. Harlow* 1946 and *C. B. Stuart* 1871.

6

On Lowell see the town histories, esp. *F. W. Coburn* 1920. On early Lowell see also the histories of early textile manufacture, e.g., *C. F. Ware* 1931, also aur notes to Ch. 5, 8. Impressions of foreign visitors in *C. Dickens* 1842; *M. Chevalier* 1839; *C. Lyell* 1845, I, p. 117. Coming from Europe they were all favorably impressed. Chevalier, who visited Lowell in 1834, felt a "sort of giddiness" at the "sight of this extemporaneous town," noted the tendency to reduce wages. On Moody, Boott, Ames, Bartlett, Francis, S. L. Dana, see D.A.B. and *F. W. Coburn* 1920. On Francis Boott see this book Ch. 6, 7. On S. L. Dana see also N.A.C.; *H. R. Bartlett* 1940 has shown the importance of S. L. Dana as a pioneer in industrial chemistry. See also *S. L. Dana* 1842. On Holyoke see *R. H. Gabriel* 1936. Among the young men of Lowell was also Alvan Clark, the future telescope builder (see

Ch. 12, 6), his marriage of Mar. 25, 1826, is the first on record in Lowell.

7

On Francis see D.A.B. and *H. F. Mills* 1892; his book is quoted in most textbooks on hydraulics, e.g. *M. Merriman* 1898, esp. p. 298. On the Boydens see D.A.B.; on their inventions see textbooks on inventions and on hydraulics, e.g. *M. Merriman* 1898; on his bequest to Harvard and the "Boyden expeditions" see *S. I. Bailey* 1931. On Storrow see D.A.B.; his son was James Storrow, in his days a well-known Boston banker and public-minded citizen.

8

The history of the Boston water supply listed as *Water supply* 1868; much material also in *J. Winsor* 1881, III, pp. 238–256. *C. Lyell* 1845, II, p. 335, describes the Croton reservoir, which he visited in 1846.

9

On patients see art. "Patients" N.A.C.; on Corliss see D.A.B.; on his engine see technical books on the steam engine, e.g. *C. S. Dow's Practical Mechanical Engineering* (Philadelphia, Stanley Inst.), 1915 ed., p. 15; popular description of Corliss's engine at the Philadelphia Centennial Exhibition in *R. M. Devens* 1881, p. 951. On E. Howe and C. Goodyear see D.A.B. and the books on invention quoted this book, Ch. 5, 2.

10

On Henry see D.A.B., N.A.C., *D. S. Jordan* 1910; on Smithsonian Institution *W. J. Rhees* 1885–1887; *G. Brown Goode* 1897; on Davenport see D.A.B. Morse's work is described in all American books on invention, his life in *C. Mabee* 1943. On Farmer see D.A.B., *A. E. Dolbear* 1897.

11

The story of Cyrus Field in D.A.B., N.A.C.; *H. M. Field* 1892; *I. F. Judson* 1896. On the early telegraph *R. Sabine* 1869; *Telegraph* 1851.

Chapter 10

The quotation from *R. W. Emerson,* 1910, III, p. 163.

1

See *G. P. Fisher* 1866. In *S. P. Fowler* 1884 a sketch of nature studies around Salem. On J. Quincy's report see *No. Am. Rev.* 4 (1816) pp. 309–319. On early entomology *J. G. Morris* 1846; *B. D. Walsh* 1864. Also *R. T. Young* 1923.

2

On Nuttall see D.A.B., *W. J. Youmans* 1896. Quotations from R. H. Dana in *Two Years Before the Mast* (Houghton Mifflin, Boston, 1911), pp. 359, 412. Also *Th. Nuttall* 1832; *W. H. Powers* 1925, *W. L. Jepson* 1926, *F. W. Pennell* 1936; *J. E. Graustein* 1951. On the Nuttall's house: *B. L. Robinson* 1911. The story of the life mask in the "Harvard Alumni Bulletin," March 1931, by M. A. Day.

3

On Rafinesque see D.A.B., *C. S. Rafinesque* 1944. Quotations from *Am. Jour. Sc.* 40 (1841) pp. 223, 329, 332; *New Flora of North America,* Part I, Introd. pp. 11, 14; Gray's opinion in *Am. Jour. Sc.* 1841. Bibliography of Rafinesque: *T. J. Fitzpatrick* 1911, *E. D. Merrill* 1943.

4

On Audubon see D.A.B., *W. J. Youmans* 1896, who quotes the Edinburgh critic Chr. North in "Noctes Ambrosianae," *Blackwood's Magazine* 30 (January 1827), Audubon's diary in *R. Buchanan* 1869; see pp. 219, 291.

5

Lyell's travels in the U.S.: *C. Lyell* 1845, 1850; see esp. 1845, pp. 81, 148.

6

E. Lurie 1960 has written an authoritative life of Agassiz. Mrs. Agassiz wrote on her husband's life and correspondence in *E. C. Agassiz* 1886, she herself is the central figure of *L. H. Tharp* 1959. There is a considerable, ever growing literature on Agassiz; e.g. *M. L. Robinson* 1939; *C. O. Peare* 1958. Letter to father in *E. C. Agassiz* 1886, I p. 98. See also D.A.B., N.A.C.; *E. Favre* 1878; *J. Marcou* 1896; *A. B. Gould* 1901; *B Jaffe* 1944, Ch. X; *W. J. Youmans* 1896; *D. S. Jordan* 1910. The first in America to ascribe the presence of boulders to the action of glaciers seems to have been a Connecticut manufacturer, Peter Dobson; see *P. Dob-*

son 1826. The French quotation from *J. Marcou* 96, I, p. 102. It states: "There must have been at the period preceding the uplifting of the Alps and the appearance of the present living beings a drop of temperature well below what it is today. And we must ascribe to this drop of temperature the formation of the immense masses of ice which must have covered the earth at all places where we find erratic blocs with rocks polished like ours. This great cold has doubtless also buried the mammoths of Siberia in ice, frozen all our lakes, and accumulated ice up to the summits of our Jura which existed before the uplifting of the Alps."

7

Agassiz's arrival in Boston in *J. Marcou* 1896, I, pp. 279–280. Agassiz describes the reasons for staying in the U.S.A. in *E. C. Agassiz* II, pp. 432–433, 436. Description of the early Agassiz household in *J. Marcou* 1896. Agassiz trip to Lake Superior in *L. Agassiz* 1850, his conclusion on pp. 399, 401. The objections to Agassiz's glacier theory reported in *G. P. Merrill* 1924, esp. pp. 621–633. See also *F. T. Thwaites* 1927.

9

The quotation from Whipple in *E. P. Whipple* 1896. See also *A. S. Packard* 1876.

10

A. Hunter *Dupree* 1959 has given us an excellent biography of Gray. See further D.A.B., N.A.C.; *D. S. Jordan* 1910; *J. L. Gray* 1893; *C. S. Sargent* 1886; *J. E. Humphrey* 1896.

11

On Torrey see *A. D. Rodgers* 1942.

12

On J. D. Dana see D.A.B., N.A.C.; *D. S. Jordan* 1910; *E. S. Dana* 1895; *C. E. Beecher* 1896; and the biography *D. C. Gilman* 1899. Also *B. D. Walsh* 1864.

14

On Pourtalès, Marcou, Guyot, Shepard, Tuckerman, see D.A.B. On Guyot also *A. Guyot* 1849, 1860; *S. C. Grant* 1905–1908; *L. D. Jones* 1929. Oakes's work listed as *W. Oakes* 1848. Lyell's remark in *C. Lyell* 1850, p. 71; pp. 56–92 give a charming description of Lyell's visit to the White Mountains in early October, 1845.

Chapter 11

Quotation from *F. Engels.* Dialectics of Nature (N.Y., Int. Publ. 1940) p. 208.

1

A good account of the whole struggle between science and religion in *A. D. White* 1896. We do not know of an investigation dealing with the social aspects of this struggle in New England. On Jefferson and science see *C. A. Browne* 1943. On R. W. Emerson and Harvard see *S. E. Morison* 1936, pp. 243–244; the address by Emerson can be found in *R. W. Emerson* 1887, pp. 117–148.

2

Many facts on religion and geology in *C. Wright* 1941. On Paley see N.A.C., written when his influence in America was still strong; also E.S.S. Paley (1743–1805) summarized the whole theological utilitarianism of the eighteenth century and prepared the way for Jeremy Bentham. His collected works were published in seven volumes (London, 1825). On Silliman's attitude see *G. P. Fisher* 1866, quotation from II, p. 134. Maclure's statement in centennial ed. *Am. Jour. Sc.* 1918, p. 74; on Huttonians and Wernerians see Ch. 6, 4. The many papers of Moses Stuart on religion and science quoted in *C. Wright* 1941 were mostly in *Biblical Repository and Quarterly Observer;* see also *M. Stuart* 1836. On T. Cooper and his attitude to religion see *B. Jaffe* 1944.

3

On Silliman at Boston 1835 *G. P. Fisher* 1866, I, pp. 347, 372; Silliman's correspondence with Hitchcock *G. P. Fisher* 1866, II, p. 144. On Hitchcock consult *E. Hitchcock* 1835; *E. Hitchcock* 1851, pp. 37–50, 104, 198, 282, 521. Also R. Stebbins's review in *Christian Examiner* 53 (1852) pp. 51–66. See *J. D. Dana* 1856.

4

See *R. Chambers* 1845, and art. in *British Biog. Cycl.* and in N.A.C. A Gray's review in *A. Gray* 1846. Letters by A. Gray in *J. L. Gray* 1893, correspondence with Darwin in *C. Darwin* 1888. The letter of September 5, 1857, is not reprinted in these books, it can be found in the Proc. Linaean Society of 1858, and has been reprinted in *W. M. Smallwood* 1941.

5

Consult for the early papers on Darwinism in the U.S. the issues of the *American Journal of Science* 1860 and later. A good account of the controversy in *B. Loewenberg* 1932–1933; also *B. Jaffe* 1944. See also *A. Gray* 1878 and the biographies of Gray in *D. S. Jordan* 1910, *C. S. Sargent* 1886. The recent books by *A. H. Dupree* 1959 and *E. Lurie* 1960 give much information on Gray's and Agassiz's position in the evolution controversy.

6

On Agassiz's position see again *B. Loewenberg* 1932–1933, and the biographies of Agassiz; see Ch. 10, 6–9. *J. Marcou* 1896 defended Agassiz's position. *E. Favre* 1878 points to Agassiz's statement of 1844 in his book on fossil fishes. On Agassiz's belief that man had not sprung from a common stock. *E. C. Agassiz* 1886, II, p. 497. Agassiz's book of 1863 originated mostly in lectures at the Lowell Institute in Boston, published in the *Atlantic Monthly: L. Agassiz* 1863. The first quotation is from the preface, the quoted metaphysical speculation from p. 318 of that book. As an example of the more popular forms of criticism to which Agassiz's opinions were subjected see *J. Fiske* 1873.

7

On Jeffries Wyman see *D. S. Jordan* 1910 and D.A.B.; he was the brother of Morrill Wyman who wrote the biography of Treadwell; both were sons of a physician who did pioneer work at the Mclean Asylum in Charlestown. Wyman's publications of 1863: *J. Wyman* 1863; his private letter in *D. S. Jordan* 1910, p. 194 (also the footnote). Wyman's study of the Chimpanzee: *J. Wyman* 1850. Agassiz's letter to Howe in *E. C. Agassiz* 1886, II, pp. 605–606. Further material on the controversy in *B. Jaffe* 1944, esp. 249. *E. Lurie* 1954 deals with Agassiz' position on races of men.

On Bowen see D.A.B. and N.A.C.; his paper on Darwinism *F. Bowen* 1861. The history of Bowen's appointment, discharge and appointment at Harvard in *S. E. Morison* 1936; Professor Morison defends the position which Bowen took in 1849. In 1849 Czarism was the main enemy of democracy, and Hungary was therefore supported in its fight against Russia by democrats of all shades.

8

On Dana's position see *B. Loewenberg* 1932–1933; *D. C. Gilman* 1899; *D. S. Jordan* 1910. Later came a period of "social Darwinism"; see *W. G. Summer* 1941, also the communications in *Science and Society* 5 (1941). Dana's cephalization theory in *D. S. Jordan* 1910, pp. 246–248; see also *W. I. Vernadsky* 1945, p. 7. Vernadsky stresses the fact that Dana's principle asserts that the evolution of living matter is proceeding in a definite direction; he compares Dana's "cephalization" with the conception of the "psychozoic era" by Joseph le Conte (1823–1901), another American geologist.

Chapter 12

1

The quotation is from Whitman's "Drum Taps." There is an extensive literature on the clipper ship period, some books factual, others more romantic. See *S. E. Morison* 1921, on which our account leans heavily; *H. I. Chapelle* 1935; *A. H. Clark* 1911; *C. C. Cutler* 1930. *A. Laing* 1944 gives interesting technical details. Some details about Enoch Train in *A. Forbes* 1930. On Donald McKay see the books on clipper ships and D.A.B. A valuable collection of prints of clipper ships in the rooms of the State Street Trust Company of Boston. Shattuck quotes S. Johnson in *L. Shattuck* 1850, p. 248; *H. I. Chapelle* 1935, p. 281, calls clipper ships "the most overadvertised type in maritime history."

2

On steam navigation see *G. H. Preble* 1883; *J. L. Ringwalt* 1888; *B. H. Meyer* 1917; *J. Flexner* 1944; also *S. E. Morison* 1921 on Massachusetts conditions. On Train see *A. Forbes* 1930.

3

On coastal towns see *S. E. Morison* 1921, esp. pp. 216, 225, and the local histories of such towns as Newburyport, Salem and Marblehead. See also *J. Winsor* 1881.

4

On Sumner see *P. V. H. Weems* 1938, also *T. H. Sumner* 1843. For modern books dealing with Sumner's method see e.g. *W. T. Skilling–R. S. Richardson* 1942; *F. G. Watson* 1942.

5

There is a history of the early years of the Coast Survey in N.A.C., also *F. Cajori* 1929 and 1930. Bache's biography has been written by *M. M. Odgers* 1947; there is also an article on A. D. Bache in *W. J. Youmans* 1896. On the Harvard Astronomical Observatory see *S. I. Bailey* 1931; on the Bonds *E. E. Holden* 1897. The history of astronomy in the U.S.A. between 1830 and 1850 in *E. Loomis* 1851, 1856; see also art. on Olmsted in *W. J. Youmans* 1896. On Hopkins see *J. H. Denison* 1935.

6

The great comet of 1843 described in *R. M. Devens* 1881, p. 424 (quotation from Devens); see also *S. I. Bailey* 1931. Jules Verne glorified the Cambridge Observatory in *Voyage de la terre à la lune,* 1866.

7

On C. H. Davis see N.A.C. and D.A.B., on Newcomb see *E. W. Brown* 1909–10, and *D. S. Jordan* 1910. On S. Walker see D.A.B. On the meteorology of this period see *W. Redfield* 1831, 1834, 1862; *J. P. Espy* 1835, 1840, 1841, 1866; *E. Loomis* 1859, 1868; *I. M. Cline* 1935; *C. Tracy* 1843, 1910.

8

On Peirce see, apart from D.A.B. and N.A.C., *H. A. Newton* 1881, *J. L. Coolidge* 1924, *R. C. Archibald* 1925; the last book with a short analysis of all Peirce's papers and books as well as a full bibliography. The story of the discovery of Neptune and Peirce's criticism in *A. Pannekoek* 1953, see also *H. H. Turner* 1904; *B. Pierce* 1846–1848. Quotation on the meeting of the Academy in art. "Peirce," N.A.C. On Peirce's criterion see *B. Peirce* 1852, 1877–1878, *B. A. Gould* 1855, *W. Chauvenet* 1886; for the criticism see *R. C. Archibald* 1925. The quotation from Byerly is also taken from Archibald.

9

On Lovering see D.A.B. and *J. Lovering* 1892; his paper on Boston and science: *J. Lovering,* 1881. On B. A. Gould see D.A.B.; *A. M. Davis* 1897; *E. Winslow* 1881–1882.

Chapter 13

Quotation from *C. A. and M. B. Beard* 1927, I, p. 741. Their words "that age" refers to the period 1840–1860 in America.

1

The description of scientific education at Harvard and Yale in 1845 leans heavily on *R. H. Chittenden* 1928, I, opening chapter. Quotation on Ticknor from *S. E. Morison* 1936, p. 233. The ditty on Euclid from *M. Rukeyser* 1942, p. 97. Conditions at Williams College in *J. H. Denison* 1935. Also *Harvard* 1926–1927.

2

Silliman quoted from *G. P. Fisher* 1866, I. See *R. H. Chittenden* 1928, I.

3

Descriptions of scientific work and inventions of the forties and fifties are found in many books on American history, e.g. *C. A. and M. B. Beard* 1927, I, pp. 738–743. The Beards philosophize: "Political democracy and natural science rose and flourished together. Whether in their inception there were deep connections, researches have not yet disclosed, but beyond question their influence upon each other has been reciprocal." This is putting the question the wrong way. The mutual influence of political democracy on science can only be discussed if the common root of both, the industrial revolution and its expansion in a capitalist economy without feudal remains, is clearly recognized. The Beards continue: "There might have been no causal relation between science and democracy"; they only see that the Jacksonian upheaval coincides with the scientific upheaval. For this period see also *B. Jaffe* 1944.

4

Quotation from Lyell *C. Lyell* 1850, II, p. 304; from S. W. Johnson *E. A. Osborne* 1913; see also *C. A. Browne* 1944, p. 280. Browne's book gives an account of Liebig's influence on the U.S.A.; material also in *Liebig and after Liebig* 1942. On some of the other European scientific centers see again *R. H. Chittenden* 1928, I. Some of Liebig's books in English translations: *J. Liebig* 1842, 1855, no date.

5

See *R. H. Chittenden* 1928, I, on the Sheffield School; on J. P. Norton see also D.A.B.; quotation concerning Norton in *Chittenden* I, p. 44. See also *J. P. Norton* 1850. On Sheffield see *Chittenden* and D.A.B.

6

On the foundation of the Harvard Scientific School see *S. E. Morison* 1936, *I. B. Cohen* 1948. On Storrow see this book Ch. 9, 6; on Horsford see *E. V. Horsford* 1846; D.A.B. and *C. I. Jackson* 1892–1893; on Eustis see D.A.B. On the engineering school *J. L. Love* 1902. General background on civil engineering *C. W. Hunt* 1897, 1902.

7

On the Rogers brothers see *E. S. Rogers* 1896, D.A.B., N.A.C. and *W. J. Youmans* 1896. On the United Irishmen in America *S. E. Morison–H. S. Commager* 1942, p. 376. On H. D. Rogers see *J. W. Gregory* 1916. Also *P. K. Rogers* 1822.

8

The correspondence of the Rogers brothers in *E. S. Rogers* 1896; some of Rogers's works listed as *W. B. Rogers* 1838, 1852, 1863. On the foundation of M.I.T. see also *F. A. Walker* 1895; *J. P. Munroe* 1888.

9

On Morrill D.A.B. and *W. B. Parker* 1924.

10

This section leans on *I. B. Cohen* 1946. See also *I. I. Hayes* 1862. On the foundation of the National Academy of Science see the biographies of the leading founders, such as Henry, Agassiz, and Gray.

Bibliography

Bibliography

ADAMS, C. R. "Benjamin Thompson, Count Rumford." *Scientific Monthly* 71 (1950) pp. 380–386.

ADAMS, H. B. *Thomas Jefferson and the University of Virginia.* Circular of Information No. 1, U. S. Bureau of Education, Washington, 1888. 300 pp.

ADAMS, J. *The works.* 10 vols. Little, Brown, Boston, 1851–1865.

ADAMS, J. *More books.* Bulletin Boston Public Library, Boston, March 1926, December 1930.

ADAMS, J. Q. *Report of the Secretary of State on weights and measures, 1819.* Report signed February 22, 1821, 135 + 100 pp. House Doc. 109, 16th Congress, 1st session.

ADAMS, J. T. *Album of American history,* II. Scribner, New York, 1942 (1783–1853).

ADAMS, S. H. *The Erie Canal.* Random House, New York, 1953, 182 pp.

AGASSIZ, E. C. *Louis Agassiz. His life and correspondence.* 2 vols. Houghton Mifflin, Boston, 1886. 794 pp.

AGASSIZ, G. R. *Letters and recollections of Alexander Agassiz.* Houghton Mifflin, Boston, 1913. 454 pp.

AGASSIZ, L. *The structure of animal life.* Six lectures delivered at the Brooklyn Academy of Music, January, February, 1862. Scribner, New York, 1866. viii + 110 pp.

AGASSIZ, L. *Methods of study in natural history.* Houghton Mifflin, Boston, 1863. vi + 319 pp.

AGASSIZ, L. *Lake Superior . . . narrative of the tour by J. E. Cabot.* Gould, Kendall & Lincoln, Boston, 1850. x + 428 pp.

AGASSIZ, L. *Introduction to the study of natural history.* New York, 1847.

AGASSIZ, L., and GOULD, A. A. *Principles of zoology. Part 1, Comparative physiology.* For the use of schools and colleges. Revised ed. Gould and Lincoln, Boston, 1854, 240 pp. (First ed. 1848.)

ALLEN, Z. *Philosophy of the mechanics of nature, and the source and modes of action of natural motive power.* Appleton, New York, 1852. vii + 797 pp.

ALLEN, Z. *The science of mechanics, as applied to the present improvements in the useful arts in Europe, and in the United States of America.* Hutchins, Providence, 1829. v + 364 pp.

APPLETON, N. *Correspondence between Nathan Appleton and John A. Lowell in relation to the early history of the City of Lowell.* Eastburn's Press, Boston, 1848. 19 pp.

ARCHER, W. H. Life and letters of Horace Wells, discoverer of anesthesia. *Journal American College of Dentists,* 11 (1944), pp. 81–210.

ARCHIBALD, R. C. *Benjamin Peirce, 1809–1880.* Mathematical Association of America, Oberlin, Ohio, 1925. iv + 30 pp.

BAGNALL, W. R. *Samuel Slater and the early development of the cotton manufacture in the U. S.* Middletown, Connecticut, 1890. 70 pp.

BAILEY, S. I. *The history and work of Harvard Observatory 1839–1927.* McGraw-Hill, New York, 1931. xiii + 301 pp.

BAKER, R. P. *A chapter in American education.* Rensselaer Polytechnic Institute, 1824–1924. Scribner, New York, 1924. viii + 170 pp.

BALCH, E. S. *Antarctica.* Allen, Lane & Scott, Philadelphia, 1902. 230 pp.

BALDWIN, S. *The first century of the Connecticut Academy of Arts and Sciences, 1799–1899.* A historical address. Tuttle, Morehouse & Taylor, New Haven, 1901. pp. xii–xxxv. From the Centennial Volume of the Transactions Connecticut Academy.

BARRETT, JOHN T. The inoculation controversy in Puritan New England. *Bulletin of Historical Medicine* 12 (1942), pp. 169-190.

BARTLETT, H. R. The development of industrial research in the U. S. 77 pp. in *Research—a national resource.* National Resources Planning Board, December 1940, II, Industrial Research.

BATCHELDER, S. *Introduction and early progress of the cotton manufacture in the United States.* Little, Brown, Boston, 1863. 108 pp.

[BATES, R. S.] *The American Academy of Arts & Sciences 1780–1940.* Published by the American Academy, 1940, 19 pp.

BATES, R. S. *Scientific societies in the U. S.* John Wiley, New York. Chapman Hall, London, 1945. v + 246 pp. Second edition, Technology Press, M.I.T., Columbia Univ. Press, New York, 1958. ix + 297 pp.

BATHE, G. *An engineer's miscellany*. Patterson and White, Philadelphia, 1938, xii + 136 pp.

BATHE, G. AND D. *Jacob Perkins: His inventions, his times and his contemporaries*. Historical Society of Pennsylvania, Philadelphia, 1943. xiv + 215 pp.

BATHE, G. AND D. *Oliver Evans, a chronicle of early American engineering*. Historical Society of Pennsylvania, Philadelphia, 1935. xviii + 362 pp.

BATTELL, J. *The Morgan horse and register*. 2 vols. U. S. Department of Agriculture Circular 199 (1926).

BEALL, O. T. AND SHRYOCK, R. H. *Cotton Mather, First Significant Figure in American Medicine*. Johns Hopkins Press, Baltimore, 1954. ix + 241 pp.

BEARD, C. A. AND MARY B. *The rise of American civilization*. 2 vols. Macmillan, New York, I (1927). 824 pp.

BEECHER, C. E. *J. D. Dana*. American Geologist 17 (1896).

BELL, W. J. *Early American Science*. Institute of Early American History and Culture, Williamsburg, Virginia, 1955. x + 85 pp.

[BELLINGSHAUSEN, F. G. VON] *The Voyage of Captain Bellingshausen to the Antarctic Seas 1819–1821*. Translated from the Russian. Edited by F. Debenham. Hakluyt Soc., London (2nd series 91–92). 1945, xxx + 260 pp., 216 pp.

BENJAMIN, A. *Elements of architecture, containing the Tuscan, Doric, Ionic and Corinthian orders, with all their details and embellishments. Also the theory and practice of carpentry . . .* 28 plates. B. B. Mussey, Boston, 1843. viii + 232 pp.

BENJAMIN, M. The early presidents of the American Association. 48th annual meeting, Columbus, Ohio, August 21–26, 1899, 63 pp. *Science* 10 (1899) pp. 625–637, 705–713, 759–766.

BENSON, A. B. James Gates Percival, student of German culture. *New England Quarterly* 2, (1929). ix + 553 pp.

BENSON, A. E. *History of the Massachusetts Horticultural Society*. Printed for the Society 1929. ix + 553 pp.

BENT, A. H. *A bibliography of the White Mountains*. Published for the Appalachian Mountain Club. Houghton Mifflin, Boston, 1912. vii + 114 pp.

BERNSTEIN, S. Robert Fulton's unpublished memoir to Pitt. *Science and Society* 8 (1944) pp. 40–63.

BERRY, R. E. *Yankee Stargazer. The life of N. Bowditch.* Whittlesey, New York 1941. xi + 234 pp.

BIDWELL, P. W. The agricultural revolution in New England. *American Historical Review* 26 (1921) pp. 683–702.

BIDWELL, P. W. Rural economy in New England at the beginning of the nineteenth century. *Transactions of Connecticut Academy* 20 (1916) pp. 319–353.

BIDWELL, P. W. AND FALCONER, J. I. *History of agriculture in the Northern United States, 1620–1860.* Carnegie Institute Publications 358, Washington, 1925. xii + 512 pp.

BIGELOW, J. *The useful arts, considered in connexion with the applications of Science.* 2 vols. Boston, March, etc. 1840. (2nd ed. Harper, New York, 1847.)

BIGELOW, J. *American Medical Botany, being a collection of the native medicinal plants of the United States.* Cummings and Hilliard, Boston. I (1817), 197 pp.; II (1818), 199 pp.; III (1820), 197 pp.

BIGELOW, J. Some account of the White Mountains of New Hampshire. *New England Journal of Medicine and Surgery* 5 (1816) pp. 321–338.

BIGELOW, J. *Florula Bostoniensis. A collection of plants of Boston and its vicinity.* 2nd ed., greatly enlarged. Cummings, Hilliard, University Press, Boston, 1824. 423 pp. Little, Brown, Boston, 1840. vi + 468 pp.

BIGELOW, J. *Elements of Technology, taken chiefly from a course of lectures delivered at Cambridge, on the application of the sciences to the useful arts.* 1st ed. 1829; 2nd ed. Hilliard, Gray, Little & Wilkins, Boston, 1831. xv + 521 pp.

BIGELOW, J. Inaugural address as Rumford professor in Harvard University. *North American Review* 3 (1816) pp. 271–283.

BINNEY, A. *Terrestrial air-breathing mollusks of the United States and the adjacent territories of North America.* Ed. by A. Gould. 2 vols. Little, Boston, 1851. I, xxix + 366 pp.; II, 362 pp.

BISHOP, J. L. *A history of American manufactures from 1608 to 1860.* 3 vols. E. Young, Philadelphia; S. Low, London, 1866. I, 642 pp.; II, 617 pp.

BLACK, A. *The story of tunnels.* McGraw-Hill, New York, 1937. xv + 245 pp.

BLACK, N. H. Certain ancient physical apparatus belonging to Harvard College. *Harvard Alumni Bulletin* 35 (1930) p. 660.

BLAKE, J. B. "Lemuel Shattuck and the Boston Water Supply." Bull. Hist. Medicine 29 (1955) pp. 554-562.

———. *Benjamin Waterhouse and the introduction of vaccination: a reappraisal. Monograph* 33, Yale University, Department of Hist. Medic., University of Pennsylvania Press, Philadelphia, 1957, 95 pp.

BLAKE, J. L. *First book in astronomy, adapted to the use of common schools.* Lincoln & Edmands, Boston, 1831. 115 pp.

BOORENSTEIN, D. J. *The Americans, The Colonial experience.* Random House, New York, 1950, 434 pp.

BORDEN, S. *A system of useful formulae, adapted to the practical operations of locating and constructing railroads.* Little, Boston, 1851. 10 + 188 pp.

BORDEN, S. An account of the trigonometrical survey of Massachusetts. *Transactions American Philosophical Society* 9 (1846) pp. 33–91. See *North American Review* 61 (1845) p. 455; *Proceedings American Philosophical Society* 2 (1841–1843) p. 60.

BOSTON SOCIETY OF NATURAL HISTORY. *Annual* I (1868–1869).

BOUVÉ, T. T. Historical sketch of the Boston Society of Natural History, with a Notice of the Linnaean Society which preceded it. *Anniversary Memoirs Boston Society of Natural History 1830–1880* (1880) pp. 14–250. Also *id.* Some reminiscences of earlier days in the history of the Society. *Proceedings of Boston Society of Natural History* 18 (1876) pp. 242–250.

BOWDITCH, H. *A Catalogue of a special exhibition of manuscripts, books, portraits and personal relics of Nathaniel Bowditch (1773–1838). Sketch of his life by H. Bowditch and essay by R. C. Archibald.* Peabody Museum, Salem, Massachusetts, 1937. iv + 40 pp.

BOWDITCH, N. An estimate of the height, direction, velocity, and magnitude of the meteor that exploded over Weston, in Connecticut, December 14, 1807. *Memoirs American Academy* 3 (2) (1815) pp. 213–237.

BOWDITCH, N. *Mécanique céleste by the Marquis de la Place . . .* translated, with a commentary, by N. Bowditch. Hilliard, Gray, etc., Boston. I (1829) xxiv + 747 pp.; II (1832) xviii + 991 pp.; III (1834) xxx + 1017 pp.; IV (1839) 168 + xxxvi + 1018 pp.

BOWDITCH, N. *The new practical navigator.* E. M. Blunt, Newburyport, 1799. 574 pp.

BOWDITCH, N. *The new American practical navigator.* E. M. Blunt. Newburyport, 1802. 594 pp. Editions of 1807, 1811, 1817, 1821, 1826, 1832, 1836, 1837, eleventh ed. continued after death of the author in 1839. (From 1868 the work was published by the Government; by 1880 the title was changed to *The American practical navigator.*)

BOWDITCH, N. On the motion of a pendulum suspended from two points. *Memoirs American Academy* 3 (2), 1815, pp. 413–436.

BOWDITCH, N. I. *A history of the Massachusetts General Hospital.* Wilson, Boston, 1851, xi + 442 pp. (2nd ed., Boston, 1872, xvii + 734 pp.)

BOWDITCH, V. Y. *Life and correspondence of Henry Ingersoll Bowditch.* 2 vols. Houghton Mifflin, Boston, 1902. vi + 337 pp.

BOWEN, F. Remarks on the latest form of the development theory. *Memoirs American Academy Arts and Sciences* 8 (1) (1861) pp. 97–122.

BOWERS, C. G. *Jefferson in power, the death struggle of the Federalists.* Houghton Mifflin, Boston, 1936. xix + 538 pp.

BOYD, T. *Poor John Fitch. Inventor of the steamboat.* Putnam, New York, 1935. 315 pp.

BRADLEE, F. B. C. *The Boston and Lowell, the Nashua and Lowell, and the Salem & Lowell Railroads.* Essex Institute. Salem, 1918, 64 pp.

BRAGG, W. The contribution of Count Rumford and M. Faraday to the modern museum of science. *Science* 72 (1930) pp. 19–23.

BRASCH, F. John Winthrop, our first astronomer. *Journal Astronomical Society Pacific,* 1916. Revised and enlarged in *Newton's first critical disciple in the American Colonies, John Winthrop.* In *Sir Isaac Newton* 1727–1927. Williams & Wilkins, Baltimore, 1928. ix + 351 pp.; esp. pp. 301–338.

BRASCH, F. The Newtonian Epoch in the American Colonies (1686–1783). *Proceedings American Antiquarian Society* (for October 1939), 1940. 21 pp.

BRASCH, F. E. The Royal Society of London and its influence upon scientific thought in the American colonies. *Scientific Monthly* 33 (1931) pp. 336–355, 448–469.

BREARLEY, H. C. *Time telling through the ages.* Doubleday, Page for R. H. Ingersoll, New York, 1919. 294 pp.

BRECK, S. *Recollections of Samuel Breck with passages from his note-books (1771–1862)*, ed. by H. E. Scudder. Porter and Coates, Philadelphia, 1877. 316 pp.

BRIDENBAUGH, C. *Cities in the wilderness; the first century of urban life in America, 1625–1742.* Ronald Press, New York, 1938. xiv + 500 pp.

BRIDENBAUGH, C. AND J. *Rebels and gentlemen: Philadelphia in the age of Franklin.* Reynal and Hitchcock, New York, 1942. xvii + 393 pp.

BRIGGS, L. VERNON. *History of shipbuilding on North River, Plymouth County, Massachusetts, 1640–1872.* Coburn Bros., Boston, 1889. xv + 420 pp.

BRISSOT, J. P. (Warville). *Nouveau voyage dans les États-Unis de l'Amérique septentrionale, fait en 1788.* Buisson, Paris, 1791. I, liii + 395 pp.

BROCKETT, L. P. *The silk industry in America.* A history: prepared for the centennial exposition. For Silk Association America, Nesbitt, New York, 1876. 237 pp.

BROOKS, VAN WYCK. *The flowering of New England.* Dutton, New York, 1936. Modern Library Edition. 550 pp.

BROWN, E. W. "Simon Newcomb." *Bulletin American Mathematical Society* 16 (1909–1910) pp. 341–355.

BROWN, R. H. *Mirror for Americans. Likeness of the Eastern Seaboard 1810.* American Geographical Society, New York, 1943. xxxii + 312 pp.

BROWN, S. C. "Count Rumford and the caloric theory of heat," *Procedures American Philosophic Society,* 93 (1949) pp. 316–325 (see also *American Journal of Physics* 18 (1940) pp. 367–373).

———. Scientific drawings of Count Rumford at Harvard, *Harvard Library Bulletin* 9 (1955) pp. 350–369.

———. *An exhibition of the scientific works of Count Rumford.* American Academy of Arts and Sciences, March 26, 1953, 30 pp. (See also *Procedures American Academy of Arts and*

Sciences, 82 (1953). pp. 266–289, and *American Scientist,* January, 1954.)

BROWN GOODE, G. *The origin of the natural scientific & educational institutions of the U.S.A.* Reprint from papers of the American Historical Association. Putnam, New York, 1890. 112 pp.

BROWN GOODE, G. *The Smithsonian Institution 1846–1896.* Smithsonian Transactions, 1897, 10 + 856 pp.

BROWN, GOODE G. The beginnings of natural history in America. Address 6th anniversary meeting of Biological Society Washington (1886). *Proceedings Biological Society,* Washington 3 (1884–1886) pp. 35–105.

BROWNE, C. A. Some relations of the New Harmony movement to the history of science in America. *Scientific Monthly,* 42, 1936, pp. 483–497.

BROWNE, C. A. The life and chemical services of Frederick Accum. *Journal of Chemical Education* 2 (1925) pp. 829–851, 1008–1039, 1140–1149.

BROWNE, C. A. *A source book of agricultural chemistry. Chemica Botanica* 8, No. 1, Waltham, Massachusetts, 1944. x + 290 pp.

BROWNE, C. A. Emerson and chemistry. *Journal of Chemical Education* 5 (1938), pp. 269–279, 391–403.

BROWNE, C. A. *Thomas Jefferson and the scientific trends of his time.* Chronica Botanica Company, reprinted from *Chronica Botanica* 8. Waltham, Massachusetts (1943), 63 pp.

BRUCE, J. L. Filling in of the Backbay and the Charles River Development. *Proceedings Boston Society* 1940, pp. 25–38.

BUCHANAN, R. *Life and adventures of Audubon, the naturalist* (1869). Everyman's library. Dent, London, Dutton, New York. xx + 335 pp.

BULFINCH, E. S. *The life & letters of Charles Bulfinch, architect.* Houghton Mifflin, Boston, 1896. xiv + 323 pp.

BURLINGAME, R. *Whittling boy. The story of Eli Whitney.* Harcourt, Brace, New York, 1941. viii + 370 pp.

BURLINGAME, R. *March of the iron men. A social history of union through invention.* Scribner, New York, 1938. xv + 500 pp.

BURRAGE, W. L. *A history of the Massachusetts Medical Society, with brief biographies of the founders and chief officers 1781–1922.* Privately printed 1923. ix + 505 pp.

BURSTYN, H. L. The Salem Philosophical Library: its history and importance for American Science. *Essex Institute Historical Collections* 96 (1960) pp. 169–206.

BUSH, G. G. *History of higher education in Massachusetts.* Bureau of Education, Circular of Information 6 (1891). Government Printing Office, Washington. 445 pp.

BYRNE, E. W. *The progress of invention in the nineteenth century.* New York, 1900, viii + 476 pp.

CAJORI, F. *The early mathematical sciences in North and South America.* Badger, Boston, 1928. 156 pp.

CAJORI, F. *The teaching and history of mathematics in the United States.* Bureau of Education Circular of Information 3 (1890) (whole, number 167), Washington, Government Printing Office, 1890. 400 pp.

CAJORI, F. *The chequered career of Ferdinand Rudolph Hassler.* Christopher, Boston, 1929. 245 pp.

CAJORI, F. A century of American geodesy. *Isis* 14 (1930) pp. 411–416.

CALLENDER, G. S. The early transportation and banking enterprises of the states in relation to the growth of corporations. *Quarterly Journal of Economics* 17 (1902–1903) pp. 111–162.

CANNON, W. B. Henry Pickering Bowditch, physiologist. *Science* 87 (1938) pp. 471–478.

CASWELL, L. B. *Brief history of the Massachusetts Agricultural College.* Semicentennial 1917. Massachusetts Agricultural College, Amherst. 72 pp.

CENTENNIAL OF CHEMISTRY. *American Chemist* 5 (1874–1875).

CENTENNIAL year 1792–1892 of the Massachusetts Society for promoting agriculture, Salem, 1892, 146 pp.

CHAMBERLAIN, E. G. Altitudes in Massachusetts. *Appalachia* 4 (1884–1886) pp. 132–145.

CHAMBERLAIN, J. E. *The Boston Transcript. A history of its first hundred years.* Houghton Mifflin, Boston, 1930. xii + 241 pp.

[CHAMBERS, R.] *Vestiges of the natural history of creation.* Wiley & Putnam, New York, 1845, 291 pp.; 8th ed. Churchill, London, 1850, vi + 319 pp.

CHAPELLE, H. I. *The history of American sailing ships.* Norton, New York, 1935. xvii + 400 pp.

CHAUVENET, W. *A treatise on the method of least squares. Being the appendix to the author's manual of spherical and practical astronomy.* Lippincott, Philadelphia, 1886, pp. 469–566. (Entered, according to act of Congress, in 1868.)

CHEVALIER, M. *Society, manners and politics in the U. S.; being a series of letters on North America.* Translated from third Paris ed., Weeks, Jordan, Boston, 1839. 467 pp., transl. T. G. Bradford.

CHICKERING, J. *A statistical view of the population of Massachusetts from 1765 to 1840.* Little & Brown, Boston, 1846. 160 pp.

CHILD, E. *The tools of the chemist. Their ancestry and American evolution. Reinhold,* New York, 1940. 220 pp.

CHINARD, G. *Honest John Adams.* Little, Brown, Boston, 1933. xii + 359 pp.

CHITTENDEN, R. H. *History of the Sheffield Scientific School.* Yale University Press, New Haven, 1928. 2 vols. I, viii + 298 pp.

CLARK, A. H. *The clipper ship era,* 1843–1869. Putnam, New York, 1911. xii + 404 pp.

CLARK, V. S. *History of manufactures in the United States, 1607–1860.* Carnegie Inst., Washington, 1916. 675 pp. (new ed. 1929).

CLARKE, E. H., a.o. *A century of American medicine 1776–1876.* Lea, Philadelphia, 1876, 3 + 366 pp.

CLEAVELAND, P. *An elementary treatise on mineralogy and geology.* 6 plates. Harvard University Press, Cambridge, 1816. xii + 668 pp.

CLINE, I. M. *A century of progress in the study of cyclones.* New Orleans, Louisiana, 1935. 29 pp.

COBURN, F. W. *History of Lowell and its people.* 3 vols. Lewis Historical Publication Company, New York, 1920. 572 pp. + 441 pp.

COFFIN, R. P. T. *Kennebec; Cradle of Americans.* Farrar and Rinehart, New York, 1937. x + 292 pp.

COHEN, I. B. *Benjamin Franklin's experiments.* A new edition of Franklin's *Experiments & Observations on Electricity.* Ed., with a critical and historical introduction. Harvard University Press, Cambridge, 1941. xxviii + 453 pp.

COHEN, I. B. Science and the revolution. *Technology Review* 47 (1945) no. 6, 4 pp.

COHEN, I. B. Science and the Civil War. *Technology Review* 48, Jan. 1946, no. 3.

COHEN, I. B. Harvard and the scientific spirit. *Harvard Alumni Bulletin,* Feb. 7, 1948, pp. 393–398.

COHEN, I. B. *Some early tools of American Science.* Harvard University Press, Cambridge, 1950, xxi + 201 pp.

———. Benjamin Franklin and the mysterious "Dr. Spence." *Journal of the Franklin Institute* 235 (1943) pp. 1–25.

COLBURN, W. *Intellectual arithmetic upon the inductive method of instruction.* Revised and enlarged edition, with appendix. Houghton Mifflin, Boston, 1891. xiii + 216 pp.

COLBURN, W. *An arithmetic on the plan of Pestalozzi, with some improvements.* Cummings & Hilliard, Boston, 1821. 143 pp.

COLBURN, Z. *A memoir of Zerah Colburn, written by himself.* Merriam, Springfield, 1833. 104 pp.

COLE, A. H. *The American wool manufacture.* Harvard University Press, Cambridge, 1926. I, xv + 392 pp.; II, viii + 328 pp.

COLMAN, H. *First report on the agriculture of Massachusetts, County of Essex 1837.* Dutton & Wentworth, Boston, 1838. 139 pp. Reports 1–4 (1837–1840).

COLMAN, H. *Address before the Agriculture and Mechanical Institute of New London & Windham Counties, Connecticut,* October 8, 1840.

COLMAN, H. *Agriculture of the United States.* Address before the American Institute, New York, April 14, 1841.

COLMAN, H. *Agriculture and rural economy from personal observation.* 2 vols., 1857. Phillips, Sampson, Boston. I, xxvi + 492 pp.; II, xxiv + 558 pp. (first ed. 1849).

COLTON, C. Q. *Anesthesia, who made and developed this great discovery.* A. C. Sherwood, New York, 1886.

[CONCHOLOGY] A sketch of the history of conchology in the United States. *American Journal of Science* (2) 33 (1862), pp. 161–180.

CONGDON, H. W. *The covered bridge.* Knopf, New York, 1946. 151 pp.

COOLIDGE, J. L. *An introduction to mathematical probability*. Oxford University Press, 1925. xii + 215 pp.

COOLIDGE, J. L. The story of mathematics at Harvard. *Harvard Alumni Bulletin* 26 (1924), pp. 372-378.

COOLIDGE, J. L. Three hundred years of mathematics at Harvard. *American Mathematical Monthly* 50 (1943) pp. 347–356.

COUPER, W. *Claudius Crozet: Soldier—Scholar—Educator—Engineer,* (1789–1864). *Historical Publishing Company,* Charlotteville, Virginia, 1936, 221 pp.

CROSBY, I. B. *Boston through the ages. The geological story of Greater Boston.* Marshall Jones, Boston, 1928. xvii + 166 pp.

CUNNINGHAM, C. E. *Timothy Dwight 1752–1817. A biography.* Macmillan, New York, 1942. viii + 403 pp.

CURTI, M. *The learned blacksmith. The letters and journals of Elihu Burritt.* Wilson-Erickson, New York, 1937. viii + 241 pp.

CURTI, M. *The growth of American thought.* Harper, New York, 1943. xx + 848 pp.

CUTLER, C. C. *Greyhounds of the sea, the story of the American clipper ship.* Putnam, New York, 1930. xxvii + 592 pp. [New ed. Halcyon, New York, 1937.]

CUTLER, M. An account of some of the vegetable productions, naturally growing in this part of America, botanically arranged. *Memoirs American Academy* 1 (1785), pp. 396–493.

DAMPIER, W. C. *A history of science and its relations with philosophy and religion.* 3rd ed. University Press, Cambridge; Macmillan, New York, 1943. xxiii + 574 pp.

DANA, E. D. James Dwight Dana. *American Journal of Science.* (3) 49 (1895) 28 pp.

DANA, J. D. *Manual of geology; treating of the principles of the science with special reference to American geological history.* Philadelphia–London, 1862. 812 pp. 4th ed., American Book Company, New York, 1895. 1087 pp.

DANA, J. D. *A system of mineralogy: including an extended treatise of crystallography.* New Haven, 1837. xiv + 580 pp. 2nd ed. 1844; 3rd ed. 1850; 4th ed. 1854, 2 vols.; 5th ed. 1868.

DANA, J. D. *Manual of mineralogy, including observations on mines, rocks, reduction of ores, and the application of the science to the arts.* New Haven, 1848. 430 pp., 2nd ed. 1857 [15th ed. 1891].

DANA, J. D. Science and the Bible; remarks on "The six days of creation," of Professor Tayler Lewis. *Bibliotheca Sacra* 13, pp. 80–129, 631–656; 14, pp. 338–413, 461–524 (1856–1857); reprinted as a pamphlet.

DANA, J. D. Address, as retiring president of the A. A. A. S., 1855. *Proceedings A. A. A. S.*, 9th meeting, Providence, Rhode Island, August 1855. (1856), pp. 1–36.

DANA, J. F. AND S. L. *Outlines of the mineralogy and geology of Boston and its vicinity.* Cummings & Hilliard, University Press, Boston, 1818, 108 pp.

DANA, S. L. *A muck manual for farmers.* Saxton, New York, 1860. vii + 312 pp. 4th ed., first ed. Bixby, Lowell, 1842, 242 pp.

DARLING, A. B. *Political changes in Massachusetts.* Yale University Press, New Haven, 1925. xii + 392 pp.

DARWIN, F. *The life and letters of Charles Darwin.* 2 vols. Appleton, New York, 1888. I (viii + 558 pp.), II (iv + 562 pp).

DAVIES, C. *The metric system, considered with reference to its introduction into the United States.* Barnes, New York, Chicago, 1871. 327 pp., with reports of J. Q. Adams and J. Herschel.

DAVIS, A. MCF. Benjamin Apthorp Gould. *Proceedings American Antiquarian Society,* April 1897.

DAVIS, N. S. *Contributions to the history of medical education and medical institutions in the United States.* Government Printing Office. Washington, 1877, 60 pp.

DAVIS, W. T. *The New England states.* 4 vols. Boston, 1897.

DEAN, J. Investigation of the motion of the earth viewed from the moon, arising from the moon's librations. *Memoirs American Academy* 3 (2), 1815, pp. 241–245.

DEANE, J. *Ichnographs from the sandstone of the Connecticut River.* 46 plates, 1863. 62 pp.

DEANE, S. *The New England farmer, or Georgical dictionary.* I. Thomas, Worcester, Massachusetts, 1790. viii + 335 pp. 3rd ed., Wells & Lilly, Boston, 1822. xi + 532 pp.

DEEVEY, E. S. A re-examination of Thoreau's *Walden. Quarterly Review of Biology* 17 (1942), pp. 1–11.

DEFEBAUGH, J. E. *History of the lumber industry of America.* 2 vols. The American Lumberman, Chicago. Vol. 2 (1907). xiii + 655 pp.

DEMAREE, A. L. *The American agricultural press 1819–1860*. Columbia University Studies in the History of American Agriculture, Columbia University Press, New York, 1941. xix + 430 pp.

DENISON, J. H. *Mark Hopkins. A Biography*. Scribner, New York, 1935. viii + 327 pp.

DENTISTRY AS A PROFESSIONAL CAREER. A brochure for the use of guidance officers and prospective students. The Council on Dental Education of the American Dental Association, Chicago, 1941. 71 pp.

DEVENS, R. M. *Our first century: being a popular descriptive portraiture of the one hundred great and memorable events of perpetual interest in the history of our country*. Nichols, Springfield, Massachusetts, 1881. 1004 pp.

DEXTER, F. B. *Sketch of a history of Yale University*. Holt, New York, 1887.

DICKENS, C. *American Notes for general circulation,* Harper, New York, 1842, 92 pp.

DICKINSON, H. W. *Robert Fulton, engineer and artist.* Lane, London, 1913. 14 + 333 pp.

DIX, D. L. *Memorial to the legislature of Massachusetts.* Munroe & Francis, Boston, 1843. 32 pp.

DOBSON, P. Remarks of Bowlders. *American Journal of Science* 10 (1826) pp. 217–218.

DOLBEAR, A. E. Moses G. Farmer as an electrical pioneer. *Electricity* (N.Y.) 13 (1897) pp. 56–58.

DOOLITTLE, W. H. *Inventions in the century;* vol. XVI of the Nineteenth Century Series, Chambers, London, Edinburgh, 1903. xxx + 495 pp.

DRAYTON, J. *Letters written during a tour through the northern and eastern states of America,* Harrison–Bowen, Charleston, South Carolina, 1794. 3 + 130 pp.

DUNCAN, L. C. *Medical men in the American Revolution* (*1775–1783*). Carlisle Barracks, Pennsylvania, Medical Field Service School, 1931. 414 pp.

DUPREE, A. H. *Science in the Federal Government. A History of Policies and Activities to 1940.* Harvard University Press, Cambridge, 1957, x + 460 pp.

———, *Asa Gray, 1810–1888,* Harvard University Press, Cambridge, Massachusetts, 1960. x + 505 pp.

DUPUY, R. E. The story of West Point; 1802–1946. The West Point tradition in American life. *Infantry Journal,* Washington, 1943. 282 pp. (first ed. under title: Where they have trod, 1940).

DWIGHT, T. *Travels; in New England and New York.* 4 vols, Dwight, New Haven, 1821–22. (London ed. 1823.)

DWIGHT, T. *A statistical account of the city of New Haven.* xi + 83 pp. *A statistical account of the towns and parishes in the state of Connecticut, published by the Connecticut Academy of Arts and Sciences.* Vol. I, No. 1. Walter & Steel, New Haven, 1811.

DYER, W. A. *Early American craftsmen.* Century, New York, 1915. xv + 387 pp.

Early history of science and learning in America. *Proceedings American Philosophical Society* 86, No. 1, September 25, 1942.

EATON, A. *Geological textbook.* Albany, 1830. 9 + 63 pp. [2nd ed. 1832.]

EATON, A. *Index to the geology of the northern states.* Albany, Northampton, Boston, 1818, 52 pp. [2nd ed. 1820.]

EATON, A. *Manual of botany for North America,* 6th ed. Steele, Albany, 1933, x + ii + 103, 401 + 137 pp. (first ed. 1817, 2nd ed. 1820, xi + 286 pp.).

———. *North American botany; comprising the native and common cultivated plants, north of Mexico. Gates, Troy,* 1840, vii + 625 pp. (8th ed. of the previous book with J. Wright).

EATON, C. *Freedom of thought in the Old South.* Duke University Press, Durham, North Carolina, 1940. xix + 343 pp.

EDDES, H. H. Josiah Barker, and his connection with shipbuilding in Massachusetts, *New England Historical and Genealogical Register* 24 (1870) pp. 297–305.

EDDY, C. *Historical sketch of the Middlesex Canal.* Dickinson, Boston, 1843. 53 pp.

EDSON, T. *Memoir of Warren Colburn.* Brown, Taggard & Chase, Boston, 1856. 27 pp. (Appeared in *American Jour. of Education,* September 1856.)

ELIOT, J. A. *An Essay on the invention of making sand iron.* Subtitle: An Essay on the invention or Art of making very good, if not the best Iron, from black sea sand. Holt, New York, 1762. 34 pp.; Fascimile reprint from original in Massachusetts Historical Society, February 1937, from copy of Ezra Stiles.

ELLIS, G. E. *Memoir of Sir Benjamin Thompson, Count Rumford.* American Academy of Arts and Sciences, Boston, 1871. 16 + 650 pp.

ELLIS, G. E. *Memoir of Jacob Bigelow, M. D., LL. D.* Cambridge, Wilson, University Press, 1880. 105 pp. (reprinted from Proceedings Massachusetts Historical Society).

EMERSON, G. B. Address delivered at Framingham, December 15, 1853. *17th Annual Report Board of Education,* Boston, 1854. pp. 29–53.

EMERSON, G. B. *Report on the trees and shrubs growing naturally in the forests of Massachusetts.* Boston, 1846. 534 pp.

EMERSON, R. W. *Journals,* ed. by E. W. Emerson and W. E. Forbes, 1833–1835. Houghton Mifflin, Boston, 1910. xvi + 575 pp.

EMERSON, R. W. *Nature, addresses, and lectures.* New and revised ed. Houghton Mifflin, Boston, 1887. iv + 372 pp.

ESPY, J. P. Contributions on storms. *Journal of the Franklin Institute.* 16 (1835) pp. 4–6; 17 (1866) pp. 386–393; etc., till 23 (1839); *American Journal of Science* 39 (1840) pp. 120–132; 40 (1841) pp. 327–332.

ESPY, J. P. *The philosophy of storms.* Bohn, London, 1841. XII, xl + 552 pp.

FAIRCHILD, H. L. The history of the American Association for the Advancement of Science. *Science* (new series) 59 (1924), pp. 365–368; *Comp. Science* 60 (1924) pp. 134–135.

FANNING, E. *Voyages and discoveries in the South Seas.* Marine Research Society, Salem, Massachusetts, 1924. 335 pp. (first publication 1833).

FARRAR, J. Review of *Ferguson's Astronomy, explained upon Sir I. Newton's principles, by D. Brewster.* Philadelphia, 1817, 2 vols. *North American Review* 6 (1818) pp. 205–224.

FARRINGTON, O. C. Catalogue of Meteorites of America, to January 1, 1909. *Memoirs National Academy of Science* 13 (1915) pp. 1–513.

FAVRE, E. Louis Agassiz. A biographical notice. Translated by M. A. Henry. *Annual Report Smithsonian Institution,* 1878, pp. 236–261.

FIELD, H. M. *The story of the Atlantic telegraph.* 2nd ed.,

Scribner, New York, 1867, viii + 435 pp. (new ed. ill. 1892, ix + 415 pp).

FISH, C. R. *The rise of the common man, 1830–1850.* A history of American life. VI. Macmillan, 1929. xix + 391 pp.

FISHER, G. P. *Life of Benjamin Silliman, M.D., LL.D., late professor of chemistry, mineralogy and geology in Yale College.* 2 vols. Scribner, New York, 1866. I, xvi + 407 pp.

FISKE, J. Agassiz and Darwinism. *Popular Science Monthly* 3 (1873) pp. 692–705.

FITZ, R. The medical school library—its first hundred years. *Harvard Library Notes* 3 (1939) pp. 262–270.

FITZGERALD, D. [On history of civil engineering in the U.S.A.] Address American Convention, Cape May, New Jersey, June 27, 1899. *Trans. American Society Civil Engineers* 41 (1899) pp. 596–617.

FITZPATRICK, T. *J. Rafinesque. A sketch of his life with bibliography. Historical Department Iowa,* Des Moines, 1911, 241 pp.

FLEXNER, J. *Steamboats come true. American inventors in action.* Viking, New York, 1944. x + 406 pp.

[FORBES, A.] *Boston, England & Boston, New England, 1630–1930.* Printed for State Street Trust Company, 1930. vii + 45 pp.

FORMAN, S. *West Point: A history of the U. S. Military Academy.* Columbia University Press, New York, 1950, vii + 255 pp.

FOWLER, S. P. Ornithology of the U. S. *Proc. Essex Inst.* 2 (1856–1866), pp. 327–334.

FOWLER, S. P. An historical sketch. *Bulletin Essex Institution* 16 (1884), pp. 141–145.

FRANCIS, J. B. *Lowell hydraulic experiments.* V. Nostrand, New York, 1871. xi + 251 pp. 3rd ed. (1st ed. 1855, 2nd ed. 1868).

FRASER, C. *The story of engineering in America.* Crowell, New York, 1928. 471 pp.

[FREEMAN, H. F., AND ABBOTT, N. G.] *The factory mutuals 1835–1935.* Manufacturers Mutual Fire Insurance Co., Providence, Rhode Island, 1935. 384 pp.

FRENCH, H. P. Country roads. *Report Committee on Agriculture,* Washington, 1866, pp. 538–567.

FUESS, C. M. *Amherst, the story of a New England college.* Little, Brown, Boston, 1935. xiii + 372 pp.

FULLER, E. *Tinkers and genius, the story of the Yankee inventors.* Hastings House, New York, 1955, 308 pp.

FÜLÖP-MILLER, R. *Triumph over pain.* Translated by E. and C. Paul. Bobbs-Merrill, Indianapolis, 1930. xii + 438 pp.

FULTON, J. F. AND THOMSON, E. H. *Benjamin Silliman 1779–1864. Pathfinder in American Science.* Schuman, New York, 1947. ix + 294 pp.

GABRIEL, R. H. *The founding of Holyoke 1848.* Newcomen Address 1936. The Newcomen Society, American Branch, 1936. 23 pp.

GALLATIN, A. Report of manufactures. Eleventh Congress, 2nd session, April 17, 1810. Doc. No. 325, *American State Papers* II (Washington 1832) pp. 425–439.

GARDINER, R. H. Memoir of B. Vaughan. *Collection of Maine Historical Society* 6 (1859) pp. 85–92.

GARDNER, R. E. *American arms and arms makers.* Heer, Columbus, Ohio, 1930. 167 pp.

GARDNER, W. *The clock that talks and what it tells. A portrait story of the maker:* Hon. Walter Folger, Jr. *Whaling Museum Publications,* Nantucket Island, Massachusetts, 1954, viii + 143 pp.

GARRISON, F. H. *An introduction to the history of medicine.* 2nd ed. Saunders, Philadelphia–London, 1917. 905 pp.

[GAS] *Lectures delivered at the Centenary Celebration of the first commercial gas company.* Held at Franklin Institute, Philadelphia, 1912. Edited and published by American Gas Institute, New York, 174 pp.

GEE, JOSHUA. *The trade and navigation of Great Britain considered.* 4th ed. and suppl. A. Betterworth, C. Hitch, S. Birk, London, 1738. xxxix + 239 pp.

GETMAN, F. H. Samuel Morey, an pioneer of science in America, *Osiris* 1 (1936) pp. 278–302.

GILMAN, D. C. *The life of James Dwight Dana.* Harper, New York, 1899. xii + 409 pp.

GOOD, H. G. To the future biographers of John Quincy Adams. *Scientific Monthly* 39 (1934) pp. 247–251.

GOOD, H. G. Amos Eaton (1776–1842). Scientist and teacher of science. *Scientific Monthly* 53 (1941) pp. 464–469.

GOODALE, A. M. Some points in the history of the Boston Manufacturing Company. *Papers Citizens' Club of Waltham, Season of 1891–1892*. Waltham, Massachusetts, 1891. pp. 3–16.

GOULD, A. B. *Louis Agassiz*. Boston, 1901.

GOULD, B. A. On Peirce's criterion for the rejection of doubtful observations with tables for facilitating its application. *Astronomical Journal* 4 (1855) pp. 81–87.

GOULD, B. A. *Untersuchungen über die gegenseitige Lage der Bahmen der zwischen Mars und Jupiter sich bewegenden Planeten*. Diss. Göttingen, 1847.

GOULD, R. T. *The marine chronometer. Its history and development*. Potter, London, 1923. xvi + 287 pp.

GRANT, S. C. With Professor Guyot on Mount Washington and Carrigain in 1857. *Appalachia* 11 (1905–1908) pp. 229–239.

GRAS, N. S. B. *A history of agriculture in Europe and America*. 2nd ed. Crofts, New York, 1940. xxvii + 496 pp.

GRAUSTEIN, F. E. Nuttall's travels into the Old Northwest. An unpublished 1810 diary. *Chronica Botanica* 14, Nr 1–2 (1951) 88 pp.

GRAY, A. Mémoires et Souvenirs de Augustin-Pyramus DeCandolle, écrits par lui-même et publiées par son fils. Geneva and Paris, 1862. *American Journal of Science* (2) 25 (1863) pp. 1–16.

GRAY, A. Review of Darwin's Theory on the Origin of Species by means of Natural Selection. *American Journal of Science* (2) 29 (1860) pp. 154–184.

GRAY, A. Discussion between two readers of Darwin's treatise on the Origin of Species, upon its natural theology. *American Journal of Science* (2) 30 (1860) pp. 226–239.

GRAY, A. Sequel to the vestiges of creation. *American Journal of Science* (2) 1 (1846) pp. 250–254.

GRAY, A. *Darwiniana: Essays and reviews pertaining to Darwinism. Appleton,* New York, 1878. xii + 396 pp.

GRAY, A. *The botanical text-book for colleges, schools and private students: comprising Part I: An Introduction to structural and physiological botany. Part II: The Principles of systematic botany*. Wiley and Putnam, New York; Little, Brown, Boston, 1842. 413 pp.

GRAY, A. *Manual of the botany of the northern United States,* 1st ed. 1848. The revised ed. of 1863 has added an entirely new part: "Garden Botany, an introduction to a knowledge of the common cultivated plants," pp. xxix-lxxxix.

GRAY, J. L. *Letters of Asa Gray*. Houghton Mifflin, Boston, 1893. 2 vols. 838 pp.

GREEN, C. M. *Eli Whitney and the birth of American technology*. Little, Brown, Boston, 1956, 215 pp.

GREEN, S. A. *History of medicine in Massachusetts.* 1881.

GREEN, S. A. *Medicine in Boston.* Winsor's memorial history of Boston 1881. IV pp. 527–549, with additional memoranda by O. W. Holmes, pp. 549–570.

GREENE, S. A taste for figures. *New England Quarterly* 26 (1953) pp. 65–77.

GREENE, J. C. *The death of Adam, evolution and its impact on Western thought.* Iowa State University Press, Ames, 1959, 388 pp.

———. Some aspects of American astronomy, 1750–1815. *Isis* 45 (1954) pp. 339–358.

GREGORY, J. W. *H. D. Rogers*. Glasgow, 1916. 38 pp.

GREGORY, O. *A treatise of mechanics, theoretical, practical and descriptive.* Vol. II, 3rd ed. London, 1815. xx + 565 pp.

GRISWOLD, C. Description of a machine, invented and constructed by David Bushnell. *American Journal of Science* 2 (1820) pp. 94–100. See also *Transactions American Philosophical Society* 4, p. 312.

GRUND, J. *An elementary treatise on geometry, simplified for beginners not versed in algebra.* Part I containing Plane Geometry. Carter, Hendee and Babcock, Boston, 2nd ed., 1830. vii + 238 pp.

GUYOT, A. *The earth and man: lectures on comparative physical geography, in its relation to the history of mankind.* Translated by C. C. Felton. Gould, Kendall & Lincoln, Boston, 1849. xviii + 310 pp.

GUYOT, A. *On the Appalachian Mountain System. American Journal of Science* 31 (1861) pp. 157–187.

HALE, E. E. *A New England boyhood.* Cassel, New York, 1893. xxiii + 267 pp. New ed. Little, Brown, Boston, 1927.

HALL, C. R. *A scientist in the early republic, Samuel Latham Mitchell, 1764–1831.* Columbia University Press, New York, 1934. v + 162 pp.

HALL, H. *Report on the ship-building of the U. S.* Report to superintendent of Census, Washington, D. C. Letter of transmissal November 30, 1882, 276 pp. (10th census, v. 8, 1884).

HALLOCK, W., AND WADE, H. T. *Outline of the evolution of weights and measures and the metric system. Macmillan,* 1906. xi + 304 pp.

HALSEY, R. H. *How the President, Thomas Jefferson, and Doctor Benjamin Waterhouse established vaccination as a public health procedure.* History of Medical series, auspices Library New York Academy of Medicine 5, 1936. 58 pp.

HAMILTON, A. *Report on the subject of manufactures.* Report of the Secretary of the Treasury, in obedience to the order of the House of Representatives of January 15, 1790. Philadelphia, 1821. vii + 325 pp., see pp. 157–274.

HAMMOND, M. B. Correspondence of Eli Whitney relative to the invention of the cotton gin. *American Historical Review* 3 (1897–1898) pp. 90–127.

HARLOW, A. F. *Steelways of New England.* Creative Age Press, New York, 1946. 461 pp.

HARLOW, A. F. *Old towpaths.* Appleton, New York, 1926. xiv + 403 pp.

HARRINGTON, T. F. *The Harvard Medical School.* Lewis, New York, 1905. 3 vols.

HARRIS, T. W. *A treatise on some of the insects injurious to vegetation.* New edition, enlarged and under supervision of L. Agassiz. Ed. by C. L. Flint, Boston, 1862. xi + 640 pp.

HARRIS, T. W. *A report on the insects of Massachusetts.* Folsom, Wells & Thurston, Cambridge, 1841. viii + 459 pp.

HART, A. B. ed. *Commonwealth History of Massachusetts.* 5 vols. States History Company, New York, 1927–1930. I (1927) 23 + 608 pp.; II (1928) 13 + 592 pp.; III (1929) 14 + 582 pp.; IV (1930) 14 + 626 pp.; V (1930) 20 + 808 pp.

HARTE, C. R. Connecticut's iron and copper. *Annual Report Connecticut Society Civil Engineers* 1944, pp. 131–166.

HARTE, C. R. Some engineering features of the old Northampton Canal. *Annual Report Connecticut Society Civil Engineers* 1933, pp. 21–53.

HARTLEY, E. N. *The ironworks on the Saugus.* University of Oklahoma Press, Norman, 1957, 328 pp.

[HARVARD] Some of Harvard's endowed professorships. *Harvard Alumni Bulletin* 29 (1926–1927) pp. 65–69, 145–150.

[HARVARD] *The Harvard Medical School 1782–1906.* 1906. x + 212 pp.

HAYES, C. B. *The American Lyceum. Its history and contribution to education.* Bulletin 12 U. S. Department of the Interior, Office of Education, 1932, xii + 72 pp.

[HAYES, I. I.] The polar expedition of Dr. Hayes. *American Journal of Science* (2) 33 (1862) pp. 263–265.

HAYNES, W., BASS, L. W. American chemical chronology. In *Our Chemical Heritage,* American Chemical Society, 89th meeting. New York, 1935. 54 pp.

HAYNES, W. *Chemical pioneers; the founders of the American chemical industry.* The story of fifteen pioneers. Van Nostrand, New York, 1939. 288 pp.

HAZLITT, G. H. A. *The watch factories of America past and present* 1888.

HEATON, C. E. The history of anesthesia and analgesia in obstetrics. *Journal of Historical Medicine* 1 (1946) pp. 567–572.

HEINDEL, R. H. Americans and the Royal Society. 1783–1937. *Science* 87 (1938) pp. 267–272.

HERRICK, E. C. Historical Sketch of the Connecticut Academy of Arts and Sciences. *American Quarterly Register* 13 (1840) pp. 23–28.

HILL, H. A. *Memoir of the Hon. Marshall P. Wilder Ph.D., LL.D.* Boston, printed for private distribution 1888, 15 pp. (from *New England Historical and General Register,* July 1858).

HINDLE, B. *The pursuit of science in revolutionary America,* 1735–1789, University of North Carolina Press, Chapel Hill, 1956, XI + 410 pp.

HINKS, A. R. On some misrepresentations of Antarctic history. *Geographical Journal* 94 (1939) pp. 309–330.

HITCHCOCK, E. *Report on the geology, mineralogy, botany &*

zoology of Massachusetts. J. S. and C. Adams, Amherst, Massachusetts, 1833. xii + 700 pp.

HITCHCOCK, E. The connection between geology and natural religion. *Biblical Repos. & Quart. Obs.* 5 (1835) pp. 113–138. Other articles in vols. 5–11; *American Journal of Science* 30 (1836) pp. 114–130.

HITCHOCK, E. *Final report on the geology of Massachusetts.* 2 vols. Northampton, 1841.

HITCHCOCK, E. *Report on a re-examination of the economical geology of Massachusetts.* Boston, 1838. 139 pp.

HITCHCOCK, E. *The religion of geology and its connected sciences.* Phillips, Boston, 1851. xvi + 511 pp.; new ed. 1859.

HOBBS, W. H. *The discoveries of Antarctica within the American sector as revealed by maps and documents.* Transactions American Philosophical Society (1939) 71 pp. [see *Isis* 32 (1940, 1947) pp. 214–218].

HODGINS, E., AND MAGOUN, F. A. *Behemoth, the story of power.* Doubleday, Doran, Garden City, New York, 1932. xviii + 354 pp.

[HOLBROOK, J.] *American Lyceum, or Society for the improvement of schools and the diffusion of useful knowledge.* Perkins and Marvis, Boston, 1829. 24 pp.

HOLDEN, E. S. *Memorials of William Cranch Bond, director of the Harvard College Observatory 1840–1859, and of his son George Phillips Bond, director of the Harvard College Observatory 1859–1865.* San Francisco and New York, 1897.

HOLLAND, J. G. *History of Western Massachusetts.* Bowles, Springfield, 1855. 10 + 520 pp.

HOLMES, O. W. *Medical Essays 1842–1882.* Houghton Mifflin, Boston, 1892. xvii + 445 pp.

HOLMES, T. J. *Cotton Mather, a bibliography of his works.* 3 vols. Harvard University Press, Cambridge, 1940. xxxvi + 1395 pp.

HOLYOKE, E. A. A meteorological journal from the year 1786 to the year 1829, inclusive. With a prefatory memoir by E. Hale. *Memoirs American Academy,* New Series 1 (1833) pp. 107–216.

HOMANS, G. C., AND MORISON, S. E. *Massachusetts on the sea 1630–1930.* Massachusetts Bay Colony Tercentenary Commission 1930. 32 pp.

HOOKER, W. J. On the botany of America. *American Journal of Science* 9 (1925) pp. 263–284.

HOOL, G. A., AND KINNE, W. S. *Steel and timber structures.* McGraw Hill, New York, 1924. 695 pp.

HOOPES, P. R. *Connecticut clockmakers of the eighteenth century.* Dodd, Mead, New York, 1930. 3 + 178 pp.

HOOPES, P. R. *Connecticut's contribution to the development of the steamboat.* Yale University Press, New Haven, 1936. 31 pp.

HORSFORD, E. V. *Chemical essays relating to agriculture.* Munroe, Boston, 1846. 68 pp.

HOWE, H. *Memoirs of the most eminent American mechanics.* Derby and Jackson, New York, 1858. 482 pp.

HOWE, H. F. *Prologue to New England.* Farrar and Rinehart, New York, Toronto, 1943, xii + 324 pp.

———. *Early explorers of Plymouth Harbor* 1525–1619. Plymouth Plantation Inc. and Pilgrim Society, Plymouth, 1953, 30 pp.

HUBBARD, O. P. *The early history of New Hampshire medical institutions.* Globe Printing House, Washington, D. C., 1880. 41 pp.

HUBERT, P. G. *American inventors.* Scribner, New York, 1896. 299 pp.

HULTMAN, E. C. The Charles River Basin. *Proceedings Boston Society,* 1940, pp. 39–48.

HUMPHREY, J. E. Botany and botanist in New England. *New England Magazine* 14 (1896) pp. 27–44.

HUMPHREYS, D. *The Miscellaneous Works.* Swords, New York, 1804. xv + 394 pp.

HUNT, C. W. The first fifty years of the American Society of Civil Engineers, 1852–1902. *Transactions American Society of Civil Engineers.* 48 (1902), pp. 220–226.

HUNT, C. W. *Historical Sketch of the American Society of Civil Engineers.* New York, 1897, 92 pp.

HUNTINGTON, J. H. The flowering plants of the White Mountains. *Appalachia* 1 (1876–78) pp. 100–106.

ILES, G. *Leading American inventors.* Holt. New York, 1912. xv + 447 pp.

INGERSOLL, E. Harvard's Botanical Garden and its botanists. *Century Magazine* 32 (1886) pp. 237–240.

JACKSON, C. L. Eben Norton Horsford. *Proceedings American Academy* 28 (1892–1893) pp. 340–346.

JACKSON, C. T. *Final report on the geology and mineralogy of the state of New Hampshire; with contributions towards the improvement of agriculture and metallurgy.* Carroll & Baker, Concord, New Hampshire, 1844. viii + 376 pp.

JACKSON, C. T. On the collection of geological specimens and on geological surveys. *American Journal of Science* 30 (1836) pp. 203–208.

JACKSON, S. I. Some ancestors of the "extension course." *New England Quarterly* 14 (1941) pp. 505–508.

JAFFE, B. *Men of science in America.* The role of science in the growth of our country. Simon and Schuster, New York, 1944. x + 600 pp., revised ed., ib., 1958, xliii + 716 pp.

JARVIS, E. *Practical physiology; for the use of schools and families.* Thomas, Cowperthwait, Philadelphia, 1847. 368 pp.

JARVIS, E. *Primary physiology, for schools.* Thomas, Cowperthwait, Philadelphia, 1848. 168 pp.

JEFFERSON, T. *Writings,* ed. H. A. Washington. Riker, New York, 1857.

JEFFERSON, T. The description of a mould-board of the least resistance, and of the easiest and most certain construction. *Transactions American Philosophical Society* 4 (1799) pp. 313–322.

JENKINS, R. A sketch of the industrial history of the Coalbrookdale District. *Transactions Newcomen Society* 4 (1923–1924) pp. 102–107.

JEPSON, W. L. The overland journey of Thomas Nuttall. *Madroño, San Francisco* 2 (1926) pp. 143–147.

JEROME, C. *History of the American clock business for the past sixty years and the life of Chauncey Jerome.* New Haven, 1860.

JOERG, W. L. G. *Brief history of polar exploration since the introduction of flying.* 2nd rev. ed. *American Geographical Society,* New York, 1930. 95 pp.

[JOHNSON, S. W.] *From the letter files of S. W. Johnson, Professor of Agricultural Chemistry in Yale University.* Ed. by Elizabeth A. Osborne. Yale University Press, New Haven, 1913.

JONES, H. M. *The future of the Academy.* President. Address, American Academy of Arts and Sciences, 1944. 9 pp.

JONES, L. C. *Arnold Guyot et Princeton.* Rec. Trav. publ. Fac. des Lettres, Neuchatel 14 (1929) 125 pp.; a shorter version in *Union College Bulletin* 23 (1930) pp. 31–65.

JORDAN, D. S. *Leading American men of science.* Holt, New York, 1910. 471 pp.

JUDSON, I. F. *Cyrus W. Field, his life and work.* Harper, New York, 1896. 332 pp.

KAEMPFFERT, W. (ed.) *A popular history of American invention.* Burt, New York. I. *Transportation, communication & power,* 1924. xvi + 577 pp.

KAPLAN, S. Lewis Temple and the hunting of the whale. *New England Quarterly* 26 (1953) pp. 78–88.

KAROLIK, M. The American way. *Atlantic Monthly,* 170 (1942) pp. 101–105.

KARPINSKI, LOUIS CHARLES. *Bibliography of mathematical works printed in America through 1850 with the co-operation for Washington libraries of Walter F. Shenton.* University of Michigan Press, Ann Arbor, 1940. xxvi + 697 pp.

KEITH, H. C., AND HARTE, C. R. *The early iron industry of Connecticut.* Reprinted from 51st Annual Report of the Connecticut Society of Civil Engineers (1935). Mead & Noel, New Haven, 69 pp.

KERFOOT, J. B. *American Pewter.* Houghton Mifflin, Boston, 1924. xxiii + 239 pp.

KEYES, C. William Maclure: father of modern geology. *American Geologist* 44 (1925) pp. 81–94.

KILBOURNE, F. W. *Chronicles of the White Mountains.* Houghton Mifflin, Boston, 1916. xxxii + 433 pp.

KILGOUR, F. G. Professor John Winthrop's notes on sun spot observations. *Isis* 29 (1938) pp. 355–361.

KILGOUR, F. G. The first century of scientific books in the Harvard College library. *Harvard Library Notes* 3 (1939) pp. 217–225.

KILGOUR, F. G. The rise of scientific activity in colonial New England. *Yale Journal Biology and Medicine* 22 (1949) pp. 123–138.

KINGSLEY, J. L. Articles on the history of Yale College. *American Quarterly Register* 1835–1836; American Bibliography Report 1841–1842.

KIRBY, R. S. *Inventors and engineers of old New Haven.* A series of six lectures given in 1938 under the auspices of the School of Engineering, Yale University. New Haven Tercentenary Publications, New Haven Colony Historical Society, New Haven, Connecticut, 1939. 111 pp.

KIRBY, R. S. Some early American civil engineers and surveyors. *Papers and Transactions Connecticut Society of Civil Engineers* 1930, pp. 26–47.

KIRBY, R. S. William Weston and his contributions to early American engineering. *Transactions Newcomen Society* 16 (1935–1936) pp. 111–127.

KIRBY, R. S., AND LAURSON, P. G. *The early years of modern civil engineering.* Yale University Press, New Haven, 1932. xvi + 324 pp.

KIRKLAND, E. C. The "Railroad Scheme" of Massachusetts. *Journal of Economic History* 5 (1945) pp. 145–171.

KITTREDGE, G. L. *The old farmer and his almanack.* Harvard University Press, Cambridge, 1920. xiv + 403 pp.

KLEIN, F. *Vorlesungen über die Entwicklung der Mathematik im 19en Jahrhundert.* I, Springer, Berlin, 1926. xvi + 385 pp.

KNIGHT, C. K. *The history of life insurance in the United States to 1870 with an introduction to its development abroad.* Thesis, University of Pennsylvania, Philadelphia, 1920. 160 pp.

KRAUS, M. Scientific relations between Europe and America in the eighteenth century. *Scientific Monthly* 55, 1942, pp. 259–272.

LAING, A. *Clipper ship men.* Duell, Sloan & Pearce, New York, 1944. 279 pp.

LAPLACE (PIERRE SIMON DE). *Traité de Mécanique Celeste* (*Oeuvres complètes,* vol. 1–5, Paris, 1878–1882, with the list of errata suggested by Bowditch).

LATHAM, J. L. *Carry on, Mr. Bowditch.* Houghton Mifflin, Boston, 1955, 251 pp.

———. *The story of Eli Whitney; invention and progress in the young nation.* Aladdin Books, New York, 1953, 192 pp.

LATHROP, W. G. *The brass industry in the United States.* A study of the origin and the development of the brass industry in the

Naugatuck Valley and its subsequent extension over the nation. Revised ed. 1926, Lathrop, Mount Carmel, Connecticut, 174 pp. (Originally thesis, Yale.)

LENIN, V. *A letter to American workers,* August 20, 1918. International Publications, New York, 1934. 22 pp.

LIEBIG, JUSTUS. *Animal chemistry, or organic chemistry in its application to physiology and pathology.* Ed. by Wm. Gregory. Cambridge, Mass.

LIEBIG, J. *Principles of agricultural chemistry.* Walton and Maberly, London, 1855, vii + 136 pp.

LIEBIG, J. *Liebig's complete works on chemistry.* Peterson, Philadelphia, no date. iv + 135 pp. + 111 + 48 + 47 + 48 pp.

Liebig and after Liebig. A Century of progress in agricultural chemistry. American Association for the Advancement of Science, Washington, D. C., 1942. viii + 111 pp.

Light Houses: Report of the Secretary of the Treasury on the Improvements in the Light House System, embracing a report from Lieut. T. A. Jenkins & Lieut. R. Bache. Washington, 1846. 272 pp.

LOEWENBERG, B. The reaction of American scientists to Darwinism. *American Historical Review* 38 (1932–1933) pp. 687–701.

LOOMIS, E. *A treatise on algebra.* Harper, New York, 1846. vii + 346 pp.

LOOMIS, E. *On certain storms in Europe & America, December 1836.* Smithsonian Contributions. Washington. 11 (1859). 26 pp.

LOOMIS, E. *A treatise on meteorology.* Harper, New York (1868). viii + 305 pp.

LOOMIS, E. *The recent progress of astronomy, especially in the United States.* Harper, New York, 1851. 257 pp. (3rd ed., mostly rewritten and much enlarged, 1856, 396 pp.)

LOOMIS, E. *Elements of natural philosophy designed for academies and high schools.* Harper, New York, 1858. xii + 351 pp.

LOREE, L. F. *The first steam ("fire") engine in America.* [Publ. for Newcomen Society] 1929, 12 pp.

LOVE, J. L. A history of engineering at Harvard University. *Harvard Engineering Journal* 1 (1902) pp. 1–29.

LOVERING, J. Boston and science. Winsor's *Memorial history of Boston,* 1881. IV, pp. 489–526.

[LOVERING, J.] Memorial of J. Lovering. *Memoirs American Academy of Science* 1892, 40 pp.

LOWELL, J. R. *The life and letters of James Gates Percival.* Literary Essays II, 1891, Houghton Mifflin, Boston, pp. 140–161.

LOWNES, A. E. The 1769 Transit of Venus and its relation to early American astronomy. *Sky and Telescope* 2 (1943) pp. 3–6.

LOWRY, L. M. H., LOWRY, C. D., JR., AND MINER, J. R. Count Rumford, a youthful meteorologist; the graveyard fort. *Isis* 27 (1937) pp. 264–266, 268–285.

LURIE, E. Louis Agassiz and the races of man. *Isis* 45 (1954) pp. 227–242

———. *Louis Agassiz, a life in science.* University of Chicago Press, Chicago, 1960, 449 pp.

LYELL, C. *Travels in North America 1841–42, with geological observations on the U. S., Canada and Nova Scotia.* Wiley & Putnam, 1845, New York. I, 250 pp.; II, 231 pp.

LYELL, C. *A second visit to the United States of North America.* Murray, London, 2nd ed., 1850. I, xii + 368 pp.; II, xii + 385 pp.

LYONS, H. G. John Winthrop (Jr.) F. R. S. *Notes & Records Royal Society, London* 3 (1941) pp. 110–115.

MABEE, C. *The American Leonardo: a life of Samuel F. B. Morse.* With an introduction by A. Nevins. Knopf, 1943, xix + 420 pp.

MCALISTER, ETHEL M. *Amos Eaton, scientist & educator 1776–1842.* University of Pennsylvania Press, Philadelphia, 1941. xiv + 587 pp.

MACDONALD, ELEANOR J. *A history of the Massachusetts Department of Public Health,* Commonwealth of Massachusetts, Department of Public Health, 1936. 43 pp. (reprint from *The Commonwealth* 23, 1936).

MACDONALD, T. H. The history and development of road building in the U. S. *Transactions American Society of Civil Engineers* 92 (1928) pp. 1181–1206.

MCGEE, W. J. Fifty years of American science. *Atlantic Monthly* 82 (1898) pp. 307–320.

MACK, H. *Carl Friedrich Gauss und die Seinen. Festschrift zu seinem 150. Geburtstage.* A. Appelhans, Braunschweig, 1927. xi + 130 pp.

McKeehan, L. W. *Yale science. The first hundred years 1701–1801.* H. Schuman, New York, 1947. vii + 82 pp.

McMaster, J. B. *A history of the people of the United States from the Revolution to the Civil War.* Appleton, New York, I (1893), xv + 622 pp.; II (1885), xx + 656 pp.

Mann, Mrs. H. *Life of Horace Mann, by his wife.* 2nd ed., Walker, Fuller, Boston, 1865. xi + 609 pp.

Mann, M. W. The Middlesex canal. *Publications Bostonian Society* 6 (1910) pp. 69–88.

[Manning, R.] *History of the Massachusetts Horticultural Society 1829–1878.* Boston, printed for the Society, 1880. v + 545 pp.

[Mansfield, J.] *Essays, mathematical and physical: containing new theories and illustrations of some very important and difficult subjects of the sciences.* Never before published. Morse, New Haven. viii + 274 pp. (1801?)

Marcou, J. *Life, letters and works of Louis Agassiz.* 2 vols. Macmillan, New York, 1896. II, vii + 318 pp.

Marshall, Helen E. *Dorothea Dix, forgotten samaritan.* University of North Carolina Press, Chapel Hill, 1937. x + 298 pp.

Martin, L. The log of Palmer's discovery of Antarctica. *Science* 37 (1938) p. 165.

Martin, E. D. *Thomas Jefferson: scientist.* Schuman, New York, 1952. x + 289 pp.

Martineau, Harriet. *Autobiography, ed. by Maria W. Chapman.* Osgood, Boston, I (1877). 594 pp.

Marx, K. *Capital.* A critical analysis of capitalist production. Translated from the third German ed. Swan Sonnerschein, Lowrey and Co., London, 1887. xxxi + 882 pp.; reprint International Publications, New York, 1939.

Mason, G. C. *Reminiscences of Newport.* Hammott, Newport, Rhode Island, 1884. 407 pp.

Massachusetts Society for Promoting Agriculture. *Rules and regulations* T. Fleet, Boston, 78 pp.

Massachusetts Society for Promoting Agriculture. Centennial year (1792–1892). Salem 1892, 146 pp.

Meisel, M. *A bibliography of American Natural History; the Pioneer Century, 1769–1865.* 3 vols. Brooklyn, 1924–1929.

Merrill, G. P. *Contributions to the history of state geological and natural history surveys.* Bulletin 109, U. S. Natural Museum, 1920.

MERRILL, G. P. *The first one hundred years of American geology.* Yale University Press, New Haven, 1924. xxi + 773 pp. (slightly changed from Annual Report Smithsonian Institute 1904). Review *Isis* 7 (1925) pp. 498–501.

MERRILL, E. D. A generally overlooked Rafinesque paper. *Proceedings American Philosophical Society* 86 (1943) pp. 72–90.

MERRIMAN, M. *A treatise on hydraulics.* Wiley, New York, 6th ed., 1898. viii + 427 pp.

MEYER, B. H. *History of transportation in the U. S. before 1860.* Carnegie Institute, Washington, 1917. 11 + 678 pp.

MICHAUX, F. A. *The North American Sylva, or a description of the forest trees of the U. S., Canada and Nova Scotia.* 3 vols. D'Hautel, Paris, 1819.

MILBANK, J. *The first century of flight in America.* Princeton University Press, 1943, Princeton, New Jersey. x + 248 pp.

Milestones 1830–1930. The Boston Society of Natural History 1830–1930. Boston, printed for the Society, 1930. xii + 117 pp.

MILLER, J. C. *Origins of the American Revolution.* Little, Brown, Boston, 1943. xiv + 519 pp.

MILLER, PERRY. *The New England mind; the seventeenth century.* Macmillan, New York, 1939, xi + 528 pp.

———. *The New England mind from Colony to Province.* Harvard University Press, Cambridge, 1953.

MILLS, A. P. Tests of wrought iron links after 100 years in service. History of the Essex-Merrimac Chain Bridge and notes on early suspension bridges. *Cornell Civil Engineer* 19 (1910–1911) pp. 251–282.

MILLS, H. F. *James Bicheno Francis.* Read to Corporation M. I. T. December 14, 1892.

MIRSKY, J. AND NEVINS, A. *The world of Eli Whitney.* Macmillan, New York, 1952, xvi + 346 pp.

MOORE, C. W. *Timing a century.* History of the Waltham Watch Company. Harvard University Press, Cambridge, 1945. xxxiv + 362 pp.

MOORE, S. *An accurate system of surveying.* Printed by T. Collier, Litchfield, Connecticut, 1796. x + 131 pp.

MORISON, S. E. *The maritime history of Massachusetts, 1783–1860*. Houghton Mifflin, Boston, 1921. xv + 401 pp.

MORISON, S. E. (ed.) *Development of Harvard University since the inauguration of President Eliot 1869–1929*. Harvard University Press, Cambridge, 1930. xc + 660 pp.

MORISON, S. E. Astronomy at Colonial Harvard. *New England Quarterly* 7 (1934) pp. 3–24.

MORRISON, S. E. *Three centuries of Harvard 1636–1936*. Harvard University Press, Cambridge, 1936, 512 pp.

MORISON, S. E. *Builders of the Bay Colony*. Houghton Mifflin, Boston, 1930, xiv + 365 pp.

MORISON, S. E. AND COMMAGER, H. S. *The growth of the American Republic*. Oxford University Press, New York, I (1942).

MORRIS, J. G. Contributions toward a history of entomology in the U. S. *American Journal of Science* (2) 1 (1846) pp. 17–27.

MORSE, J. *Elements of geography*. 4th ed., Thomas & Andrews, Boston, 1804. xii + 143 pp.

MORSE, J. K. *Jedediah Morse: A champion of New England orthodoxy*. Columbia Studies in American Cultures. Columbia University Press, New York, 1939, ix + 179 pp.

MORSE, J. T. *Life and letters of O. W. Holmes*. 2 vols. Houghton Mifflin, Boston, 1896.

MORSE, S., AND ANDERSON, J. *Book of the White Mountains*. Minton, Balch, New York, 1930. 300 pp.

MUNROE, J. P. The beginning of the Massachusetts Institute of Technology. *Technology Quarterly* 1 (1888) pp. 285–297.

MUNSTERBERG, M. The Bowditch Collection in the Boston Public Library. *Isis* 34 (1938) pp. 140–142.

NASON, E. H. *Old Hallowell on the Kennebec*. Burleigh & Flynt, Augusta, Maine, 1909. xii + 359 pp.

NAVIER, M. *Rapport à M. Becquey et mémoire sur les ponts suspendus*. Imp. Roy., Paris, 1823. xxiv + 228 pp.

NAYLOR, M. V. Sylvester Graham 1794–1851. *Annals of Medical History* 4 (1942) pp. 236–240.

NEWCOMB, S. Abstract science in America 1776–1876. *North American Review* 122 (1876) Centennial Number, pp. 88–123.

NEWELL, L. C. Benjamin Hale, professor of chemistry and college president. *Journal of Chemical Education* 2 (1925) pp. 457–458.

NEWELL, L. C., AND DAVIS, T. L. *Notable New England chemists.* Pamphlet prepared by the Division of History of Chemistry, American Chemical Society, Boston, 1928. 16 pp.

NEWTON, H. A. Benjamin Peirce. *American Journal of Science* (3) 22 (1881) pp. 167–178.

NEWTON, R. H. *Town and Davis, Architects. Pioneers in American revivalist architecture 1812–1870.* Columbia University Press, New York, 1942. xx + 334 pp.

NORTON, J. T. *Elements of scientific agriculture, or the connection between science and the art of practical farming.* E. H. Pease, Albany, 1850. x + 208 pp.

NUTTALL, TH. *A manual of the ornithology of the United States and Canada.* Hilliard & Brown, Cambridge, 1832. vi + 683 pp.

OAKES, W. *Scenery of the White Mountains.* Crosby, Nichols, Boston, 1848.

ODGERS, M. M. *Alexander Dallas Bache: scientist and educator* 1806–1867. University of Pennsylvania Press 1947, vii + 223 pp.

OLIVER, J. W. *History of American technology.* Ronald Press, New York, 1956, viii + 676 pp.

PACKARD, A. S. A century's progress in American zoology. *American Naturalist* 10 (1876) pp. 591–598.

PACKARD, F. R. *History of medicine in the U.S.* Hoeber, New York, 1931. 1323 pp.

PAINE, T. *The complete writings,* ed. by P. S. Foner. Citadel Press, New York, 1945, 2 vols.

PALFREY, J. G. *History of New England.* Vol. I. Little, Brown, Boston, 1858. xvii + 636 pp.

PANNEKOEK, A. The Discovery of Neptune. *Centaurus* 3 (1953) pp. 126–137, [*S. Lilley, ed.*] part of "Essays on the social history of science."

PARKER, W. B. *The life and public services of Justin Smith Morrill.* Houghton Mifflin, Boston, 1924. viii + 378 pp.

PARRINGTON, V. L. *Main currents in American thought.* An interpretation of American literature from the beginnings to 1920. 3 vols. (1927–1930) Harcourt, Brace, New York, I (xvii + 413 pp.); II (xxii + 493 pp.); III (xxxix + 429 pp).

PARSONS, W. B. The Cape Cod Canal. *Transactions American Society of Civil Engineers* 82 (1918) pp. 1–143.

PEARE, C. O. *A scientist of two worlds: Louis Agassiz.* Lippincott, Philadelphia, 1958. 192 pp.

PEIRCE, B. Bowditch's translations of the *Mécanique céleste. North American Review* 48 (1839) pp. 143–180.

PEIRCE, B. Criterion for the rejection of doubtful observations. *Astronomical Journal* 2 (1852) pp. 161–163.

PEIRCE, B. On Peirce's criterion. *Proceedings American Academy* 13 (1877–1878) pp. 348–351.

PEIRCE, B. Notice of the computations of Mr. Sears C. Walker. *Proceedings American Academy* 1 (1846–1848) pp. 57–68.

PEIRCE, B. *System of analytic mechanics.* Little, Brown, Boston, 1855. xxxix + 436 pp. (New ed. New York, 1872.)

PENNELL, F. W. Travels and scientific collections of Thomas Nuttall. Bartonia, *Proceedings Philadelphia Botany Club* 18 (1936). 64 pp.

PERCIVAL, J. G. *Poetical works.* 2 vols. Ticknor & Fields, Boston, 1859. I (lxii + 402 pp.); II (vii + 517 pp.).

PERCIVAL, J. G. *Report on the geology of the state of Connecticut.* Osborn and Baldwin, New Haven, 1842. 495 pp.

PERKINS, V. C. A mining boom in Maine. *New England Quarterly* 14 (1941) pp. 437–456.

PERRIN, P. G. *The life and works of T. G. Fessenden* 1771–1837. University Press, Orono, Maine, 1925, 206 pp.

PERRY, A. *Memorial to Zachariah Allen, 1795–1882.* J. Wilson, University Press, Cambridge, 1883. 182 pp.

Pharmacopoeia of the Massachusetts Medical Society, The. Larkin, Boston, 1808. x + 272 pp.

PHILLIPS, J. D. The Salem shipbuilding industry before 1812. *American Neptune* 2 No. 4 (1942) pp. 1–11.

PICKERING, C. *Chronological history of plants: man's record of his own existence illustrated through their names, uses and companionships.* Little, Brown, Boston, 1879. xvi + 1222 pp.

PIDDINGTON, H. *The sailor's horn-book for the law of storms.* 4th ed. William & North, London, Edinburgh, 1864. xviii + 408 pp.

PITKIN, T. *A statistical view of the commerce of the United States of America, its connection with agriculture and manufactures, etc.* (1st ed. Hartford, 1816); 2nd ed. Eastburn, New York, 1817. xii + 445 + vii pp. (Another ed. 1835.)

POOR, H. V. *History of the railroads and canals of the United States of America.* Schultz, New York, 1860. viii + 612 pp.

POORE, B. P. History of the agriculture of the U. S. *Report of Commission on Agriculture* 1866, Washington, pp. 498–527.

POPE, THOMAS. *A treatise on bridge architecture, in which the superior advantages of the flying, pendent lever bridge are fully proved.* New York, for the author by A. Niven, 1811. xxxii + 288 pp.

POST, S. S. *Treatise on principles of civil engineering as applied to the construction of wooden bridges.* 1859.

POUSSIN, G. T. *Chemins de fer américains.* Carilian-Goeury, Paris, 1836. xviii + 271 pp.

POWERS, W. H. Some facts in the life of Thomas Nuttall. *Science* 62 (1925) pp. 389–391.

[PRATT, T. W.] (obituary), Minutes of meetings. *American Society of Civil Engineering,* December 1, 1875, *Proceedings of American Society of Civil Engineers* 1–2 (1873–76) pp. 332–335.

PREBLE, G. H. *A chronological history of the origin and development of steam navigation 1543–1882.* Hamersly, Philadelphia, 1883, xix + 483 pp.

QUINCY, J. *History of Harvard University,* Cambridge, Owen, 1840. xiv + 612 pp.

RAFINESQUE, C. S. *A life of travels and researches in North America and South Europe.* Turner, Philadelphia, 1836. Reprinted by *Chronica Botanica,* Vol. 8, No. 2, Waltham, Massachusetts (1944) pp. 292–360.

RAISZ, E. Outline of the history of American cartography. *Isis* 26 (1937) pp. 373–391.

READ, D. *Nathan Read.* Hurt & Houghton, New York, 1870.

REDFIELD, W. Prevailing storms of the Atlantic Coast and the Northeastern States. *American Journal of Science* 20 (1831) pp. 17–51.

REDFIELD, W. Hurricanes and storms of the U. S. and W. Indies. *American Journal of Science* 25 (1834) pp. 114–121; 26 (1834) pp. 122–135.

[Redfield, W. C.] Correspondence of Sir Wm. Reid and W. C. Redfield. *American Journal of Science* (2) 34 (1862) pp. 442–443.

Reilly, J., and O'Flynn, N. Richard Kirwan, an Irish chemist of the eighteenth century. *Isis* 13 (1930) pp. 298–319.

Rhees, W. J. *The Smithsonian Institution. Origin and history.* 2 vols. I (1835–1887).

Rice, N. P. *Trials of a public benefactor, as illustrated in the discovery of etherization.* Pudney & Russell, New York, 1859. xx + 460 pp.

Rice, W. N. The contributions of America to geology. *Science* 25 (1907) pp. 161–175.

Richman, I. B. *Rhode Island.* Houghton Mifflin, Boston, 1905. x + 395 pp.

Ricketts, P. C. *History of the Rensselaer Polytechnic Institute 1824–1894.* Wiley, New York, 1895. ix + 193 pp.

Ringwalt, J. L. *Development of transportation systems in the U. S.* Railway World Office, Philadelphia, 1888. 398 pp.

Roberts, C. *The Middlesex Canal 1793–1860.* Harvard University Press, Cambridge, 1938. xii + 250 pp.

Robinson, M. L. *Runner of the mountain tops. The life of L. Agassiz.* Random House, New York, 1939, vii + 290 pp.

Robinson, B. L. The removal of an old landmark. *Harvard Graduates Magazine* 75 (1911) pp. 418–421.

Robison, J. *System of mechanical philosophy, with notes by D. Brewster.* J. Murray, Edinburgh. 4 vols. and vol. of plates. II (1822). x + 708 pp.

Rodgers, A. D. *"Noble fellow," William Starling Sullivant.* Putnam, New York, 1940. xxii + 361 pp.

Rodgers, A. D. *John Torrey. A story of North American botany.* Princeton University Press; London (Milford), Oxford University Press, 1942. xii + 352 pp.

Rogers, Emma S. *Life and letters of William Barton Rogers, edited by his wife.* Houghton Mifflin, Boston. 2 vols. 1896. I (viii + 427 pp.); II (vi + 451 pp.).

Rogers, P. K. *An introduction to the mathematical principles of natural philosophy.* Shepherd & Pollard, Richmond, 1822. 144 pp.

ROGERS, W. B. *An elementary treatise on the strength of materials.* Thompkins & Noel, Charlottesville, 1838. 50 pp.

ROGERS, W. B. *Elements of mechanical philosophy, for the use of the junior students of the University of Virginia.* Thurston, Torry & Emerson, Boston, 1852. 339 pp.

ROGERS, W. B. Electric illumination at Boston, — photometric powers of the light. Letter to Professor B. Silliman, Jr., Boston, August 19, 1863. *American Journal of Science* (2) 36 (1863) pp. 307–308.

ROPER, R. C. Thomas Paine: scientist-religionist. *Scientific Monthly* 58 (1944) pp. 101–111.

ROTH, G. B. The "original Morton inhaler" for ether. *American Medical History* 4 (1932) pp. 390–397.

RUKEYSER, M. *Willard Gibbs.* Doubleday, Doran, Garden City, New York, 1942. xi + 465 pp.

RUMFORD, COUNT (B. THOMPSON). *The complete works.* 4 vols. Boston, American Academy of Arts and Sciences. 1 (1870) 493 pp.; II (1873) 570 pp.; III (1874) 504 pp.; IV (1875) 842 pp.

[RUMFORD] The Rumford Bicentennial. *Procedures of the American Academy of Arts and Sciences* 82, No. 7 (Dec. 1953) pp. 249–368.

RUMFORD An inquiry concerning the source of the heat which is excited by friction. *Philosophical Transactions* 88 (1798) pp. 80–102; also *Complete works* I, pp. 469–492.

SABINE, R. *History and progress of the electric telegraph.* 2nd ed. Norstrand, New York, 1869. xiv + 280 pp.

SANBORN, F. B. *New Hampshire.* American Commonwealth Series, Houghton Mifflin, Boston, 1904. x + 354 pp.

SARGENT, C. S. Portions of the journal of André Michaux, botanist, written during his travels in the U. S. and Canada, 1785–1796. *Proceedings American Philosophical Society* 26 (1889) pp. 1–145.

SARGENT, C. S. Asa Gray. *New York Sun,* January 3, 1886. 16 pp.

SARGENT, P. E. *A handbook of New England.* 2nd ed. P. E. Sargent, Boston, 1917. 895 pp.

SARTON, G. The whaling museums of New Bedford and Nantucket. *Isis* 16 (1931) pp. 115–123.

SCHAFER, J. *Social history of American agriculture.* Macmillan, 1936. 302 pp.

SCHLESINGER, A. M., JR. *The age of Jackson.* Little, Brown, Boston, 1945. xiv + 577 pp.

SEDGWICK, W. T., AND TYLER, H. W. *A short history of science.* Rev. ed. Macmillan, New York, 1939. x + 512 pp.

SEITZ, D. C. Thomas Paine, bridge builder. *Virginia Quarterly Review* 3 (1927) pp. 571–584.

SHAFER, H. B. *The American Medical Profession, 1783 to 1850.* Columbia University Press, New York, 1936, 271 pp. (thesis).

SHATTUCK, L. The vital statistics of Boston; containing an abstract of the bills of mortality for the last twenty-nine years, and a general view of the population and health of the city at other periods of its history. Extract from *American Journal of Medical Science,* April 1841. Lea & Blanchard, Philadelphia, 1841. xliii pp.

SHATTUCK, L. *Report of a general plan for the promotion of public and personal health, devised, prepared and recommended by the Commissioners appointed under a resolve of the legislature of Massachusetts, relating to a sanitary survey of the state.* Dutton & Wentworth, Boston, 1850. 544 pp.

SHEPARD, CHARLES. *Treatise on mineralogy.* H. Howe, New Haven, 1832. xi + 256 pp.

SHEPARD, O. *The journals of Bronson Alcott.* Little, Brown, Boston, 1938. xxx + 559 pp.

SHOEMAKER, E. C. *Noah Webster, pioneer of learning.* Columbia University Press, New York, 1936, x + 347 pp. (thesis).

SILLIMAN, B. *First principles of chemistry, for the use of colleges and schools.* Loomis-Peck, Philadelphia, 1847. 492 pp.

SILLIMAN, B. American contributions to chemistry. *The American Chemist* 5 (1874–1875), pp. 70–114, 195–209.

SIMONS, L. G. *Introduction of algebra into American schools in the eighteenth century.* Washington, Government Printing Office, 1924. vi + 80 pp.

SIMONS, L. G. The influence of French mathematicians at the end of the eighteenth century upon the teaching of mathematics in American Colleges. *Isis* 15 (1931) pp. 104–123.

SIMONS, L. G. Bibliography of early American textbooks on

algebra published in the colonies and the U. S. through 1850. *Scripta mathematica*, New York, 1936. 68 pp.

SIMONS, L. G. Short stories in colonial geometry. *Osiris* I (1936) pp. 585–605.

SINGER, C. *A short history of medicine*. Oxford University Press, New York, 1928. xxiv + 368 pp.

SKILLING, W. T., AND RICHARDSON, R. S. *The practical essentials of pretraining navigation*. Holt, New York, 1942. v + 113 pp.

SMALLWOOD, W. M. *Natural history and the American mind*. Columbia University Press, New York, 1941. xiii + 455 pp. No. 8 of Columbia Studies in American Culture.

SMALLWOOD, W. M. How Darwinism came to the United States. *Scientific Monthly* 52 (1941), pp. 342–349.

SMILES, S. *Lives of the engineers, with an account of their principal works; comprising also a history of inland communication in Britain*. J. Murray, London. I, 1861 (xvi + 484 pp.); II, 1861 (xiv + 502 pp.)

SMITH, A. *An inquiry into the nature and causes of the wealth of nations, 1776*. (Available in Everyman's Library ed., 2 vols.)

SMITH, D. E. Thomas Jefferson and mathematics. *Scripta mathematica* I (1932) pp. 3–14.

SMITH, D. E., AND GINSBURG, J. *History of mathematics in America before 1900*. Carus Mathematical Monographs 5, 1945. 209 pp.

SMITH, E. A. *Life and letters of Nathan Smith*. 1914.

SMITH, E. F. *Chemistry in America*. Chapters from the history of the science in the United States. Appleton, New York, 1914. xiii + 356 pp.

SMITH, HARRIETTE K. *History of the Lowell Institute*. Boston, Lamson, 1898. 105 pp.

SPEARS, J. R. *Captain Nathaniel Brown Palmer. An old-time sailor of the sea*. Macmillan, New York, 1922. xii + 252 pp.

STANFORD, A. *Navigator. The Story of Nathaniel Bowditch*. Morrow, New York, 1927. xi + 308 pp.

STEARNS, R. P. Colonial fellows of the Royal Society of London, 1661–1788. *Osiris 8* (1948) pp. 73–121.

STEINMAN, D. B., AND WATSON, S. R. Bridges and their builders, Putnam, New York, 1941. xvi + 379 pp.

STEPHENSON, O. W. The supply of gunpowder in 1776. *American Historical Review* 30 (1924–1925), pp. 271–281.

STERN, B. *Society and medical progress.* Princeton University Press, Princeton, New Jersey, 1941. xvii + 264 pp.

STEVENS, N. E. America's first agricultural school. *Scientific Monthly* 13 (1921) pp. 531–540.

STEVENS, W. O. *Nantucket. The far-away island.* Dodd, Mead, New York, 1936. xi + 313 pp.

STEVENSON, D. *Sketch of the civil engineering of North America.* Weale, London, 1838. xv + 320 pp.

STILES, E. *The United States elevated to glory and honour.* Sermon, New Haven, 1783, 99 pp.

STILES, E. *The literary diary of Ezra Stiles,* ed. by F. B. Dexter. 3 vols. New York, 1901. I, 665 pp.; II, 573 pp.; III, 648 pp.

STILLMAN, C. K. The development of the builders' half-hull model in America. *Publication of the Marine Historical Association,* Mystic, Connecticut. Vol. I, No. 7, July 10, 1933, pp. 107–115.

STOKES, A. P. Memorials of eminent Yale men. 1914.

STORER, D. H. *Synopsis of the fishes of North America.* Publications by American Academy of Arts & Sciences, Cambridge, 1846, 298 pp.

STORROW, C. S. *A treatise on water-works for conveying and distributing supplies of water.* Hilliard, Gray, Boston, 1835. xii + 242 pp.

STRUIK, D. J. Mathematicians at Ticonderoga. *Scientific Monthly* 82 (1956) pp. 236–240.

———. American Science between 1780 and 1830. *Science* 129 (1959) pp. 1100–1106, also *ibid.,* 130 (1959), p. 190.

STUART, C. B. *The naval dry docks of the U. S.* 4th ed. Van Nostrand, New York, 1870.

STUART, C. B. *Lives and work of civil and military engineers of America.* Van Nostrand, New York, 1871. 343 pp.

[STUART, M.] Remarks on a "Critical examination of some passages in Genesis I." *American Journal of Science* 30 (1836) pp. 114–130.

SUMMER, W. G. Social Darwinism. *New England Quarterly* 1941,

457; see also "Social Darwinism," *American Journal of Science* 12 (1907) pp. 695–716.

SWAN, W. U. Early visual telegraphs in Massachusetts. *Proceedings Bostonian Society* 1933, pp. 31–47.

SWANK, J. M. *History of the manufacture of iron in all ages, and particularly in the U. S. from colonial times to 1891.* 2nd ed. American Iron & Steel Association, Philadelphia, 1892. xix + 554 pp.

TAYLOR, E. G. R. Notes on John Adams and contemporary mapmakers. *Geographical Journal* 97 (1941) pp. 182–184.

TAYLOR, J. A. *History of denistry.* Philadelphia 1922.

[TELEGRAPH] *Book of the telegraph.* D. Davis, Boston, 1851. 44 pp.

TERRY, H. *American clock making. Its early history.* Waterbury, Connecticut, 1872.

TERRY, J. History of clock making in our country. Supplement to *Scientific American* 27 (1889) June 15.

THARP, L. H. *Adventurous alliance. The story of the Agassiz family of Boston.* Little, Brown, Boston, 1959, xiii + 354 pp.

THOMPSON, H. *The age of invention. A chronicle of mechanical conquest.* Yale University Press, New Haven, 1921. 267 pp.

THOMPSON, J. A. *Count Rumford of Massachusetts.* Farrar & Rinehart, New York, 1935. xvi + 275 pp.

THOMPSON, L. *Young Longfellow (1807–1843).* Macmillan, New York, 1938. xxiv + 443 pp.

THOMPSON, R. L. *Wiring a continent. The history of the telegraph industry in the United States 1832–66.* Princeton University Press, 1948. xviii + 544 pp.

THOREAU, H. *A week on the Concord and Merrimack rivers. Writings I,* Houghton Mifflin, Boston, 1896. xvii + 531 pp.

THOREAU, H. *Natural history of Massachusetts.* Writings IX, Houghton Mifflin, Boston (1893), pp. 127–162.

THURSFIELD, H. Smallpox in the American War of Independence. *Annals of Medical History* 2 (1940) pp. 312–318.

THWAITES, F. T. The development of the theory of multiple glaciation in North America. *Transactions Wisconsin Academy of Science, Arts* 23 (1927) pp. 41–164.

TOCQUEVILLE, A. DE. *Democracy in America.* Trans. by H. Reeve, ed. by F. Bowen. 2 vols. 7th ed. Allyn, Boston, 1882. I (xxiii + 559 pp.); II (xiv + 499 pp.) first publication of translation 1862, after French ed. of 1850, original French ed. published 1834–1835.

TOWN, I. *Description of Ithiel Town's improvement in the construction of wood and iron bridges.* S. Converse, New Haven, 1821. Other pamphlets of 1825, 1839.

TRACY, C. On the rotary action of storms. *American Journal of Science* 45 (1843) pp. 65–72.

TRACY, C. *Mechanics of the earth's atmosphere.* Smithsonian Collection, Washington, D. C., 1910.

TREDGOLD, T. *The steam engine: its invention and progressive improvements.* New ed., enlarged by W. S. B. Woolhouse. 2 vols. J. Wealey, London, 1838. I (xix + 500 pp. + 250 pp.); II (125 plates).

TRUE, A. C. *A history of agricultural experimentation and research in the United States 1607–1925.* Miscellaneous Publication 251, U. S. Dept. of Agriculture, U. S. Printing Office, Washington, 1937. vi + 321 pp.

TUCKER, L. President Thomas Clap and the rise of Yale College. *The Historian* 19 (1956) pp. 66–81.

TURBERVILLE, A. S. *Johnson's England. An account of the life and manners of his age.* Clarendon Press, Oxford, II (1933). ix + 404 pp.

TURNER, F. J. *The United States 1830–'50. The Nation and its sections.* Holt, New York, 1935. xiv + 602 pp.

TURNER, H. H. *Astronomical discovery.* Arnold, London, 1904. 225 pp.

TYLER, ALICE F. *Freedom's ferment. Phases of American social history to 1860.* University of Minnesota Press, Minneapolis, 1944. x + 608 pp.

TYLER, M. C. *The literary history of the American Revolution.* 2 vols. Facsimile Library Barnes & Noble, New York, 1941. Copyright 1897. Vol. II (xix + 527 pp).

UPHAM, C. W. Memoir of the Reverend John Prince. *American Journal of Science* 31 (1837) pp. 201–222.

VAN DOREN, C. *Benjamin Franklin.* Viking, New York, 1938. xix + 845 pp.

VAN KLOOSTER, H. S. 125 years of chemistry at Rensselaer Polytechnic Institute. *Journal Chemical Education* 26 (1949) pp. 346–361.

———. The beginnings of laboratory instruction in chemistry in the U. S. A. *Chymia* 2 (1949) pp. 1–15.

VERNADSKY, W. I. The biosphere and the noösphere. *American Scientist* 33 (1945) pp. 1–12.

VIETS, H. R. *A brief history of medicine in Massachusetts.* Houghton Mifflin, Boston, 1930. x + 194 pp.

VOSE, G. L. *A sketch of the life and works of G. W. Whistler, civil engineer.* Lea & Shepard, Boston, 1887. 45 pp.

VOWLES, H. P., AND M. W. Jacob Perkins, 1766–1849. A study in American ingenuity and intrepid pioneering. *Mechanical Engineering* 53 (1931) pp. 785–790.

WADE, M. *Francis Parkman. Heroic historian.* Viking, New York, 1942. xii + 466 pp.

WAITE, F. C. Birth of the first independent proprietary medical school in New England, at Castleton, Vermont, in 1818. *Annals of Medical History* 7 (1935) pp. 242–252.

WALKER, B. The Middlesex Canal. *Contributions of the Old Residents' Historical Association,* Lowell, Massachusetts. 3 (1883–1886), pp. 273–308.

WALKER, F. A. Memoir of William Barton Rogers, 1804–1882. *Biographical Memoirs National Academy of Science* 3 (1895) pp. 1–13.

WALKLING, A. A. Henry Ingersoll Bowditch. *Annals Medical History* 5 (1933) pp. 428–437.

WALSH, B. D. On certain entomological speculations of the New England school of naturalists. *Proceedings Entomological Society of Philadelphia* 3 (1864).

WARD, J. H. *The life and letters of James Gates Percival.* Ticknor & Fields, Boston, 1866. xiii + 583 pp.

WARE, C. F. *The early New England cotton manufacture.* Houghton Mifflin, Boston, 1931. 349 pp.

WARFEL, H. R. *Noah Webster. Schoolmaster to America.* Macmillan, New York, 1936. ix + 460 pp.

WATER SUPPLY. *History of the introduction of pure water into the city of Boston, with a description of its Cochituate water works*. Compiled by a member of the water board. Mudge, Boston, 1868. xxi + 299 pp.

WATERHOUSE. Letter from Dr. Benjamin Waterhouse to Dr. James Tilton, Cambridge, March 24, 1815. *Proceedings Massachusetts Historical Society* Dec. 1920–Jan. 1921, pp. 159–165.

WATERHOUSE, B. *The Botanist, being the botanical part of a course of lectures on natural history, delivered in the University at Cambridge. Together with a discourse on the principle of vitality by Benjamin Waterhouse, M.D.* Buckingham, Boston, 1811.

WATERSTON, R. C. Memoir of G. B. Emerson. *Proceedings Massachussetts Historical Society*. 20 (1882–1883) pp. 232–259.

WATSON, E. History of agricultural societies on the modern Berkshire system. In *History of the rise, progress and existing condition of the western canals in the state of New York.* Albany, 1820.

WATSON, F. G. Elements of celestial navigation. *Sky & Telescope* 1 (1942) pp. 7–10.

WATSON, W. C. *Men and times of the revolution; or, memoirs of Elkanah Watson, including journals of travels in Europe and America, from 1777 to 1842*. By W. C. Watson, ed. New York, 1856. 460 pp.

WAYMAN, D. *Edward Sylvester Morse. A biography.* Harvard University Press, Cambridge, 1942. xvi + 457 pp.

WEBBER, S. *Institutes of natural philosophy,* by W. Enfield. J. Thomas & E. T. Andrews. Boston. xii + 428 + 20 pp., 1802; 2nd Am. ed., 1811. xvi + 420 pp.

WEBSTER, A. Development of the watch industry. *Papers Citizens' Club of Waltham, Season of 1891–1892,* Waltham, Massachusetts, 1891, pp. 17–31.

WEBSTER, J. W. *Manual of chemistry on the basis of Professor Brande's . . . compiled from the works of Brande, Henry, Berzelius, Thompson and others.* Richardson & Lord, Boston, 1826. xii + 603 pp.

WEEDEN, W. B. *Economic and social history of New England, 1620–1789.* 2 vols. Houghton, Mifflin, Boston, 1890, xvi + 964 pp.

WEEMS, P. V. H. Captain Thomas Hubbard Sumner of Sumner

line fame. *U. S. Naval Institute Proceedings* 64 (1938) pp. 61–65.

WEINBERGER, B. W. John Greenwood — pioneer American dental scientist. *Dental items of interest,* November 1943, 16 pp.

WELLS, H. *An essay on teeth; comprising a brief description of their formation, diseases, and proper treatment.* Case, Tiffany, Hartford, 1838.

WELLS, H. *A history of the discovery of the application of nitrous oxide gas, ether, and other vapors, to surgical operations.* Wells, Hartford, 1847.

WESTCOTT, TH. *The life of John Fitch, the inventor of the steamboat.* Lippincott, Philadelphia, 1878.

WHEATLAND, H. Sketch of the social and philosophical libraries, which in 1810 were purchased by the Salem Athenaeum. *Proceedings Essex Institute* 2 (1856–1860) pp. 140–146.

WHEILDON, W. W. *Memoir of Solomon Willard, architect and superintendent of the Bunker Hill Monument.* Prepared and printed by Monument Association, 1865. 272 pp.

WHIPPLE, E. P. *Recollections of eminent men.* Houghton Mifflin, Boston, 1896. xviii + 397 pp. (copyright 1886).

WHIPPLE, G. C. *State sanitation.* A review of the work of the Massachusetts State Board of Health. 2 vols. Harvard University Press, Cambridge, 1917; Vol. I, viii + 377 pp.

WHIPPLE, S. *A work upon bridge building.* Utica, New York, 1847.

WHITTLESEY, C. Life of John Fitch. *Sparks,* Vol. 16, 1854.

WHITE, A. D. *A history of the warfare of science with theology.* 2 vols. Appleton, 1896.

WHITE, G. S. *Memoir of Samuel Slater connected with a history of the rise and progress of the cotton manufacture in England and America.* 2nd ed., Philadelphia, 1836. 448 pp.

WILDER, M. P. The horticulture of Boston and vicinity. *Winsor's memorial history of Boston,* 1881. IV, pp. 607–640.

[WILKES, C.] Centenary celebration. The Wilkes exploring expedition of the U. S. Navy, 1838–1842, and symposium on American Polar Exploration, February 23–24, 1940. *Proceedings American Philosophical Society* 82 (1940) pp. 517–950.

WILLIS, B. American geology 1850–1900. *Proceedings American Philosophical Society* 86, No. 1 (1942) pp. 34–44.

WILSON, J. W. The first natural history lectures at Brown University, 1786, by Dr. Benjamin Waterhouse. *Annals Medical History* 4 (1942) pp. 390–398.

WINSLOW, E. Sketch of Professor B. A. Gould. *Popular Science Monthly* 20 (1881–82) pp. 683–687.

[WINSOR, J., ED.] *The memorial history of Boston, 1630–1880.* 4 vols. Osgood, Boston, 1881. Vol. IV (x + 713 pp.).

WINTHROP, JOHN. *Two lectures on comets.* Boston, 1759.

WOOD, F. J. *The turnpikes of New England, and evolution of the same through England, Virginia and Maryland.* Marshal Jones, Boston, 1919. xvii + 461 pp.

WOOD, N. A practical treatise on railroads and interior communication in general, 1825. 315 pp. 1st American ed. Philadelphia, 1832.

WOODBURY, R. S. *History of the gear-cutting machine: a historical study in geometry and machines.* Technology Press, Cambridge, Massachusetts, 1958. 135 pp.

———. *History of the grinding machine: a historical study in tools and precision instruments.* Technology Press, Cambridge, 1959, 6 +191 pp.

———. *History of the milling machine: a study in technical development.* Technology Press, Cambridge, 1960, 107 pp.

WOODWARD, C. R. *Ploughs and politicians. C. Read of New Jersey.* Rutgers University Studies in History 2 (1941). 468 pp.

WOOLF, H. Eighteenth-century observations of the transit of Venus. *Annals of Science* 9 (1953) pp. 176–190.

WRIGHT, H. *Sweeper in the sky.* Macmillan, New York, 1949.

WRIGHT, C. The religion of geology. *New England Quarterly* 14 (1941) pp. 335–358.

WRIGHT, ELIZUR. A theory of fluxions. *American Journal of Science* 14 (1828) pp. 330–350.

WRIGHT, P. G. AND E. O. *Elizur Wright, the father of life insurance.* University of Chicago Press, Chicago, 1937. xi + 380 pp.

WYMAN, J. A description of two additional crania of the En-géena (Troglodytes gorilla, Savage) from Gaboon, Africa. *American Journal of Science* 2 (9) 1850, pp. 34–45.

WYMAN, J. Review of "Monograph of the Aye-Aye" by Richard Owen. *American Journal of Science* (2) 36 (1863) pp. 294–299.

WYMAN, M. *Memoir of Daniel Treadwell.* Wilson, Cambridge, 1888, pp. 325–524. Reprint from *Memoirs American Academy Arts and Sciences* 11.

YOUMANS, W. J. *Pioneers of science in America.* Appleton, New York, 1896. viii + 508 pp. (Reprints with additions from *Popular Science Monthly.*)

YOUNG, R. T. *Biology in America.* Boston, 1923.

ZIRKLE, C. Some forgotten records of hybridization and sex in plants 1716–1739. *Journal of Heredity,* Washington, D. C., 23 (1932) pp. 432–438.

ZIRKLE, C. *The beginnings of plant hybridization.* University of Pennsylvania Press, Philadelphia, 1935. xiii + 231 pp.

Supplementary Bibliography (1991)

BROWN, C. M. *Benjamin Silliman, a life in the young republic.* Princeton University Press, Princeton, New Jersey, 1989. 363 pp.

BRUCE, P. V. *The launching of modern American science, 1846–1876.* Cornell University Press, Ithaca, New York, 1987. 446 pp.

CAMERON, E. H. *Samuel Slater, father of American manufactures.* Wheelwright, Freeport, Maine, 1960. 260 pp.

[DANIELS, G. H., ed.] *Nineteenth-century American science: a reappraisal.* Northwestern University Press, Evanston, Illinois, 1972. 274 pp.

DIBNER, B. *The Atlantic cable.* Bundy Library, Norwalk, Connecticut, 1959. 96 pp.

FURNIER, M., GINGRAS, Y., and KEEL, O. *Science et médecine au Québec. (Perspectives sociohistoriques.)* Institut Québecois de recherches sur la culture, Quebec, 1987. 210 pp.

GIFFORD, G. E. Ichthyologist dean. *Harvard Medical Alumni Bulletin* 39 (1964), pp. 22–27 (on D. H. Storer).

GORTARI, E. DE. *La ciencia en la historia de México.* Fondo de cultura económica, Mexico City, Buenos Aires, 1963. 461 pp.

GRAUSTEIN, J. E. *Thomas Nuttall, naturalist: explorations in America, 1808–1841.* Harvard University Press, Cambridge, Massachusetts, 1967. xiii + 481 pp.

GREENE, J. C. *American science in the age of Jefferson.* Iowa State University Press, Ames, Iowa, 1984. 484 pp.

JONES, B. Z., and BOYD, L. G. *The Harvard College Observatory: the first four directorships, 1839–1914.* Harvard University Press, Cambridge, Massachusetts, 1971.

KOHLSTEDT, S. G. *The formation of the American scientific community: the American Association for the Advancement of Science, 1848–60.* University of Illinois Press, Urbana, Illinois, 1976.

REINGOLD, N., ed. *Science in nineteenth-century America.* Hill & Wang, New York, 1964.

REINGOLD, N. *Science in America since 1820.* N. Watson, New York, 1976.

[RODGERS, J., GROSSO, T., and JORDON, W. M., eds.] *Boston to Buffalo in the footsteps of Amos Eaton and Edward Hitchcock.* American Geophysical Union, Washington, D.C., 1989. 89 pp.

ROSSITER, M. W. *The emergence of agricultural science: Justus Liebig and the Americans, 1840–1880.* New Haven, Yale University Press, 1975.

The *Dictionary of Scientific Biography* (Scribner, New York), which started with Vol. I in 1970 and is being continued beyond its *Index* volume XVI (1980), contains biographies of several of the persons mentioned in our book, with bibliographies. Examples: Agassiz, Bowditch, Dana, Eaton, Gray, Hitchcock, Silliman.

Index

A CATALOG OF SELECTED

DOVER BOOKS

IN ALL FIELDS OF INTEREST

A CATALOG OF SELECTED DOVER BOOKS IN ALL FIELDS OF INTEREST

DRAWINGS OF REMBRANDT, edited by Seymour Slive. Updated Lippmann, Hofstede de Groot edition, with definitive scholarly apparatus. All portraits, biblical sketches, landscapes, nudes. Oriental figures, classical studies, together with selection of work by followers. 550 illustrations. Total of 630pp. 9⅛ × 12¼. 21485-0, 21486-9 Pa., Two-vol. set $29.90

GHOST AND HORROR STORIES OF AMBROSE BIERCE, Ambrose Bierce. 24 tales vividly imagined, strangely prophetic, and decades ahead of their time in technical skill: "The Damned Thing," "An Inhabitant of Carcosa," "The Eyes of the Panther," "Moxon's Master," and 20 more. 199pp. 5⅜ × 8½. 20767-6 Pa. $4.95

ETHICAL WRITINGS OF MAIMONIDES, Maimonides. Most significant ethical works of great medieval sage, newly translated for utmost precision, readability. Laws Concerning Character Traits, Eight Chapters, more. 192pp. 5⅜ × 8½. 24522-5 Pa. $4.50

THE EXPLORATION OF THE COLORADO RIVER AND ITS CANYONS, J. W. Powell. Full text of Powell's 1,000-mile expedition down the fabled Colorado in 1869. Superb account of terrain, geology, vegetation, Indians, famine, mutiny, treacherous rapids, mighty canyons, during exploration of last unknown part of continental U.S. 400pp. 5⅜ × 8½. 20094-9 Pa. $7.95

HISTORY OF PHILOSOPHY, Julián Marías. Clearest one-volume history on the market. Every major philosopher and dozens of others, to Existentialism and later. 505pp. 5⅜ × 8½. 21739-6 Pa. $9.95

ALL ABOUT LIGHTNING, Martin A. Uman. Highly readable non-technical survey of nature and causes of lightning, thunderstorms, ball lightning, St. Elmo's Fire, much more. Illustrated. 192pp. 5⅜ × 8½. 25237-X Pa. $5.95

SAILING ALONE AROUND THE WORLD, Captain Joshua Slocum. First man to sail around the world, alone, in small boat. One of great feats of seamanship told in delightful manner. 67 illustrations. 294pp. 5⅜ × 8½. 20326-3 Pa. $4.95

LETTERS AND NOTES ON THE MANNERS, CUSTOMS AND CONDITIONS OF THE NORTH AMERICAN INDIANS, George Catlin. Classic account of life among Plains Indians: ceremonies, hunt, warfare, etc. 312 plates. 572pp. of text. 6⅛ × 9¼. 22118-0, 22119-9, Pa. Two-vol. set $17.90

ALASKA: The Harriman Expedition, 1899, John Burroughs, John Muir, et al. Informative, engrossing accounts of two-month, 9,000-mile expedition. Native peoples, wildlife, forests, geography, salmon industry, glaciers, more. Profusely illustrated. 240 black-and-white line drawings. 124 black-and-white photographs. 3 maps. Index. 576pp. 5⅜ × 8½. 25109-8 Pa. $11.95

ILLUSTRATED DICTIONARY OF HISTORIC ARCHITECTURE, edited by Cyril M. Harris. Extraordinary compendium of clear, concise definitions for over 5,000 important architectural terms complemented by over 2,000 line drawings. Covers full spectrum of architecture from ancient ruins to 20th-century Modernism. Preface. 592pp. 7½ × 9⅝. 24444-X Pa. $15.95

THE NIGHT BEFORE CHRISTMAS, Clement Moore. Full text, and woodcuts from original 1848 book. Also critical, historical material. 19 illustrations. 40pp. 4⅝ × 6. 22797-9 Pa. $2.50

THE LESSON OF JAPANESE ARCHITECTURE: 165 Photographs, Jiro Harada. Memorable gallery of 165 photographs taken in the 1930's of exquisite Japanese homes of the well-to-do and historic buildings. 13 line diagrams. 192pp. 8⅜ × 11¼. 24778-3 Pa. $10.95

THE AUTOBIOGRAPHY OF CHARLES DARWIN AND SELECTED LETTERS, edited by Francis Darwin. The fascinating life of eccentric genius composed of an intimate memoir by Darwin (intended for his children); commentary by his son, Francis; hundreds of fragments from notebooks, journals, papers; and letters to and from Lyell, Hooker, Huxley, Wallace and Henslow. xi + 365pp. 5⅜ × 8. 20479-0 Pa. $6.95

WONDERS OF THE SKY: Observing Rainbows, Comets, Eclipses, the Stars and Other Phenomena, Fred Schaaf. Charming, easy-to-read poetic guide to all manner of celestial events visible to the naked eye. Mock suns, glories, Belt of Venus, more. Illustrated. 299pp. 5¼ × 8¼. 24402-4 Pa. $7.95

BURNHAM'S CELESTIAL HANDBOOK, Robert Burnham, Jr. Thorough guide to the stars beyond our solar system. Exhaustive treatment. Alphabetical by constellation: Andromeda to Cetus in Vol. 1; Chamaeleon to Orion in Vol. 2; and Pavo to Vulpecula in Vol. 3. Hundreds of illustrations. Index in Vol. 3. 2,000pp. 6⅛ × 9¼. 23567-X, 23568-8, 23673-0 Pa., Three-vol. set $41.85

STAR NAMES: Their Lore and Meaning, Richard Hinckley Allen. Fascinating history of names various cultures have given to constellations and literary and folkloristic uses that have been made of stars. Indexes to subjects. Arabic and Greek names. Biblical references. Bibliography. 563pp. 5⅜ × 8½. 21079-0 Pa. $8.95

THIRTY YEARS THAT SHOOK PHYSICS: The Story of Quantum Theory, George Gamow. Lucid, accessible introduction to influential theory of energy and matter. Careful explanations of Dirac's anti-particles, Bohr's model of the atom, much more. 12 plates. Numerous drawings. 240pp. 5⅜ × 8½. 24895-X Pa. $5.95

CHINESE DOMESTIC FURNITURE IN PHOTOGRAPHS AND MEASURED DRAWINGS, Gustav Ecke. A rare volume, now affordably priced for antique collectors, furniture buffs and art historians. Detailed review of styles ranging from early Shang to late Ming. Unabridged republication. 161 black-and-white drawings, photos. Total of 224pp. 8⅜ × 11¼. (Available in U.S. only) 25171-3 Pa. $13.95

VINCENT VAN GOGH: A Biography, Julius Meier-Graefe. Dynamic, penetrating study of artist's life, relationship with brother, Theo, painting techniques, travels, more. Readable, engrossing. 160pp. 5⅜ × 8½. (Available in U.S. only) 25253-1 Pa. $4.95

SUNDIALS, Albert Waugh. Far and away the best, most thorough coverage of ideas, mathematics concerned, types, construction, adjusting anywhere. Over 100 illustrations. 230pp. 5⅜ × 8½. 22947-5 Pa. $5.95

PICTURE HISTORY OF THE NORMANDIE: With 190 Illustrations, Frank O. Braynard. Full story of legendary French ocean liner: Art Deco interiors, design innovations, furnishings, celebrities, maiden voyage, tragic fire, much more. Extensive text. 144pp. 8⅞ × 11¾. 25257-4 Pa. $10.95

THE FIRST AMERICAN COOKBOOK: A Facsimile of "American Cookery," 1796, Amelia Simmons. Facsimile of the first American-written cookbook published in the United States contains authentic recipes for colonial favorites—pumpkin pudding, winter squash pudding, spruce beer, Indian slapjacks, and more. Introductory Essay and Glossary of colonial cooking terms. 80pp. 5⅜ × 8½. 24710-4 Pa. $3.50

101 PUZZLES IN THOUGHT AND LOGIC, C. R. Wylie, Jr. Solve murders and robberies, find out which fishermen are liars, how a blind man could possibly identify a color—purely by your own reasoning! 107pp. 5⅜ × 8½. 20367-0 Pa. $2.50

ANCIENT EGYPTIAN MYTHS AND LEGENDS, Lewis Spence. Examines animism, totemism, fetishism, creation myths, deities, alchemy, art and magic, other topics. Over 50 illustrations. 432pp. 5⅜ × 8½. 26525-0 Pa. $8.95

ANTHROPOLOGY AND MODERN LIFE, Franz Boas. Great anthropologist's classic treatise on race and culture. Introduction by Ruth Bunzel. Only inexpensive paperback edition. 255pp. 5⅜ × 8½. 25245-0 Pa. $6.95

THE TALE OF PETER RABBIT, Beatrix Potter. The inimitable Peter's terrifying adventure in Mr. McGregor's garden, with all 27 wonderful, full-color Potter illustrations. 55pp. 4¼ × 5½. (Available in U.S. only) 22827-4 Pa. $1.75

THREE PROPHETIC SCIENCE FICTION NOVELS, H. G. Wells. *When the Sleeper Wakes, A Story of the Days to Come* and *The Time Machine* (full version). 335pp. 5⅜ × 8½. (Available in U.S. only) 20605-X Pa. $6.95

APICIUS COOKERY AND DINING IN IMPERIAL ROME, edited and translated by Joseph Dommers Vehling. Oldest known cookbook in existence offers readers a clear picture of what foods Romans ate, how they prepared them, etc. 49 illustrations. 301pp. 6⅛ × 9¼. 23563-7 Pa. $7.95

SHAKESPEARE LEXICON AND QUOTATION DICTIONARY, Alexander Schmidt. Full definitions, locations, shades of meaning of every word in plays and poems. More than 50,000 exact quotations. 1,485pp. 6½ × 9¼.
22726-X, 22727-8 Pa., Two-vol. set $31.90

THE WORLD'S GREAT SPEECHES, edited by Lewis Copeland and Lawrence W. Lamm. Vast collection of 278 speeches from Greeks to 1970. Powerful and effective models; unique look at history. 842pp. 5⅜ × 8½. 20468-5 Pa. $12.95

PLANTS OF THE BIBLE, Harold N. Moldenke and Alma L. Moldenke. Standard reference to all 230 plants mentioned in Scriptures. Latin name, biblical reference, uses, modern identity, much more. Unsurpassed encyclopedic resource for scholars, botanists, nature lovers, students of Bible. Bibliography. Indexes. 123 black-and-white illustrations. 384pp. 6 × 9. 25069-5 Pa. $8.95

FAMOUS AMERICAN WOMEN: A Biographical Dictionary from Colonial Times to the Present, Robert McHenry, ed. From Pocahontas to Rosa Parks, 1,035 distinguished American women documented in separate biographical entries. Accurate, up-to-date data, numerous categories, spans 400 years. Indices. 493pp. 6½ × 9¼. 24523-3 Pa. $10.95

THE FABULOUS INTERIORS OF THE GREAT OCEAN LINERS IN HISTORIC PHOTOGRAPHS, William H. Miller, Jr. Some 200 superb photographs capture exquisite interiors of world's great "floating palaces"—1890's to 1980's: *Titanic, Ile de France, Queen Elizabeth, United States, Europa,* more. Approx. 200 black-and-white photographs. Captions. Text. Introduction. 160pp. 8⅞ × 11¾. 24756-2 Pa. $9.95

THE GREAT LUXURY LINERS, 1927-1954: A Photographic Record, William H. Miller, Jr. Nostalgic tribute to heyday of ocean liners. 186 photos of Ile de France, Normandie, Leviathan, Queen Elizabeth, United States, many others. Interior and exterior views. Introduction. Captions. 160pp. 9 × 12. 24056-8 Pa. $10.95

A NATURAL HISTORY OF THE DUCKS, John Charles Phillips. Great landmark of ornithology offers complete detailed coverage of nearly 200 species and subspecies of ducks: gadwall, sheldrake, merganser, pintail, many more. 74 full-color plates, 102 black-and-white. Bibliography. Total of 1,920pp. 8⅞ × 11¼. 25141-1, 25142-X Cloth. Two-vol. set $100.00

THE SEAWEED HANDBOOK: An Illustrated Guide to Seaweeds from North Carolina to Canada, Thomas F. Lee. Concise reference covers 78 species. Scientific and common names, habitat, distribution, more. Finding keys for easy identification. 224pp. 5⅜ × 8½. 25215-9 Pa. $6.95

THE TEN BOOKS OF ARCHITECTURE: The 1755 Leoni Edition, Leon Battista Alberti. Rare classic helped introduce the glories of ancient architecture to the Renaissance. 68 black-and-white plates. 336pp. 8⅜ × 11¼. 25239-6 Pa. $14.95

MISS MACKENZIE, Anthony Trollope. Minor masterpieces by Victorian master unmasks many truths about life in 19th-century England. First inexpensive edition in years. 392pp. 5⅜ × 8½. 25201-9 Pa. $8.95

THE RIME OF THE ANCIENT MARINER, Gustave Doré, Samuel Taylor Coleridge. Dramatic engravings considered by many to be his greatest work. The terrifying space of the open sea, the storms and whirlpools of an unknown ocean, the ice of Antarctica, more—all rendered in a powerful, chilling manner. Full text. 38 plates. 77pp. 9¼ × 12. 22305-1 Pa. $4.95

THE EXPEDITIONS OF ZEBULON MONTGOMERY PIKE, Zebulon Montgomery Pike. Fascinating first-hand accounts (1805-6) of exploration of Mississippi River, Indian wars, capture by Spanish dragoons, much more. 1,088pp. 5⅜ × 8½. 25254-X, 25255-8 Pa. Two-vol. set $25.90

A CONCISE HISTORY OF PHOTOGRAPHY: Third Revised Edition, Helmut Gernsheim. Best one-volume history—camera obscura, photochemistry, daguerreotypes, evolution of cameras, film, more. Also artistic aspects—landscape, portraits, fine art, etc. 281 black-and-white photographs. 26 in color. 176pp. 8⅜ × 11¼. 25128-4 Pa. $13.95

THE DORÉ BIBLE ILLUSTRATIONS, Gustave Doré. 241 detailed plates from the Bible: the Creation scenes, Adam and Eve, Flood, Babylon, battle sequences, life of Jesus, etc. Each plate is accompanied by the verses from the King James version of the Bible. 241pp. 9 × 12. 23004-X Pa. $9.95

WANDERINGS IN WEST AFRICA, Richard F. Burton. Great Victorian scholar/adventurer's invaluable descriptions of African tribal rituals, fetishism, culture, art, much more. Fascinating 19th-century account. 624pp. 5⅜ × 8½. 26890-X Pa. $12.95

FLATLAND, E. A. Abbott. Intriguing and enormously popular science-fiction classic explores the complexities of trying to survive as a two-dimensional being in a three-dimensional world. Amusingly illustrated by the author. 16 illustrations. 103pp. 5⅜ × 8½. 20001-9 Pa. $2.50

THE HISTORY OF THE LEWIS AND CLARK EXPEDITION, Meriwether Lewis and William Clark, edited by Elliott Coues. Classic edition of Lewis and Clark's day-by-day journals that later became the basis for U.S. claims to Oregon and the West. Accurate and invaluable geographical, botanical, biological, meteorological and anthropological material. Total of 1,508pp. 5⅜ × 8½.
21268-8, 21269-6, 21270-X Pa. Three-vol. set $26.85

LANGUAGE, TRUTH AND LOGIC, Alfred J. Ayer. Famous, clear introduction to Vienna, Cambridge schools of Logical Positivism. Role of philosophy, elimination of metaphysics, nature of analysis, etc. 160pp. 5⅜ × 8½. (Available in U.S. and Canada only) 20010-8 Pa. $3.95

MATHEMATICS FOR THE NONMATHEMATICIAN, Morris Kline. Detailed, college-level treatment of mathematics in cultural and historical context, with numerous exercises. For liberal arts students. Preface. Recommended Reading Lists. Tables. Index. Numerous black-and-white figures. xvi + 641pp. 5⅜ × 8½.
24823-2 Pa. $11.95

HANDBOOK OF PICTORIAL SYMBOLS, Rudolph Modley. 3,250 signs and symbols, many systems in full; official or heavy commercial use. Arranged by subject. Most in Pictorial Archive series. 143pp. 8⅛ × 11. 23357-X Pa. $6.95

INCIDENTS OF TRAVEL IN YUCATAN, John L. Stephens. Classic (1843) exploration of jungles of Yucatan, looking for evidences of Maya civilization. Travel adventures, Mexican and Indian culture, etc. Total of 669pp. 5⅜ × 8½.
20926-1, 20927-X Pa., Two-vol. set $11.90

DEGAS: An Intimate Portrait, Ambroise Vollard. Charming, anecdotal memoir by famous art dealer of one of the greatest 19th-century French painters. 14 black-and-white illustrations. Introduction by Harold L. Van Doren. 96pp. 5⅜ × 8½.
25131-4 Pa. $4.95

PERSONAL NARRATIVE OF A PILGRIMAGE TO ALMANDINAH AND MECCAH, Richard Burton. Great travel classic by remarkably colorful personality. Burton, disguised as a Moroccan, visited sacred shrines of Islam, narrowly escaping death. 47 illustrations. 959pp. 5⅜ × 8½. 21217-3, 21218-1 Pa., Two-vol. set $19.90

PHRASE AND WORD ORIGINS, A. H. Holt. Entertaining, reliable, modern study of more than 1,200 colorful words, phrases, origins and histories. Much unexpected information. 254pp. 5⅜ × 8½. 20758-7 Pa. $5.95

THE RED THUMB MARK, R. Austin Freeman. In this first Dr. Thorndyke case, the great scientific detective draws fascinating conclusions from the nature of a single fingerprint. Exciting story, authentic science. 320pp. 5⅜ × 8½. (Available in U.S. only) 25210-8 Pa. $6.95

AN EGYPTIAN HIEROGLYPHIC DICTIONARY, E. A. Wallis Budge. Monumental work containing about 25,000 words or terms that occur in texts ranging from 3000 B.C. to 600 A.D. Each entry consists of a transliteration of the word, the word in hieroglyphs, and the meaning in English. 1,314pp. 6⅞ × 10.
23615-3, 23616-1 Pa., Two-vol. set $35.90

THE COMPLEAT STRATEGYST: Being a Primer on the Theory of Games of Strategy, J. D. Williams. Highly entertaining classic describes, with many illustrated examples, how to select best strategies in conflict situations. Prefaces. Appendices. xvi + 268pp. 5⅜ × 8½. 25101-2 Pa. $6.95

THE ROAD TO OZ, L. Frank Baum. Dorothy meets the Shaggy Man, little Button-Bright and the Rainbow's beautiful daughter in this delightful trip to the magical Land of Oz. 272pp. 5⅜ × 8. 25208-6 Pa. $5.95

POINT AND LINE TO PLANE, Wassily Kandinsky. Seminal exposition of role of point, line, other elements in non-objective painting. Essential to understanding 20th-century art. 127 illustrations. 192pp. 6½ × 9¼. 23808-3 Pa. $5.95

LADY ANNA, Anthony Trollope. Moving chronicle of Countess Lovel's bitter struggle to win for herself and daughter Anna their rightful rank and fortune—perhaps at cost of sanity itself. 384pp. 5⅜ × 8½. 24669-8 Pa. $8.95

EGYPTIAN MAGIC, E. A. Wallis Budge. Sums up all that is known about magic in Ancient Egypt: the role of magic in controlling the gods, powerful amulets that warded off evil spirits, scarabs of immortality, use of wax images, formulas and spells, the secret name, much more. 253pp. 5⅜ × 8½. 22681-6 Pa. $4.50

THE DANCE OF SIVA, Ananda Coomaraswamy. Preeminent authority unfolds the vast metaphysic of India: the revelation of her art, conception of the universe, social organization, etc. 27 reproductions of art masterpieces. 192pp. 5⅜ × 8½.
24817-8 Pa. $5.95

CHRISTMAS CUSTOMS AND TRADITIONS, Clement A. Miles. Origin, evolution, significance of religious, secular practices. Caroling, gifts, yule logs, much more. Full, scholarly yet fascinating; non-sectarian. 400pp. 5⅜ × 8½.
23354-5 Pa. $6.95

THE HUMAN FIGURE IN MOTION, Eadweard Muybridge. More than 4,500 stopped-action photos, in action series, showing undraped men, women, children jumping, lying down, throwing, sitting, wrestling, carrying, etc. 390pp. 7⅞ × 10⅝.
20204-6 Cloth. $24.95

THE MAN WHO WAS THURSDAY, Gilbert Keith Chesterton. Witty, fast-paced novel about a club of anarchists in turn-of-the-century London. Brilliant social, religious, philosophical speculations. 128pp. 5⅜ × 8½. 25121-7 Pa. $3.95

A CEZANNE SKETCHBOOK: Figures, Portraits, Landscapes and Still Lifes, Paul Cezanne. Great artist experiments with tonal effects, light, mass, other qualities in over 100 drawings. A revealing view of developing master painter, precursor of Cubism. 102 black-and-white illustrations. 144pp. 8¾ × 6⅜. 24790-2 Pa. $6.95

AN ENCYCLOPEDIA OF BATTLES: Accounts of Over 1,560 Battles from 1479 B.C. to the Present, David Eggenberger. Presents essential details of every major battle in recorded history, from the first battle of Megiddo in 1479 B.C. to Grenada in 1984. List of Battle Maps. New Appendix covering the years 1967–1984. Index. 99 illustrations. 544pp. 6½ × 9¼. 24913-1 Pa. $14.95

AN ETYMOLOGICAL DICTIONARY OF MODERN ENGLISH, Ernest Weekley. Richest, fullest work, by foremost British lexicographer. Detailed word histories. Inexhaustible. Total of 856pp. 6½ × 9¼.
21873-2, 21874-0 Pa., Two-vol. set $19.90

WEBSTER'S AMERICAN MILITARY BIOGRAPHIES, edited by Robert McHenry. Over 1,000 figures who shaped 3 centuries of American military history. Detailed biographies of Nathan Hale, Douglas MacArthur, Mary Hallaren, others. Chronologies of engagements, more. Introduction. Addenda. 1,033 entries in alphabetical order. xi + 548pp. 6½ × 9¼. (Available in U.S. only)
24758-9 Pa. $13.95

LIFE IN ANCIENT EGYPT, Adolf Erman. Detailed older account, with much not in more recent books: domestic life, religion, magic, medicine, commerce, and whatever else needed for complete picture. Many illustrations. 597pp. 5⅜ × 8½.
22632-8 Pa. $8.95

HISTORIC COSTUME IN PICTURES, Braun & Schneider. Over 1,450 costumed figures shown, covering a wide variety of peoples: kings, emperors, nobles, priests, servants, soldiers, scholars, townsfolk, peasants, merchants, courtiers, cavaliers, and more. 256pp. 8⅜ × 11¼. 23150-X Pa. $9.95

THE NOTEBOOKS OF LEONARDO DA VINCI, edited by J. P. Richter. Extracts from manuscripts reveal great genius; on painting, sculpture, anatomy, sciences, geography, etc. Both Italian and English. 186 ms. pages reproduced, plus 500 additional drawings, including studies for *Last Supper, Sforza* monument, etc. 860pp. 7⅞ × 10¾. (Available in U.S. only) 22572-0, 22573-9 Pa., Two-vol. set $31.90

SIR HARRY HOTSPUR OF HUMBLETHWAITE, Anthony Trollope. Incisive, unconventional psychological study of a conflict between a wealthy baronet, his idealistic daughter, and their scapegrace cousin. The 1870 novel in its first inexpensive edition in years. 250pp. 5⅜ × 8½. 24953-0 Pa. $6.95

LASERS AND HOLOGRAPHY, Winston E. Kock. Sound introduction to burgeoning field, expanded (1981) for second edition. Wave patterns, coherence, lasers, diffraction, zone plates, properties of holograms, recent advances. 84 illustrations. 160pp. 5⅜ × 8¼. (Except in United Kingdom) 24041-X Pa. $3.95

INTRODUCTION TO ARTIFICIAL INTELLIGENCE: SECOND, ENLARGED EDITION, Philip C. Jackson, Jr. Comprehensive survey of artificial intelligence—the study of how machines (computers) can be made to act intelligently. Includes introductory and advanced material. Extensive notes updating the main text. 132 black-and-white illustrations. 512pp. 5⅜ × 8½. 24864-X Pa. $8.95

HISTORY OF INDIAN AND INDONESIAN ART, Ananda K. Coomaraswamy. Over 400 illustrations illuminate classic study of Indian art from earliest Harappa finds to early 20th century. Provides philosophical, religious and social insights. 304pp. 6⅝ × 9⅝. 25005-9 Pa. $11.95

THE GOLEM, Gustav Meyrink. Most famous supernatural novel in modern European literature, set in Ghetto of Old Prague around 1890. Compelling story of mystical experiences, strange transformations, profound terror. 13 black-and-white illustrations. 224pp. 5⅜ × 8½. (Available in U.S. only) 25025-3 Pa. $6.95

PICTORIAL ENCYCLOPEDIA OF HISTORIC ARCHITECTURAL PLANS, DETAILS AND ELEMENTS: With 1,880 Line Drawings of Arches, Domes, Doorways, Facades, Gables, Windows, etc., John Theodore Haneman. Sourcebook of inspiration for architects, designers, others. Bibliography. Captions. 141pp. 9 × 12. 24605-1 Pa. $7.95

BENCHLEY LOST AND FOUND, Robert Benchley. Finest humor from early 30's, about pet peeves, child psychologists, post office and others. Mostly unavailable elsewhere. 73 illustrations by Peter Arno and others. 183pp. 5⅜ × 8½. 22410-4 Pa. $4.95

ERTÉ GRAPHICS, Erté. Collection of striking color graphics: *Seasons, Alphabet, Numerals, Aces* and *Precious Stones*. 50 plates, including 4 on covers. 48pp. 9⅜ × 12¼. 23580-7 Pa. $7.95

THE JOURNAL OF HENRY D. THOREAU, edited by Bradford Torrey, F. H. Allen. Complete reprinting of 14 volumes, 1837–61, over two million words; the sourcebooks for *Walden*, etc. Definitive. All original sketches, plus 75 photographs. 1,804pp. 8½ × 12¼. 20312-3, 20313-1 Cloth., Two-vol. set $125.00

CASTLES: THEIR CONSTRUCTION AND HISTORY, Sidney Toy. Traces castle development from ancient roots. Nearly 200 photographs and drawings illustrate moats, keeps, baileys, many other features. Caernarvon, Dover Castles, Hadrian's Wall, Tower of London, dozens more. 256pp. 5⅜ × 8¼. 24898-4 Pa. $6.95